PROPERTY OF:
DAVID O. McKAY LIBRARY
BYU-IDAHO
REXBURG ID 83460-0405

Roland Keller

Identification of tropical woody plants in the absence of flowers

A field guide (2nd edition)

Birkhäuser Verlag
Basel · Boston · Berlin

Author:

Roland Keller
Primerose 6
CH-1007 Lausanne
Switzerland

A CIP catalogue record for this book is available from the
Library of Congress, Washington D.C., USA

Bibliographic information published by Die Deutsche Bibliothek
Die Deutsche Bibliothek lists this publication in the Deutsche Nationalbibliografie;
detailed bibliographic data is available in the Internet at <http://dnb.ddb.de>.

ISBN 3-7643-6453-X Birkhäuser Verlag, Basel – Boston – Berlin

The publisher and editor can give no guarantee for the information on drug dosage and administration contained in this publication. The respective user must check its accuracy by consulting other sources of reference in each individual case.

The use of registered names, trademarks etc. in this publication, even if not identified as such, does not imply that they are exempt from the relevant protective laws and regulations or free for general use.

This work is subject to copyright. All rights are reserved, whether the whole or part of the material is concerned, specifically the rights of translation, reprinting, re-use of illustrations, recitation, broadcasting, reproduction on microfilms or in other ways, and storage in data banks. For any kind of use, permission of the copyright owner must be obtained.

© 2004 Birkhäuser Verlag, P.O. Box 133, CH-4010 Basel, Switzerland
Part of Springer Science+Business Media
Printed on acid-free paper produced from chlorine-free pulp. TCF ∞
Cover design: Micha Lotrovsky, CH-4106 Therwil, Switzerland
Printed in Germany
ISBN 3-7643-6453-X

9 8 7 6 5 4 3 2 1   www.birkhauser.ch

Acknowledgements

I wish to thank the following for their assistance and their friendship,

Enrique Araujo Salazar (Pelotas, Brasil), Daniel Atuany (Manusela National Park, Moluccas), Claude Edelin (Montpellier, France), Peter Endress (Zürich, Suisse), Jacques Fournet (Pointe-à-Pitre, Guadeloupe), Pierre Hainard (Crissier, Suisse), Francis Hallé (Montpellier, France), René Hebding (Saint-Jean-Cap-Ferrat, France), Pak Ijun (Bogor, Java), Anton Leeuwenberg (Wageningen, Nederland), Pierre Lombion (Morne-à-l'Eau, Guadeloupe), Paul Maas (Utrecht, Nederland), Jean-François and Caroline Molino (Bukittinggi, Sumatra), Jeanine Raharilala (Tsimbazaza, Madagascar), Mamisoa Rapanoelina (Diego Suarez, Madagascar), Elio Sanoja (Upata, Venezuela), N. Sasidharan (Peechi, India), Francisco Antonio da Silva (Florianopolis, Brasil), Rodolphe Spichiger (Genève, Suisse), M.M.J. van Balgooy (Leiden, Nederland) Jasper Zanco (Florianopolis, Brasil), without whom this identification key would simply have been mere speculation.

I am indebted to the Société Académique Vaudoise who generously helped finance this edition. The drawing up was carried out under the auspices of the Institut de Botanique Systématique et de Géobotanique at Lausanne University.

Adrian Bell contributed to the translation of the French version into English and made numerous technical and scientific suggestions, Finally many thanks to Jean-Charles Delacrétaz, Caroline and Peter Nazroo, for their technical and linguistic assistance.

à Zoé et Lucien

CONTENTS

Foreword XI

Introductory remarks 1

Part I: Key and glossary

Key 9

Glossary, notes and illustrations 123
1. Geographical distribution 124
2. Outer bark and lenticels 126
3. Macroanatomy I: inner bark, rays and exudates 128
4. Macroanatomy II: internal phloem, wood and pith 130
5. Climbing systems 132
6. Unit of extension, monopodium and sympodium 134
7. Architectural models (trunk monopodial) 136
8. Architectural models (trunk sympodial) 138
9. Phyllotaxy and torsion 140
10. Heterophylly 142
11. Shape of stems 144
12. Axillary buds and prophylls 146
13. Stipules and interpetiolar ridge 148
14. Leaf: blade and rachis 150
15. Leaf: petiole and petiolule 152
16. Leaf folding and aestivation 154
17. Venation I 156
18. Venation II 158
19. Glands and translucent dots 160
20. Epidermal structures of leaves 162
21. Key to architectural models 164

Part II: The principal families of tropical woody Dicotyledons illustrated by means of their vegetative characters

A classification of the families identified by the key 169

1. Arecaceae 172
2. Hernandiaceae, Illiciaceae, Canellaceae, Lauraceae 173

3.	Monimiaceae, Annonaceae	178
4.	Magnoliaceae, Myristicaceae, Piperaceae, Chloranthaceae	179
5.	Menispermaceae, Aristolochiaceae, Meliosmaceae, Lardizabalaceae, *Clematis*	184
6.	Proteaceae, Dilleniaceae, Polygonaceae	185
7.	Nyctaginaceae, Phytolaccaceae, Olacaceae, Opiliaceae, Loranthaceae	190
8.	Hamamelidaceae, Fagaceae, Zygophyllaceae, Staphylaceae, Melianthaceae, Greyiaceae, Rosaceae, *Eryobotrya*	191
9.	Ulmaceae, Urticaceae, Moraceae	196
10.	Rhamnaceae, Vitaceae, Leeaceae	197
11.	Theaceae, Actinidaceae, Marcgraviaceae, Pellicieraceae, Bonnetiaceae, Myrsinaceae, Theophrastaceae	202
12.	Ericaceae, Cyrillaceae, Ebenaceae, Sapotaceae	203
13.	Lecythidaceae, Combretaceae, Myrtaceae	208
14.	Lythraceae, Vochysiaceae, Melastomataceae	209
15.	Clusiaceae, Ochnaceae, Quiinaceae	214
16.	Flacourtiaceae, Capparaceae, Passifloraceae	215
17.	Violaceae, *Pangium*, Bixaceae, Tetramelaceae, Caricaceae	220
18.	Dipterocarpaceae, Elaeocarpaceae, Malvaceae *s.l.*	221
19.	Euphorbiaceae, Putranjivaceae	226
20.	Leguminosae	230
21.	Chrysobalanaceae, *Prunus*, Oxalidaceae, Connaraceae	234
22.	Celastraceae, Irvingiaceae, Humiriaceae, Linaceae, Erythroxylaceae	235
23.	Sapindaceae, Meliaceae, Simaroubaceae	240
24.	Anacardiaceae, Burseraceae, Rutaceae	241
25.	Thymeleaceae, Dichapetalaceae, Malpighiaceae	246
26.	Polygalaceae, Xanthophyllaceae, Brunelliaceae, Cunoniaceae, Rhizophoraceae	247
27.	Pittosporaceae, Aquifoliaceae, Araliaceae, Cornaceae, Alangiaceae, Anisophylleaceae	252
28.	Icacinaceae, Asteraceae	253
29.	Convolvulaceae, Boraginaceae, Solanaceae	258
30.	Bignoniaceae, Oleaceae, Verbenaceae, Acanthaceae	259
31.	Loganiaceae, Gentianaceae, Apocynaceae	264
32.	Rubiaceae	265

Bibliography 271

Index of the genera and families 275

Captions to photos 289

Foreword

A preface is often the way used to justify an opinion, in the first edition of this book, one emphasised how difficult it is to find flowers or fruits in a tropical forest. Of course, this situation has not changed and continues to be the principal reason for embarking on the construction of a vegetative key.

Since seven years, the key established has been widely used and indeed, introduced by me on personally led field courses and subsequently I have come to appreciate that identification can be difficult even in the presence of permanently visible features. The central principal is that plants should be examined from several different angles with each level of observation being benchmarked against the correct corresponding criteria. Some of these levels of observation and the criteria of examination are not included in conventional plant sciences syllabus and therefore tree architecture and the morphology and characteristics of bark are unlikely to have been studied by a recently qualified graduate.

Although the keys are now some twenty per cent broader than those included in the first edition, identification at family level remains the focal point whereas the largest families are examined in greater detail. A key has been added for the Euphorbiaceae and despite this family's reputation for being difficult to identify iota main groups of genera can be quickly identified by means of this key. Similar sub keys will certainly soon be developed for other families which appear several times in the main key (for example, Clusiaceae and Apocynaceae). The Rubiaceae could be also easily classified in greater detail whereas the identification of the different genera of Lauraceae remains problematic.
The organisation of families in the second part of this book has been reviewed taking account recent advances in genotyping and thus allowing, for better or worse, new classifications in areas where knowledge had been based on scant or outdated taxonomical data.

One can ask also oneself: is it still worthwhile building keys on paper which force the user to follow a predefined path when a computerised multi-entry system would offer the chance of choosing to observe certain characters rather than others which (often through a question of cultural baggage) are considered difficult to observe. Replying to this question pragmatically it is suggested that:

- A fair number of taxonomical characteristics are definitive and must always be considered (if one stays within a given register, for example the vegetative, reproductive or even that of pollen morphology). All that can be done is to defer the more complex observations and enter the key with the more obvious characteristics. Except from choosing between key G and H, this is precisely the method adopted in this guide and to remove any doubt it suffices to consult the general key and the start of the other keys. Given that certain characteristics must be considered, they would also have to be taken into account with a multi-entry computerised key.

- Carrying a bulky portable computer in the field is difficult and such equipment is easily damaged by the high humidity of tropical undergrowth.

- A book generally presents facts in a concise way and the contents can be easily and quickly checked ... one cannot flick through the pages of a PC! This is certainly true regarding such rarely addressed characteristics such as barks and rhitidomes, though computerisation certainly holds promise as regards knowledge of regional flora on at the species level.

Taxonomical identification is not carried out in the same way in the tropics as in the West. Knowledge of flora originating from the tropical belt is still either incomplete or in the process of being refined and hence herbaria are still a principal source of reference. Therefore a relevant vegetative key enables the user to gain time, imagine hunting through a huge herbarium with a sterile specimen in one hand and having no idea as to which family it belongs.

First and foremost this guide is dedicated to botanists who are not particularly specialised in any one flora but who wish to have a certain autonomy and independence of study. It would only be half joking to say that nowadays it is more and more difficult to find the right taxonomist in the right place at the right time!

Introductory remarks - how to use this book

Structure of the book

It remains the same as in the previous edition, the first part comprises the *keys* and a *glossary* which explains and illustrates the terms and concepts used in this section. The second part gives a *general overview of the principal families of ligneous tropical plants*. The two parts can be consulted independently, although the glossary also applies to the second part. When identifying plants, however, the reader will probably find it advantageous to refer initially to part I, Part II (family overview) then being used to check the result provided by the key.

Part I: Keys and glossary

The procedure to be followed is roughly the same as that employed with any dichotomic key. The identification process involves two or three keys, always starting with the *General key*. The result obtained here indicates the next key (*A* to *V*). A third key (*W, X, Y*, or *Z*) is sometimes necessary to identify the family. The difference with regard to a "classical" key lies in the fact that only vegetative characters are used and that particular care has been taken in the glossary to explain and illustrate the essential terms.

Part II: Illustrations of the important families of tropical woody Dicotyledons

The second part of this book presents further details of many families dealt with in the key. These families are divided into thirty-two groups the numbers of which correspond to those mentioned in the keys, such as (9A) for the ULMACEAE, (11B) for the MYRSINACEAE, etc. It helps the user to compare a specimen with a typical representation of the family to which it may belong. In this way, wrong identifications should be detected in most cases. Families not illustrated are not designated by numbers (families of the Gymosperms, Monocotyledons except Palms, and some very small Dicotyledonous families of lowland tropical forests, such as Huaceae, Rhabdodendraceae, etc.).

Botanical knowledge required

This book requires an elementary knowledge of morphology and only a basic knowledge of anatomy. It may seem difficult to use the key at first, as it refers to certain little-known concepts and characters, but it becomes easier with practice. The characters referred to can be seen with the naked eye or with a hand lens, and all are illustrated in the glossary.

Precautions to be taken in the observations of certain characters

It is better to observe the following characters in the field, before bringing the samples back to camp:

- Exudates: These should be carefully checked for by making a tangential cut in the trunk and by sectioning leaves and young twigs.

- Rhytidome characters: these are concerning the peripherical, dead tissues of trunks and branches (as the scaly rhytidome of the Myrtaceae - Combretaceae and the thin slits of the Euphorbiaceae).

- Bark characaters: these refer to the inner living bark, and have to be observed in branches or trunks sufficiently thick (the existence of networked structures must be carefully searched for).

- The monopodial or sympodial character of the branches and, if possible, of the trunk.

- The existence of stems with truly distichous phyllotaxy.

- The odour of the crushed leaves and, for those species without latex, the taste.

- The mechanical properties of the twigs, such as flexibility and rigidity.

In the camp the following characters have to be carefully examined (with a hand lens or against light, if necessary):

- Presence or absence of stipules, these can be very reduced.

- Presence or absence of glands and translucent dots on the leaves.

- Indumentum characters such as the presence of stellate or appressed hairs.

- Leaves with an mbossed or a protruding main vein.

Field of application

This key applies principally to Phanerogams in tropical lowland forests, however it should be useful, to some extent, in forests of warm temperate zones. The identification of the family to which a plant belongs can be guaranteed in 90% of cases covering the woody forest species in the great tropical land masses of America, but somewhat less for Africa and Asia. About one hundred and eighty families are concerned. As Palms (Arecaceae) are a very important group their desciption has received some attention in the present edition, see also [1]* for a comprehensive study of this group. Arborescent Ferns are excluded.

In order to specify more precisely the range of application, it must be pointed out that:

1. The key applies to *ramified woody* plants, however, some monocaulous trees (CORNER's model) are identifiable.

2. Certain very infrequently encountered or highly endemic families have been deliberately excluded as a result of a lack of observation in the field, these are listed in p. 124.

3. Individuals from certain families remain difficult to identify; extra caution is needed for the Euphorbiaceae, Flacourtiaceae, Violaceae and the different families of the order Sapindales (especially Meliaceae vs. Sapindaceae and Anacardiaceae vs. Burseraceae).

4. Some families extending beyond the Tropics can be quite readily identified (where woody species are concerned) in temperate or subtropical regions: Aceraceae, Anacardiaceae, Annonaceae, Apocynaceae, Berberidaceae, Buxaceae, Caprifoliaceae, Ericaceae, Fagaceae, Hippocastanaceae, Hydrangeaceae, Leguminosae, Moraceae, Myrtaceae, Oleaceae, Platanaceae, Proteaceae, Rosaceae (Pomoideae and Prunoideae), Rubiaceae, Rutaceae, Solanaceae, Theaceae, Thymelaeaceae, Tiliaceae, Ulmaceae, Vitaceae.

---

* Numbers in square brackets refer to bibliography.

Annexe keys

The keys W, X, Y and Z have been constructed for the Euphorbiaceae *s.l.*, the Malvaceae *sensu lato*, the Sapindales and the Leguminosae. These four taxa unite a very large number of species and are ecologically important in tropical forests. A more comprehensive identification of the taxa they comprise is therefore very useful.

- *Key W* (for Euphorbiaceae and Pandaceae) allows identification to the subfamily or tribe level, or to a group of related genera (e.g. *Hura*, *Sapium*)
- *Key X* (for Malvaceae) allows the identification to the tribe level (e.g. Durioneae), or to a group of related genera (e.g. *Bombax* and *Pochota*), or to the genus (e.g. *Triplochiton*).
- *Key Y* enables the identification of most of families of the order Sapindales.
- *Key Z* enables, in most cases, the identification of a woody Leguminosae to tribe level.

Prospects for other keys to the tribal or generic level

The individuals of many other families can be identified to the tribal or even generic level, especially if architectural characters are involved. Apart from Euphorbiaceae (*key W*), Leguminosae (*key Z*) and the subfamilies of the Malvaceae (*key X*), the Apocynaceae, Clusiaceae, Icacinaceae, Loganiaceae, Monimiaceae, Thymeleaceae and Rubiaceae could serve as examples of families which identification, without flowers or fruits, is possible to tribe level. Hence the present key might encourage additional work in vegetative identification.

Conventions

Syntax

When an item consists of more than one proposition, these propositions are separated by a full stop. A full stop could be replaced by "and" without any change in the significance of the two propositions: thus "and" is implied; "or" is always non-exclusive.

For example: 3   Branches with spiral phyllotaxy. Resin not orange-coloured.
              3* Branch phyllotaxy not spiral or orange-coloured resin.

thus items **3** and **3***, each with two propositions, are complementary.

Illustrations

Part I contains the figures illustrating the glossary, these figures are numbered from **F1** to **F21**, the drawings marked with small letters. The letter-number combinations in the key refer to these illustrations. For example: **F13 d** = Figure **13**, drawing **d**.

Part II contains the plates illustrating the families, these plates are numbered from **P1A**, **P1B** to **P32A**, **P32B**, the drawings marked with small letters. For example: **P9B c** = Plate **9B**, drawing **c**. These illustrations can be used to check the results given by the key. However, they are generally not referred to in the key.

Optional characters

The propositions concerning optional taxonomical characters are printed between brackets, but if one of them is confirmed this increases the probability of an exact identification. Propositions that are not in brackets should be sufficient for identification.

For example (see *key B*): "(Crushed leaves with resinous smell.)" is a character that may not be detected for some Anacardiaceae (with the relevant characters given by key the at that point).

Abbreviations

cult. = cultivated for its agronomic or timber value, hence possibly outside of its area of origin.
ornam. = cultivated as an ornamental, hence possibly outside of its area of origin.
f. s. spp. = for some species.
L. = leaves or leaflets.
Rhy/Per = rhytidome or periderm.
UE = unit of extension.
V. = veins or venation; V I/II/III = primary / secondary / tertiary veins.

Geographical distribution

The abbreviations for the geographic distribution are:

AM = America (tropical or subtropical); AF = Africa (tropical or subtropical); AS = Asia (tropical or subtropical, incl. New Guinea); EU = Europe; MA = Madagascar; Comoro Islands or Mascarenes; NG = New Guinea; AU = Australia (mainly Queensland); OC = Oceania (incl. New Caledonia). C = central; S = south; W = west. Paleotrop. = wide paleotropical distribution, i.e. AF, AS, AU or OC. Pantrop. = wide tropical distribution, i.e. at least AM, AF, AS.

When a genus is cited as an example, and the geographical distribution is not mentioned, then this taxon exists at least in AM, AF and AS (pantropical distribution).

An exotic species such as a Bilimbi tree cultivated in an American garden could fail to be recognized as Oxalidaceae as the key specifies that this plant grows in Asia. However, in most cases it is posssible to ignore the geographic origins of a taxon and still achieve a correct identification of the family.

Architectural models

The names of the models (see plates *F7, F8, F21*) are always written in capital letters. As these names are very often in brackets (the architectural model thus becoming an optional character), the identification of a model is rarely necessary in order to identify a family (the model is, however, easily identifiable when dealing with a young tree). Figures *F6, F7* and *F8* and associated texts introduce the aspects of architecture that are needed to follow the key. A simplified key to architectural models is given in plate *F21*. Plates ***P*** and associated texts give examples of architectural models for many genera.

Trunk ramification

Occurrence of **immediate** or **delayed** ramification in trunk is mentioned at bottom of caption pages of plates ***P***. These two character states are taxonomically sound, i.e. they are fairly constant at tribe or family level. Trunks are axes which have at most the ability to exhibit immediate ramification, however ramification in branches, especially last order branches, tend to be delayed for all trees. Thus some families (as Lauraceae, Myristicaceae) are of the 'immediate' type, while other (as Menispermaceae, Chrysobalanaceae) are of the 'delayed' type.

Genera cited as examples

The names of identified families are given in capital letter, bold. For each land mass where a family is represented (with the relevant characters given by the key at that point) an effort has been made to cite at least one genus for each main land mass, as in the following case from *key X*:

STERCULIOIDEAE
Theobroma (AM), *Scaphopetalum* (AF), *Heritiera* (AF,AS)

This means that *Theobroma* is exclusively American, *Scaphopetalum* is exclusively African and that *Heritiera* is paleotropical.

If no region is mentioned after the name of a genus, this means that the latter has a pantropical distribution (i.e. at least in America, Africa and Asia), as in the following case from *key C*:

P9A **ULMACEAE**
Gironniera (AS), *Celtis, Trema*

In a few cases no genus is cited as an example; in this case the name of the family is printed in ordinary type. This indicates that plants with the characters described probably exist, but lack of sufficient observations prevents a definitive conclusion (for example in *key D* for stipulated, simple-leafed Araliaceae).
P27A ARALIACEE

Taxonomic information can sometimes be insufficient to be able to cite an example of genus for one of the large land masses where the family might exist, e.g. in *key D*:
  20* Branches sympodial (F6 f,g,h). (PREVOST).   P19 **EUPHORBIACEAE**
e.g. *Aporusa* (AS,OC)
and one must refrain from deducing that Euphorbiaceae conforming to the PREVOST's model are absent in Africa.

If a genus is mentioned without being preceded by "e.g.", this genus alone is identified at this point in the key, as for example: P25A **THYMELAEACEAE**
Gonystylus (AS,OC)

When a genus is mentioned by the key as an example for all the characters given, this does not imply that the characters exist in all species of the genus. Nevertheless, the characters stipulated are in general fairly typical of a genus. This convention also applies to the glossary.

Sources

A prototype key was established after several journeys (totalling a period of eight months) in French Guiana, Guadeloupe, Malaysia and Indonesia (Sumatra and the Bogor botanical garden). Visits to the botanical gardens in Singapore, the Villa Les Cèdres (St-Jean-Cap-Ferrat, France), the Villa Thuret (Antibes, France) and the Villa Hanbury (Ventimiglia, Italia) also resulted in numerous contributions.

The key was subsequently tested thoroughly for six months in the Bogor botanical garden and in places very distant from the previous ones: Manusela national park of Seram (Indonesia), forest reserve of Campo (Cameroon), forests of the Sierra Imataca, Gran Sabana and upper Orinoco (Venezuela), and in the national park of the Montagne d'Ambre (Madagascar).

The present key is an improved version of the first edition after other tests in Venezuela, in South Brasil (Isle of Santa Catarina), and in South India (Kerala), Francis Hallé gave other examples of architectural models and helped to correct some inaccuracies of the first edition. Bibliography of *'The state of the art in vegetative identification'* [2] was also the source of useful informations.

The observations were made on living plants belonging to more than a thousand genera of one hundred and sixty families, and on herbarium specimens (Conservatoire Botanique de Genève, Fundación Instituto Botánico de Venezuela, Forest Research Institute of Malaysia). Compilation of other floras or field guides have also contributed to data recording (Gentry's *Field Guide for Northwestern South America*, Aubréville's *Flore Forestière de la Côte d'Ivoire, Tree Flora of Malaya*, etc.). The recording and interpretation of information taken in the field was done with the aid of notes and drawings; characters observed in herbaria were noted on file cards. All the line drawings and the photos are the work of the author.

PART I

KEYS and GLOSSARY

General key

				Key	Page
1	Tree or shrub (or plant weakly prostrate,, F5 e).				
	2 Distichous or spiral phyllotaxy, F9 (maybe scale-leaves subtending green stems).				
		3 Venation pinnate, palmate or campylodromous, F17,18.			
			4 Leaves simple.		
			5 Latex, resin or coloured exudate (or copious gum) in bark, trunk or leaves (see text of F3).		
			6 Leaves stipulate (including minute stipules, F13).	A	13
			6* Leaves not stipulate (even the youngest leaves).	B	15
			5* No latex, resin, or coloured exudate.		
			7 Leaves stipulate, F13. (Stipules can be vestigial).		
			8 Some twigs with phyllotaxy perfectly distichous, F9 h.	C	18
			8* All stems with more or less spiral phyllotaxy or trunk not branched.	D	27
			7* Leaves not stipulate (even the youngest leaves).		
			9 Venation palmate or leaves trinerved, F18 c,d,e.	E	34
			9* Venation pinnate, F18, f,g,h.		
			10 Some twigs with perfectly distichous phyllotaxy, F9 h.	F	37
			10* All stems with more or less spiral phyllotaxy or trunk not branched.		
			11 Leaves grouped (F9 b,c), (possibly short shoots, F11 p), or branches distinctly sympodial (F6 e,f). If not branched treelet: CHAMBERLAIN's model (F8 c).	G	42

		Key	Page

      11* Developed, not scaly, leaves evenly distributed on twigs. Branches monopodial (F6 a) or, if sympodial, then including some units of extensions arranged in a monopodial sequence (F6 g). Possibly a not branched treelet (CORNER's model).   **H**   48

  4* Leaves compound, F14.

    12 Leaves stipulate, F13.   **J**   54

    12* Leaves not stipulate (even the youngest leaves).   **K**   56

 3* Venation parallelodromous, F17 f, or leaves as in, F15 a1,a2,a3.   **L**   59

2* Opposite or whorled phyllotaxy, F9, (at least in branches). (Maybe scales-leaves subtending green stems).

  13 Leaves simple.

    14 Latex, resin or coloured exudate (or copious gum) in bark, trunk or leaves.   **M**   61

    14* No latex, resin, or coloured exudate.

      15 Leaves stipulate, F13. (Stipules can be vestigial).   **N**   63

      15* Leaves not stipulate (even the youngest leaves).

        16 Nodes with an interpetiolar ridge, F13 y.   **O**   68

        16* Nodes without interpetiolar ridge.   **P**   73

  13* Leaves compound, F14.   **Q**   78

1* Liana, weakly prostrate plant, F5 e, (hemi)epiphyte or parasitic plant.

  17 Latex, resin or coloured exudate (or copious gum) in bark, trunk or leaves.   **R**   80

  17* No latex, resin or coloured exudate (see text of F3).

    18 Leaves simple, not deeply divided or leaves absent (plant with green stems).

      19 Phyllotaxy distichous or spiral.

		Key	Page
20	Venation palmate, F18 c, trinerved, F18 d, supra-trinerved, F18 e, or campylodromous, F17 e.	**S**	82
20*	Venation pinnate, F18, f,g,h, parallelodromous F17 f, or leaves absent.	**T**	85
19*	Phyllotaxy opposite or whorled (at least in branches).	**U**	91
18*	Leaves compound or leaves simple and deeply divided.	**V**	95

Annexe keys

	Key	Page
Key for the principal groups of genera of the EUPHORBIACEAE *s.l.*	**W**	98
Key for the principal groups of genera of the MALVACEAE *s.l.*	**X**	106
Key for the families of the self-supporting SAPINDALES.	**Y**	110
Key for the tribes of the woody LEGUMINOSAE.	**Z**	115

Key A

1 Leaves entire.
  2 Stipule encircling the whole stem or near so, leaving an annular scar (F13 m).
    3 No resin. Latex white, pale yellow, pink, brown or black. (Venation densely reticulate, F 18a1).
      Exudate uncoloured, or brownish, or black: Cecropioideae, e.g. *Cecropia, Coussapoa, Pourouma*. Exudate white, yellow or pink: Moroideae, e.g. *Artocarpus, Brosimum, Ficus, Perebea, Pseudolmedia*. See p. 198 for examples of architectural models. **P9B MORACEAE**

    3* Wood or bark resinous.
      AS. Monopodial branches with distichous phyllotaxy (F9 h).
      P18A DIPTEROCARPACEAE
      Dipterocarpus (AS)

  2* Stipule and its scar different.
    4 Venation brochidodromous (F17 a), densely reticulate (F 18a1). Petiole not pulvinate. (RAUH: *Artocarpus*; TROLL: *Brosimum, Streblus, Trophis*; ROUX or COOK: *Castilla*, spiny shoots: *Chlorophora, Maclura*). **P9B MORACEAE**
      e.g. *Brosimum, Castilla, Pseudolmedia* (AM)
      Trophis (AM,MA,AS), *Artocarpus, Streblus* (AS)

    4* Venation or petiole different. See 5a, 5b and 5c:
      5a Resin (see text of *F 3*) more or less yellow or bark resinous.
        6 AM. Branches erect with spiral phyllotaxy (F9 b,c) or orange resin. (Venation scalariform, F17 k; translucent dots). **P15A CLUSIACEAE**
        Kielmeyeroideae, e.g. *Mahurea* (AM)

        6* AS, OC. Branches with distichous phyllotaxy (F9 h). Bark or wood resinous. (V II abruptly curved near the margin, F18 p; V. camptodromous or scalariform, F17 k). **P18A DIPTEROCARPACEAE**
        e.g. *Anisoptera, Shorea* (AS)

      5b Cut bark producing a coloured exudate.
        7 Copious red exudate in trunk. Lenticels retaining a circular form during the thickening of the rhytidome (F2 j). (Trunk with flat buttresses).
        P20 LEGUMINOSAE
        e.g. *Inocarpus* (NG,OC)

        7* Exudate yellow, orange or red (in small quantity). Lenticels different or none. (Foliar glands; twigs flexible). **P19 EUPHORBIACEAE**
        e.g. *Croton*, key W

      5c Latex (see *F3*) white or exudate faintly coloured.
        8 Rhy/Per forming thin longitudinal slits (F2 e,f, a transient character). (Petiole distally pulvinate, F15 p, or stellate or peltate hairs, or L. glandular f. spp.). **P19 EUPHORBIACEAE**
        Numerous genera, *key W*

        8* AM, AS, OC. Rhy/Per different. Petiole without pulvinus. Neither stellate hairs nor glands. Venation pinnate. Latex white (rarely coloured). (AUBREVILLE: e.g. *Palaquium*; RAUH: e.g. *Palaquium*).

                  P12B **SAPOTACEAE**
                    e.g. *Ecclinusa* (AM)
             Mimusops, Palaquium, Payena (AS, OC)

1* Leaves serrulate or lobate.
  9 Stipule encircling the whole stem, leaving an annular scar (F13 m).
    (Leaves almost palmately compound (F14 a) for some Cecropioideae: *Cecropia* spp,
    Musanga; RAUH).   P9B **MORACEAE**
          e.g. *Artocarpus* (AS, cult.), *Cecropia* (AM), *Musanga* (AF)

  9* Stipule and scar different.
    10 Exudate orange. Leaves palmately lobate (F14 a), without glands. Hairs simple.
      (Rhy/Per without longitudinal slits; bark fibrous, F3 a). P17B **BIXACEAE**
                           Cochlospermum
    10* Characters different.
      11 Venation brochidodromous (F17 a), densely reticulate (F 18a1). (Latex
      white, yellow, red or brown, TROLL).   P9B **MORACEAE**
                        e.g. *Clarisia* (AM)

      11* Venation different. (Latex white, yellow, red or uncoloured; Rhy/Per
        forming thin longitudinal slits (F2 e,f); L. glandular, e.g. F19 h: *Sapium*, or
        not glandular).
        12 Latex white.   P19 **EUPHORBIACEAE**
                   e.g. *Manihot* (AM, cult.), *Sapium*
                         see *key W*
        12* Latex other than white.
            (Old leaves turning orange frequent in *Croton*; latex opalescent
            and trunk spiny: *Hura*; latex red and petiole with distal glands,
            e.g. *Pausandra*)   P19 **EUPHORBIACEAE**
              e.g. *Hura* (AM, ornam.), *Pausandra* (AM), *Croton*
                         see *key W*

Key B

1 Venation pinnate (F18 f,g,h) or leaves supratrinerved (F18 e).
  2 Bark with red, reddish or orange in colour (sometimes appearing after a moment) or astringent exudate. (Examine stems old enough).
    3 Exudate astringent (harsh or constricting).
Branches disposed in tiers (F7 d), with distichous phyllotaxy (F9 h), (MASSART). **P4A MYRISTICACEAE**
e.g. *Virola* (AM), *Pycnanthus* (AF)
Myristica (AF,AS,AU, cult.)

    3* Exudate not astringent.
      4 Petiole distally enlarged (F15 m). Young internodes angular (F11 j). (ROUX, camptodromous venation). **P7B OLACACEAE**
e.g. *Coula* (AF, cult.)

      4* Petiole not enlarged.
        5 (ROUX, no lenticels). **P12B SAPOTACEAE**
e.g. *Chrysophyllum* (e.g. AM)

        5* Latex reddish. (MASSART, Fimbrial vein, F17 q, lenticels circular). **P31B APOCYNACEAE**
Aspidosperma (AM)

  2* Exudate in bark not orange-red in colour and not astringent.
    6 Latex mostly in bark, white, coloured or turning brown or black (in some cases hours are needed for latex to become black). V II abruptly curved near the margin (F18 p). (Crushed leaves with resinous smell). **P24A ANACARDIACEAE**
e.g. *Anacardium* (AM, cult.), *Mangifera* (AS, cult.)

    6* Latex or venation different or exudate resinous.
      7 Some twigs with distichous phyllotaxy (F9 h). Periderm retaining a green shade on numerous internodes. Young Internodes angular (F11 j). **P7B OLACACEAE**
e.g. *Heisteria* (AM,AF)

      7* Characters different.
        8 Latex (see F3) not resinous (in some cases uncoloured but becoming opalescent). (For some APOCYNACEAE, no latex in the trunk).
          9 Branches with modular architecture or seeming bi- or trifurcate (F6 e, F8 f,g,h).
            10 Branches sympodial by apposition (F6 e). (Latex white).
              11 Leaves entire. (Leaf folding conduplicate, F16 b1). **P12B SAPOTACEAE**
e.g. *Manilkara, Pradosia* (mainly AM), *Sideroxylon*

              11* Leaves serrulate. (Rhy/Per forming thin longitudinal slits, F2 e,f). **P19 EUPHORBIACEAE**

            10* Branches not sympodial by apposition.

Leaves entire. Latex white. (L. folding plane-curved, F16 a;. KORIBA, LEEUWENBERG, e.g. *Cerbera, Himatanthus, Plumeria.* PREVOST, e.g. *Geissospermum.*

          P31B **APOCYNACEAE**
      e.g. *Himatanthus, Geissospermum* (AM)
      Plumeria (AM, ornam.), *Cerbera* (AS, OC)

9* Branches monopodial, or sympodial as in (F6 f,g).
  12 Lenticels circular or latex turning brownish. Typical shiny, pale green aspect of the underside of L. (MASSART, RAUH; intra-marginal vein present, F18 m; f. s. spp.).

          P31B **APOCYNACEAE**
          e.g. *Aspidosperma* (AM)

  12* Lenticels different or none. Latex not turning brownish. Underside of L. different (RAUH, ROUX, TROLL). See 13a, 13b, and 13c:
    13a Tree. Venation brochidodromous (F17 a) or scalariform (F17 k). Apices with brown-red indumentum; latex white, very rarely red).
      TROLL: e.g. *Micropholis*; ROUX: e.g. *Chrysophyllum*; RAUH: e.g. *Pouteria, Chrysophyllum.*
      AUBREVILLE: e.g. *Faucherea, Manilkara.*

          P12B **SAPOTACEAE**
       e.g. *Chrysophyllum, Pouteria*
    Manilkara, Micropholis (AM), *Faucherea* (MA)

    13b Shrub or small tree. Rhy/Per forming longitudinal slits (F2 e,f); Latex white. V. brochidodromous. (Twigs flexible; vestigial stipules; f. s. spp.). (Petiole distally pulvinate: e.g. *Actinostemon*; extratropical: *Euphorbia*).

          P19 **EUPHORBIACEAE**
     e.g. *Actinostemon* (AM), *Euphorbia*, Key W

    13c Shrub (with bell shaped flowers). Basal secondary veins grouped (F18 h). P29A **CONVOLVULACEAE**
              e.g. *Ipomoea*

8* Resin or gum (see *F3*).
 14 Resin.
  15 AM. Bark resinous. Resin more or less yellow, pink, or uncoloured.
    Venation more or less scalariform (F17 k): e.g. *Caraipa* (AM). V. not scalariform; branches sympodial. Numerous parallel V II (F17 g): Kielmeyeroideae.

          P15A **CLUSIACEAE**
        e.g. *Caraipa, Kielmeyera* (AM)

15* AS, AU, OC. Whitish sticky resin exuding from cut twigs and petioles. Apical bud scaly. Leaves clustered in pseudowhorls.
P13B **MYRTACEAE**
e.g. *Tristania* (AS,AU,OC)

14* Gum.

16 Gum sticky. Venation camptodromous (F17 c,d).
P13B **COMBRETACEAE**
e.g. *Anogeissus* (AF,AS)

16* AF, AS, AU, OC. Gum aqueous. Venation brochidodromous (F17 a).
P27A **PITTOSPORACEAE**
e.g. *Pittosporum* (Paleotrop., ornam.)

1* Venation palmate (F18 c) or leaves trinerved (F18 d). Rhy/Per forming thin longitudinal slits (F2 e,f), (this character is transient) or apex of petiole or base of lamina glandular (F19 j,k). (Latex uncoloured f. s. spp.; minute stipules?).
P19 **EUPHORBIACEAE**
e.g. *Jatropha* (cult.), key W

Key C

1 Stipular or petiolar scar encircling the whole stem (F13 m,n,s,t,u).
   2 Persisting cylindrical stipule (ochrea), (F13 s,t,u).
      3 Venation pinnate. (F18 f,g,h). (Short twigs (F 11 p), e.g. *Ruprechtia*).
                                                                      P6B POLYGONACEAE
                                                      e.g. *Coccoloba, Ruprechtia, Triplaris* (AM)

      3* V. palmate (F18 c,d). Stipule trumpet-shaped. (F13 u). External wood furrowed, network of fibers in bark (F3 j).   **PLATANACEAE**
                                                   Platanus (North hemisphere)
   2* Stipule different.
      4 Stipule enveloping the terminal bud, narrow and elongated (F13 n) or flattened and adnate to the petiole (F13 h).
         5 Stipule not adnate to petiole. See a), b), and c):
            6a AF, AS. Leaves glabrous.   **P22A IRVINGIACEAE**
                                                  e.g. *Irvingia* (AF,AS)

            6b MA. Buds covered with appressed hairs (F20 e).
                MA. TROLL. False longitudinal veins for some species (F16 f).
                                                  SARCOLAENACEAE
                                            e.g. *Leptolaena, Sarcolaena* (MA)

                AS. (TROLL). Leaves without such veins.   **P9A ULMACEAE**
                                                    e.g. *Gironniera* (AS)

                AS. (TROLL). Twigs distinctly lenticellate.
                                                  P21A CHRYSOBALANACEAE
                                                         e.g. *Atuna* (AS)

            6c MA. Leaves aromatic. No appressed hairs.   **DIEGODENDRACEAE**
                                                               Diegodendron (MA)
         5* Stipule adnate to the petiole.
            7 AS. Leaves trinerved and palmately lobate. Stipules flattened, one appressed to the other.   **P8A HAMAMELIDACEAE**
                                                             Exbucklandia (AS)

            7* AF, MA, AS. Venation pinnate. Plant more or less herbaceous. Stems kneed at the nodes.   **P10B LEEACEAE**
                                                       Leea (AF,MA,AS, ornam.)

   4* Stipule enveloping the terminal bud, large, not narrow and elongated, (F13 m), not flattened.
      8 Leaves trinerved (F17 h, F18 d) or venation palmate (F18 c).
         9 Leaves palmately lobate. (MASSART).
            AF. Bark with network of fibres (F3 a,b,c,d).
                                                  P18B STERCULIACEAE
                                               Triplochiton (AF), see *Key X*
         9* Leaves different.

10 AS. Leaves faintly trinerved, entire. (TROLL). Stipules relatively large. **P8A HAMAMELIDACEAE**
Maingaya (AS)

10* Leaves (faintly) trinerved (F17 h), entire or serrulate. Stipules small. Venation camptodromous, densely reticulate (F 18a1), (with transparent dots for some *Celtis*). (ROUX, TROLL). **P9A ULMACEAE**
Gironniera (AS), *Chaetachme* (AF,MA), *Celtis*

8* Venation pinnate (F18 f,g,h).
  11 V II forming broken-line loops (F17 a2). (Bark with network of fibres (F3 a,b,c) or aromatic, f. s. spp.). (ROUX, TROLL). **P4A MAGNOLIACEAE**
e.g. *Magnolia* (AM,AS), *Elmerrillia* (AS)

  11* AS. V II forming regularly curved loops. (F17 a1). (ROUX, MASSART; resin in old branches). **P18A DIPTEROCARPACEAE**
Dipterocarpus (AS)

1* Stipular or petiolar scar not completely encircling the stem.
  12 Bark with **network** of fibres (F3 a,b,c,d).
    13 Leaves trinerved (F18 d) or venation palmate (F18 c).
      14 Base of internode swollen above the node (F11 g). Hairs not stellate. (Self-supporting form of a liana). (Leaves serrulate). **P10B VITACEAE**
e.g. *Tetrastigma* (AS,AU)

      14* Base of internode not enlarged or indumentum of stellate or peltate hairs (F20 k,m,n,p).
        15 Petiole pulvinate (F15 p) or hairs stellate or peltate. **P18B MALVACEAE** *s.l*
key X

        15* Petiole not pulvinate. Hairs simple or none. See a), b) and c):
          a) Tree growing in open places. Branches plagiotropic. (ROUX, TROLL; L. trinerved, F17 h). **P9A ULMACEAE**
e.g. *Trema*

          b) AS. Leaves almost white underneath. Bark network with very elongated meshes (F3 c). Leaves serrulate. Venation densely reticulate. (TROLL). **P9A URTICACEAE**
e.g. *Leucosyke* (AS,OC)

          c) AM. Leaves strongly asymmetric. (TROLL). **P18A MUNTINGIACEAE**
Muntingia (AM)

    13* Venation pinnate (F18 f,g).
      16 Bark that is pulled off in fibrous strips (somewhat sticky). Stipules minute. **P13A LECYTHIDACEAE**
e.g. *Eschweilera* (AM)

      16* Bark different. L. with numerous, thin parallel V II. **P15B OCHNACEAE**
e.g. *Ouratea* (AM)

12* **Bark different** (with broad strips of fibres in some OCHNACEAE).
  17 Numerous thin and parallel V II (F17 g) or V II arched, delimiting areas containing numerous intercostal V II perpendicular to the midrib (see P7B f).
    18 Branches with rhythmic growth marked by scale-leaves (F6 a). Leaves entire or denticulate. V I, V II protruding above the lamina (F18 r). (V I longitudinally striate, F18 r; intrapetiolar stipules, F13 b; f. s. spp.).
         P15B **OCHNACEAE**
         e.g. *Ouratea*

    18* AS. Intramarginal vein (F18 k,m). Branches with weak rhythmic growth. Leaves entire (aromatic). (ROUX). P18A **DIPTEROCARPACEAE**
         Dryobalanops, *Cotylelobium* (AS)
  17* Venation different.
    19 Petiole distally pulvinate (F15 p,q), (long enough for this character to be perceptible), or petiole distally enlarged (F15 m).
      20 Petiole articulate with the lamina (F15 q) or leaves bilobate. (Lenticels becoming transversely elongated, F2 k, venation palmate; f. s. spp.).
         P20 **LEGUMINOSAE**
         e.g. *Flemingia* (AF,AS), *Bauhinia* (key Z)

      20* Petiole not articulate with lamina. Leaves not bilobate.
        21 AS. Venation camptodromous with V II running parallel with the margin (F17 d). (ROUX, Venation embossed on upper side of blade, F 18s). V. scalariform, F17 k: e.g *Shorea*, *Marquesia*. Hairs in groups, these like stellate hairs, F20 h: e.g. *Shorea*. MASSART, V II parallel: e.g. *Vateria*. Domatia (F20 t): e.g. *Hopea*. P18A **DIPTEROCARPACEAE**
         e.g. *Shorea*, *Anisoptera*, *Hopea*, *Vateria* (AS)
        21* Characters different.
          22 Rhy/Per forming thin longitudinal slits (F2 e,f).
              (TROLL; quite supple young twigs).
         P19 **EUPHORBIACEAE**
         e.g. *Antidesma* (AF,AS)

          22* Rhy/Per different.
            23 Models of ROUX, COOK or TROLL. Venation pinnate. No glands. (Brittle young twigs; branches monopodial). P16A **FLACOURTIACEAE**
         e.g. *Hydnocarpus* (AS)

            23* AM. Brazil. Leaves trinerved Underside of lamina with 2 glands at base of the midrib. Serial buds (F12 c). **PERIDISCACEAE**
         Peridiscus (AM)

    19* Petiole not distally enlarged or very short in relation with blade length or with its base enlarged (F15 n).
      24 Leaves glandular or stem glandular near the petiole (see *F19*). Caution! Domatia (F20 t) are not considered here.

25 Leaves glandular.
    26 AF. One gland on the midrib (F19 t) at the base of the lamina.   P18A **DIPTEROCARPCEAE**
Marquesia, Monotes (AF)
26* Characters different.
    27 Petiolar glands (F19 f,g,h) or margin glandulate (F19 m,q). If glands as in F19 m: *Chrysobalanus, Parinari*. If no appressed hairs and glands scattered below the lamina: *Hirtella* (AM).
    28 Branches stout, not flexible. Leaf folding conduplicate (F16 b). Lenticels becoming transversely elongated (F2 k). (Bark with smell of bitter almond f. s. spp.). (TROLL, RAUH).
P21A **CHRYSOBALANACEAE**
e.g. *Licania* (AM, AF), etc.
P21A **ROSACEAE**
Prunus (EU,AS)

    28* Leaf folding involute (F16 d). (Bark with orange sclerenchymatous inclusions, F4 t; base of lamina glandular, F19 q; margin of lamina hyalin; f. s. spp.).   P16A **FLACOURTIACEAE**
e.g. *Banara, Laetia* (AM)

  27* Glands distributed otherwise (F19 j,k,n,p,w). See 29a, 29b, and 29c:
    29a AS. Glands scattered below the lamina or arranged near the margin in two symmetrical loops. Branches monopodial. Venation camptodro-mous (F17 c,d). (ROUX).
P18A **DIPTEROCARPACEAE**
e.g. *Shorea, Vatica* (AS)

    29b AS. Glands scattered below the lamina. Branches monopodial.   P21A **ROSACEAE**
e.g. *Prunus* (AS)
    29c Characters different.
        30 Buds with indument of appressed hairs (F20 e). Glands close to the midrib (F19 w). Lenticels readily observable (in longitudinal rows). (Young internodes angular; ROUX, MANGENOT).
P25B **DICHAPETALACEAE**
e.g. *Tapura* (AM), *Dichapetalum*

        30* Glands at the base of the lamina or on the main vein (F19 j,n,p). (Lenticels faintly marked; Rhy/Per forming thin longitudinal slits, F2 e,f).
P19 **EUPHORBIACEAE**
e.g. *Aporusa* (AS,OC), key W

25* Stem with glands close to the petiole (F19 a).
  31 AF. Plant with hollow internodes inhabited by ants. Stipules glandular. (ROUX). P16B **PASSIFLORACEAE**
Barteria (AF)

  31* Glands in stipular position (F19 a). (See also *key F*).
P26A **POLYGALACEAE**
e.g. *Securidaca* (AM,AF)

24* Petiole, lamina and stem without glands.
  32 External wood furrowed (F3 f,g).
    AM, AS. Branch collars (F2 g,h). P8A **FAGACEAE**
e.g. *Lithocarpus* (AS)

    Furrowed wood uncommon in the following families:
    Leaves entire. (ROUX; bark quite fibrous and weakly aromatic; petiole slightly enlarged and twisted, F15 m).
P18A **DIPTEROCARPACEAE**
e.g. *Shorea* (AS)

    Old branches stout and not flexible. Wood becoming very hard in old trunks. Branches sympodial. (Twigs lenticellate, TROLL). (Bark red-brown below the rhytidome; f. s. spp.; stipules inserted on petiole: e.g. *Licania*).
P21A **CHRYSOBALANACEAE**
e.g. *Licania* (AM), *Parinari*

    Petiole distally pulvinate (F15 p). (V. palmate or glands f. s. spp.). P19 **EUPHORBIACEAE**
e.g. *Aporusa* (AS), see *Key W*

  32* External wood not furrowed.
    33 Growth and ramification rhythmic readily observable due to short internodes on monopodial portions of twigs. Twigs not phyllomorphic. See 34a, 34b and 34c:
      34a Stipules intrapetiolar (F13 b). Leaves entire. V. not densely reticulate (F 18 a2). (MASSART, RAUH or sympodial trunk). (Two false longitudinal veins, F16 f). P22B **ERYTHROXYLACEAE**
Erythroxylum

      34b Stipules rarely intrapetiolar. Midrib or V II protruding above the lamina. V II or V III thin and numerous. (MASSART; V I striate, F18 r; leaves denticulate; f. s. spp.). P15B **OCHNACEAE**
e.g. *Ochna* (AF,AS), *Ouratea*

      34c Stipules not intrapetiolar and venation different.
      35 V II protruding on upper side of blade (F 18r).

36 Paleotrop. Stipular scars more or less as broad as the stem. Base of UE with scale-leaves. (TROLL).   P17A **VIOLACEAE**
Rinorea (MA,AS)

36* AM. Base of UE with scale-leaves. V II quite rectilinear and parallel, ending in a intramarginal vein (F18 k,m). (Young leaf folding involute, F16 d).
P22B **IXONANTHACEAE**
Ochthocosmus (AM)

35* V II embossed on upper side (F 18s). (Crushed leaves giving out an almond smell).
P21A **ROSACEAE**
e.g. *Prunus* (AS)

33* Growth of branches weakly rhythmic or branches sympodial or branches resembling compound leaves (~ phyllomorphic branches). See **37**.

37 Branches with UE forming monopodial series (F6 a,b,g) or L. with translucent dots.
38 AS. Leaves entire. Petiole bending distally (F15 m). (ROUX; bark somewhat fibrous). (Venation camptodromous and scalariform (F17 k): e.g. *Isoptera, Shorea*, or brochidodromous, e.g. *Hopea*).   P18A **DIPTEROCARPACEAE**
e.g. *Hopea, Isoptera, Shorea* (AS)

38* Leaves serrulate or petiole different.
  39 Venation more or less scalariform (F17 j,k).
    40 Venation protruding at upper side of blade (F18 r).
      41 Leaf folding involute (F16 d). V. pinnate (F18 f,g).
L. serrulate, e.g. *Casearia*. (Lenticels crossed by a longitudinal slit, F2 m; margin of lamina hyaline; bark as in F4 t; f. s. spp.).
P16A **FLACOURTIACEAE**
e.g. *Laetia, Ryania* (AM), *Casearia*

      41* Young L. folding not involute.   P18A **DIPTEROCARPACEAE**
e.g. *Shorea* (AS)

    40* Venation embossed at upper side of blade (F18 s).
      42 Leaves trinerved (F18 d).
        43 V. densely reticulate (F 18a1).
          44 Young internodes not angular. (Short twigs transformed into spines: *Celtis*; L. serrulate, e.g. *Trema*). (Bark that comes out in fibrous strips; asymmetrical L. f. s. spp.).
P9A **ULMACEAE**
e.g. *Celtis, Trema*, See also MORACEAE (in dry season)

          44* Leaves entire (Lenticels becoming transversely elongated, F2 k).   P20 **LEGUMINOSAE**
Bauhinia

43* Venation not densely reticulate (F 18a2). Bark not fibrous. (L. serrulate; Young internodes angular; stipule modified into a spine, F13 w and V. plicate, F16 b2: *Ziziphus*).
P10A **RHAMNACEAE**
e.g. *Ventilago* (AF,MA,AS), *Ziziphus*

42* Venation pinnate (F18 e, f,g).
45 AS. Venation densely reticulate.   P9A **ULMACEAE**

45* AM. V. not densely reticulate. Leaves of young tree serrulate, with long hairs (ROUX; trunk sympodial in the adult tree). Young internodes grooved or angular (F11 b,h).
GOUPIACEAE
Goupia (AM)

39* Venation not at all scalariform.
46 Leaves serrulate with V II secant (F17 p).   P8B **ROSACEAE**
e.g. *Sorbus* (North temperate)

46* Leaves different.
47 Stipular scar narrow (or minute).
48 Rhy/Per forming thin longitudinal slits (F2 e,f). Leaf folding not involute.
49 Leaves entire. (Branches like compound L.: *Phyllanthus*; young internodes angular: e.g. *Drypetes*). (Twigs flexible).   P19 **EUPHORBIACEAE s.l.**
Key W

49* Leves serrulate. Shrub (TROLL). Buds hairy. Twigs flexible.   P17A **VIOLACEAE**
Hybanthus (AM)

48* Rhy/Per different or L. folding involute. See 50a, 50b, and 50c:
50a Rhy/Per different. Leaf folding involute (F16 d) and L. with a hyaline margin. (ROUX, TROLL, sclerenchymatous orange inclusions in bark, F4 t; twigs brittle; petiole or lamina glandular; translucent dots; f. s. spp.). (Including LACISTEMATACEAE).
P16A **FLACOURTIACEAE**
e.g. *Lacistema* (AM), *Hydnocarpus* (AS), *Casearia*

50b Rhy/Per without longitudinal slits, many lenticellate. Leaf folding conduplicate (F16 b1). Leaves entire: *Distylium*. L. serrulate: *Hamamelis*.
P8A **HAMAMELIDACEAE**
e.g. *Hamamelis*, *Distylium* (AM, AS)

50c AM. L. spiny-serrulate, with a hyaline margin. Periderm retaining a green shade on numerous internodes. (TROLL).
P20 **LEGUMINOSAE**
Lecointea, *Zollernia* (AM), *key Z*

        47*   Stipular scar broad in relation to the stem. (TROLL).
                                                                                   P17A VIOLACEAE
                                                                                                  Rinorea (AF?,AS)
37*   Branches distinctly sympodial (F6 f). Leaves without translucent dots.
      51   Branches plagiotropic. Basal part of some apical branches becoming erect to form the trunk (TROLL, F8 k). See 52a, 52b and 52c:
          52a   Leaves entire, not mucronate. Lenticels well developed. V. pinnate. (V. scalariform, F17 k: *Parinari*). (Bark and wood very hard in old trunks). (Intrapetiolar stipules f. s. spp.).   **P21A CHRYSOBALANACEAE**
                                            e.g. *Couepia* (AM), *Hirtella* (AM,AF), *Licania* (mainly AM)

          52b   Leaves entire. Midrib well developed, ending in an indentation or a mucro (F14 c). (Stipules spiny, F13 w; TROLL; f. s. spp.).
                                                                  P16B CAPPARIDACEAE
                                                                                                        e.g. *Capparis*
          52c   Not as in a) or b):
              53   Young twigs flexible.   **P19 EUPHORBIACEAE**
                                                        e.g. *Chaetocarpus* (AM), *Savia* (AM,AF,MA), see *key W*

              53*   Young twigs brittle. L. with a hyaline margin. (TROLL; sclerenchymatous inclusions in bark, F4 t).
                    L. trinerved, serrulate. Stipules foliaceous: *Prockia* (AM). L. entire, trinerved. Stipules small: *Lunania* (AM). Spines: e.g. *Xylosma*.
                                                                      P16A FLACOURTIACEAE
                                                                     e.g. *Scolopia* (Paleotrop.), *Homalium*
                                                                         Aphloia (AF,MA), **APHLOIACEAE**

      51*   Branches different (see *F7, F8*), (ROUX, PETIT, MASSART, FAGERLIND) or trunk modules at first erect, later plagiotropic (MANGENOT, F8 j).
          54   Indumentum of appressed or very numerous hairs.
              55   Buds or underside of lamina with appressed hairs (F20 e). Branches sympodial. (ROUX, MANGENOT; self-supporting form of a liana f. s. spp.). (Glands really absent?).   **P25B DICHAPETALACEAE**
                                                                       e.g. *Tapura* (AM,AF), *Dichapetalum*

              55*   AM, AS. Hairs short, erect (F20 f), not appressed. Branches sympodial. Young leaves almost white underneath.   **TRIGONIACEAE**
                                                                 Euphronia (AM), *Trigoniastrum* (AS)
      54*   Indumentum different.
          56   Sheathing petiolar base (F15 a1).
                  Branches consisting of short modules, each with only one scale-leaf and one developed leaf. Nodes swollen (F11 h). All parts aromatic. (PETIT).   **P4B PIPERACEAE**
                                                                                                    Piper
          56*   Petiolar base different.
              57   Trunk with branches quite regularly spaced (F7 a,b,c).
                  58   AF. Hairs stellate (F20 k,m). Venation densely reticulate. Bark and L. smelling garlic (MANGENOT).   **HUACEAE**
                                                                                  e.g. *Afrostyrax* (AF)

58* Not in AF or characters different. See 59a, 59b, 59c:

    59a Young twigs angular (F11 h). V. not densely reticulate (F 18 a2). No hyaline margin (ROUX, twigs supple, Rhy/Per not forming thin longitudinal slits, F2 e,f). **P19 EUPHORBIACEAE**
e.g. *Drypetes*, see *Key W*

    59b Leaves without laticiferous threads, but with a hyaline margin. V. densely reticulate.(ROUX, L. serrulate f. s. spp.). **P16A FLACOURTIACEAE**
e.g. *Casearia*

    59c AM. Trunk sympodial. Twigs brittle. Laticiferous threads in leaves. (F4 u). Rhy/Per forming thin longitudinal slits, F2 e,f. Leaves (sub)glabrous. **GOUPIACEAE**
Goupia (AM)

57* Branches more or less arranged in tiers (F7 d,e,f,g, F8 h).

  60 Petiole enlarged at its base (F15 n). (Laticiferous threads in L., F4 u). **P22B HUMIRIACEAE**
e.g. *Sacoglottis* (AM,AF)

  60* Petiole different. Blade different. (Maybe also EUPHORBIACEAE: see key W).

    61 Leaves glabrous. Young leaf involute (F16 d). Branches sympodial. (Bark with thin pinkish-brown layer beneath the rhytidome, or laticiferous threads in leaves f. s. spp.). **P22A CELASTRACEAE**
e.g. *Maytenus* (AM species), *Siphonodon* (AS,AU)

    61* Underside of blade pale-green and shiny. (Sympodial trunk and branches, branches plagiotropic, lenticels and stipular / foliar scars well marked, but small in *Leonia*). **P17A VIOLACEAE**
e.g. *Leonia* (AM), *Rinorea* (AM,AF)

Key D

1   Stipular or petiolar scar encircling the whole stem (F13 h,j,m,q,s,t) or base of leaves sheathing the stem (F13 r, F15 a1).
      2   Stipule cylindrical, more or less truncate (ochrea: F13 s,t).   P6B **POLYGONACEAE**
                                                                            e.g. *Coccoloba* (AM)
      2*   Stipule different, cylindrical-conical (F13 m) or flattened (F13 h,j).
         3   Base of petioles sheathing the stem.
             4   Bark aromatic. (Wood soft; stipule adnate to petiole, F13 r).
                                                                                  P27A **ARALIACEAE**
             4*   Bark not aromatic.
                 5   AF, MA, AS. Stipule adnate to the petiole (F13 h). Shrub or herb. Stem kneed at the nodes. (L. with glandular teeth).   P10B **LEEACEAE**
                                                                                    Leea (AF,MA,AS)
                 5*   Woody plant with bark soon suberising. Stipule-like petiolar expansions (F13 j,q) or petiole canaliculate (F15 b). (L. serrulate; stems or leaves with rigid, scabrous hairs).   P6B **DILLENIACEAE**
                                                                       e.g. *Dillenia* (AS,OC,AU)
         3*   Base of leaves different.
             6   Leaves trinerved (F18 d) or palmate venation (F18 c).
                 7   Bark with network of fibres (F3 a,b,c,d).
                       Leaves entire or serrulate (cordate or peltate f. s. spp.).
                                                         P18B **MALVACEAE** *s. l.*
                                               e.g. *Hibiscus* (AM, AS), *key X*
                 7*   Bark different.
                     8   Leaves glandular. (Base of lamina glandular, F19 k).
                                                            P19 **EUPHORBIACEAE**
                                               e.g. *Macaranga* (paleotrop.), *key W*
                     8*   Leaves not glandulate.
                          V. densely reticulate (F 18a1). All twigs with spiral phyllotaxy. (L. more or less lobate: *Cecropia, Musanga*; L. entire: *Coussapoa, Pourouma*). (Leaves of young individual of different shape to that of adult leaves).   P9A **MORACEAE**
                                                    e.g. *Cecropia* (AM), *Musanga* (AF)
             6*   Venation pinnate.
                 V II irregularly curved (F17 a2). (Bark with network of fibres or aromatic f. s. spp.). (FAGERLIND).   P4A **MAGNOLIACEAE**
                                                                   e.g. *Magnolia* (AS)
1*   Stipule different. Petiole not sheathing.
      9   Bark with network of fibres (F3 a,b,c,d). (Sometimes long, broad, coloured strips).
         10   Petiole distally enlarged (F15 m,p) or leaves (sometimes slightly) trinerved (F18 d) or venation palmate (F18 c).
             11   Petiole very short or distally enlarged in a pulvinus.
                 12   AM. Underside of lamina with stellate orange hairs. Shrub with erect twigs and terminal flowering. Petiole relatively long. (If glands, or leaves aromatic: e.g. *Croton*, see *key W*).   P17B **BIXACEAE**
                                                                         Bixa (AM)
                 12*   Characters different.

13 Leaves trinerved or V. palmate (if pinnate, then scaly hairs). (Stellate or peltate hairs, F20 m,n,p; L. glabrous: e.g. *Catostemma*, AM). **P18B MALVACEAE** *s.l.*
See *key X*

13* MA, AS, OC. Venation pinnate. Hairs simple.
P18A ELAEOCARPACEAE
Elaeocarpus (MA,AS,OC)

11* Petiole not pulvinate and not very short. See a), b) and c):
a) Meshes distinctly longer than broad (F3 c). Venation densely reticulate. Stipules asymmetrically disposed, keeled. (AUBREVILLE f.s. spp.).
P9A URTICACEAE
e.g. *Urera* (AM), *Dendrocnide* (AS,OC)

b) Mostly in AM, in savannahs. Leaves palmatilobate with a long petiole. (Really no orange exudate?). **P17B BIXACEAE**
Cochlospermum

c) Characters different (this group includes numerous species). (Stipule symmetrical). **P18B MALVACEAE** *s.l.*
See *key X*

10* Petiole not enlarged and venation pinnate.
13 MA, AS, AU, OC. Branches plagiotropic by apposition (F6 e). (Buttressed tree). **P18A ELAEOCARPACEAE**
e.g. *Elaeocarpus* (mainly AS,OC)

13* AM or branches different.
14 AS. Stipules not intrapetiolar (RAUH).
15 Weak rhythmic growth. **P14A DIPTEROCARPACEAE**
e.g. *Vatica* (AS)
15* Strong rhythmic growth (Series of cataphylls).
P8B ROSACEAE
e.g. *Eryobotrya* (AS, ornam.)

14* AM. Intrapetiolar stipules (RAUH). Underside of L. pale-green and shiny. **P17A VIOLACEAE**
e.g. *Paypayrola* (AM)

9* Bark different.
16 Petiole distally enlarged (F15 m,p,q).
17 Branches plagiotropic by apposition (F6 e).
18 AS, OC. (Rhy/Per without longitudinal slits; AUBREVILLE).
P18A ELAEOCARPACEAE
Elaeocarpus (e.g. MA,AS)

18* Rhy/Per forming thin longitudinal slits (F2 e,f), this character is transient. (AUBREVILLE). **P19 EUPHORBIACEAE**
e.g. *Baccaurea* (AS,OC), *Richeria* (AM), *key W*

17* Branches different.
19 Some branches plagiotropic.
20 Branches monopodial. (ROUX). **P16A FLACOURTIACEAE**
e.g. *Hydnocarpus* (AS)

20* Branches sympodial (F6 f,g,h). (PREVOST).
　　　　　　　　　　　　　　　　　　P19 **EUPHORBIACEAE**
　　　　　　　　　　　　　　　　e.g. *Aporusa* (AS,OC), *key W*

19* All branches orthotropic or monocaulous plant.
　21 Existence of UE in monopodial series (F6 a,g), no foliar glands.
　　22 AS. V. scalariform (F17 k).
　　　　(RAUH, laticiferous threads in leaves, F4 u).
　　　　　　　　　　　　　　　　　P22A **CELASTRACEAE?**
　　　　　　　　　　　　　　　　　　　e.g. *Bhesa* (AS)

　　22* Venation not scalariform or not in AS. See 23a, 23b, 23c, and 23d:
　　　　23a Young twigs rigid. (RAUH; bark that smells bitter almond; lenticels traversed by a thin longitudinal slit f. s. spp.).　　P18A **ELAEOCARPACEAE**
　　　　　　　　e.g. *Sloanea* (AM), *Elaeocarpus* (mainly AS,OC)

　　　　23b Twigs supple. Rhy/Per with thin longitudinal slits (F2 e,f).　　P19 **EUPHORBIACEAE**
　　　　　　　　　　　　　　　　e.g. *Mareya* (AF)), *key W*

　　　　23c Twigs brittle. Young internodes grooved. Rhythmic growth. Crushed bark or leaves that smell bitter almond.　　P16A **FLACOURTIACEAE**
　　　　　　　　　　　　　　　e.g. *Lindackeria* (AM,AF)

　　　　23d AF. Underside of L. with thick hairs.
　　　　　　　　　　　　　　　　　　MEDUSANDRACEAE
　　　　　　　　　　　　　　　　　　Medusandra (AF)

　21* Branches distinctly sympodial (F6 f) or foliar glands.
　　24 Rhy/Per forming thin longitudinal slits, (F2 e,f), this character transient. (LEEUWENBERG, SCARRONE, L. glandular, twigs flexible).　　P19 **EUPHORBIACEAE**
　　　　e.g. *Macaranga, Mallotus* (Paleotrop.), *Alchornea, Croton*
　　　　　　　　　　　　　　　　　　　　key W

　　24* Midrib extended into an apical mucro (F14 c). Vertical serial buds (F12 a,b). L. not glandular.
　　　　　　　　　　　　　　　　　P16B **CAPPARIDACEAE**
　　　　　　　　　　　　　　　　　e.g. *Steriphoma* (AM)

16* Petiole not distally enlarged or very short in relation with blade length or petiolar base enlarged.
　25 Foliar glands. (Orange stellate hairs: *Hymenocardia* (AF,AS), EUPHORBIACEAE).
　　26 Petiole glandular (F19 f,g,h,j).
　　　27 Branches plagiotropic. (Self-supporting form of a liana; ROUX).
　　　　　　　　　　　　　　　　　P16B **PASSIFLORACEAE**
　　　　　　　　　　　　　　　　　　e.g. *Passiflora*

27* Branches orthotropic. Tree or shrub.
   P16A FLACOURTIACEAE
   e.g. *Idesia* (North AS)

26* Lamina glandular (F19 m,p,q).
  28 Base of lamina glandular (F19 m,n,p,q) or glands at the junction with the petiole (F19 k) or glands scattered on the lamina.
    29 Leaves aromatic (*Croton*) or no appressed hairs or Rhy/Per forming thin longitudinal slits (F2 e,f). (Quite supple young twigs).
    P19 EUPHORBIACEAE
    key W

    29* Characters different.
      30 Glands on the upper side of the lamina (F19 p).
        Base of lamina with 2 glands.
        P16A FLACOURTIACEAE
        e.g. *Hasseltia* (AM)

      30* Glands on the underside of the lamina or on its margin.
        31 Buds with an indumentum of appressed hairs (F20 e).
          32 AF. Glands below and at the very base of lamina (F19 m).
          P25B MALPIGHIACEAE
          Acridocarpus (AF,MA)

          32* AM (Guianan mountains). Base of lamina with two glands. Rhythmic growth marked by scale-leaves.
          P18A DIPTEROCARPACEAE
          Pakaraimea (AM)

        31* Buds without appressed hairs.
          33 Glands on the margin (F19 q).
          P10A RHAMNACEAE
          e.g. *Colubrina* (mainly AM)

          33* Base of the lamina glandular (Latex really absent?, see also *keys A* and *B*).
          P19 EUPHORBIACEAE
          key W

25* Leaves not glandular (possibly translucent dots).
  34 Ramified plant. Succession of UE forming monopodial series, strong rhythmic growth (but branches can be sympodial). (Short shoots, F11 p, for many temperate ROSACEAE, e.g. *Pyrus*).
    35 Venation pinnate (F18 f,g,h).
      36 Venation IV-V not visible. (Leaves serrulate, stipules minute).
      P27A AQUIFOLIACEAE
      Ilex

      36* Venation IV-V readily visible.
        37 Leaves somewhat grouped along short internodes). Branches monopodial. (RAUH). Stipules minute.
        P13A LECYTHIDACEAE
        Couroupita (AM)

37* Leaves not grouped.
    38 AM. Numerous thin parallel V II (F17 g). (SCARRONE). Main V. protruding on the upper side of the lamina. **P15B OCHNACEAE**
e.g. *Poecilandra* (AM)
    38* Venation different.
        39 AS. Elongated scale-leaves. (RAUH; Laticiferous threads, F4 u).
P22B IXONANTHACEAE
Ochthocosmus (AM), *Ixonanthes* (AS)

        39* AM. Underside of leaves pale-green and shiny. **P17A VIOLACEAE**
e.g. *Amphirrhox*, *Paypayrola* (AM)

35* Venation palmate (F18 c) or leaves trinerved (F18 d).
North AM, AS. Branches monopodial. (Leaves palmately lobate). **P8A HAMAMELIDACEAE**
Liquidambar (AM), *Altingia* (AS)

34* Trunk unbranched or branches with growth not strongly rhythmic or branches distinctly sympodial.
    40 Some branches plagiotropic.
        41 Br. monopodial or sympodial by substitution (F6 f).
            42 (Rhy/Per forming longitudinal slits, F2 e,f; PREVOST). **P19 EUPHORBIACEAE**
e.g. *Aporusa* (AS)

            42* L. with a hyaline margin. Twigs brittle with many lenticels. (TROLL). **P16A FLACOURTIACEAE**
e.g. *Flacourtia* (Paleotrop., cult.)

        41* Branches sympodial by apposition (F6 e).
Rhy/Per without thin longitudinal slits (L. becoming reddish while ageing; buttressed tree). (AUBREV.).
P18A ELAEOCARPACEAE
Elaeocarpus (mainly AS,OC)

    40* Branches not plagiotropic (orthotropic or drooping) or not branched treelet. See **43**.

43 Plant monocaulous or venation with numerous thin and parallel V II orV III. (V I, V II protruding on the upper side of the lamina, F18 r). See a), b) and c):
a) AF. Plant monocaulous (CORNER). Leaves large, cuneate, elongated, shortly petiolate. V II well spaced. (Inflorescences on the upper side of the lamina).
P16A FLACOURTIACEAE
e.g. *Phyllobotryon* (AF)

b) Numerous parallel V II or V III. (L. serrulate, V II not numerous: e.g. *Godoya*). V I or V II protruding on the upper side of the lamina (F18 r). (Plant monocaulous f. s. spp.).
P15B OCHNACEAE
e.g. *Cespedesia* (AM), *Lophira* (AF), *Euthemis* (AS)

c) AS. Plant monocaulous. Leaves elongated with V II well spaced. (Rhy/Per with thin longitudinal slits, F2 e,f). **P19 EUPHORBIACEAE**
e.g. *Agrostistachys* (AS)

43* Plant ramified. Venation different.
    44 AF. Large petiolar scars. Leaves entire (base of lamina cuneate). Petiole quite long. **P19 EUPHORBIACEAE**
e.g. *Uapaca* (AF,MA), *Macaranga* (Paleotrop.)

    44* Characters different.
        45 V II running parallel to the margin (F17 d) or external wood furrowed (F3 f,g).
        46 AM, AS. External wood furrowed.
(Axillary buds with several scale leaves; branch collars, F2 g,h; L. serrulate or lobate f. s. spp. of temperate zones). **P8A FAGACEAE**
e.g. *Quercus* (AM,AS)

        46* External wood not furrowed. Caution! some SAPOTACEAE might not produce latex during the dry season (see *key B*).
        47 AS. **P18A DIPTEROCARPACEAE**
e.g. *Parashorea* (AS)

        47* AM. (No resin?). **P15A CLUSIACEAE**
Mahurea (AM)

    45* Venation and external wood different.
        48 Young leaf involute (F16 d).
        49 AS, AU, OC. Buds with appressed hairs (F20 e). (RAUH, buttressed tree). **P18A ELAEOCARPACEAE**
Elaeocarpus (AS,AU,OC)

        49* Buds different.
           50 Petiolar base enlarged (F15 n). L. entire, glabrous, shiny. (Petiole short; brittle leaves, laticiferous threads, F4 u).
P22B LINACEAE
e.g. *Roucheria, Hebepetalum* (AM)
P22B IXONANTHACEAE
e.g. *Ixonanthes* (AS)

           50* Petiole different. Leaves glabrous. (Leaves with laticiferous threads, F4 u, f s. spp.). **P22A CELASTRACEAE**
e.g. *Maytenus* (Paleotrop. species), *Celastrus* (AS)

    48* Young leaves not involute or this character not observable.
        51 Branches distinctly sympodial (F6 f).
        52 AM, AF. Underside of leaves pale-green and shiny.
P17A VIOLACEAE
e.g. *Rinorea* (e.g. AF), *Leonia* (AM)

        52* Leaves different (sometimes shiny for CELASTRACEAE). See 53a, 53b, and 53c:
           53a Twigs brittle. Young internodes angular or grooved. Leaves faintly trinerved. Crushed bark or leaves that

smell bitter almond. P16A **FLACOURTIACEAE**
e.g. *Caloncoba* (AF)

53b Twigs flexible. Stipules vestigial. Leaves shiny, faintly crenate with venation not very visible (ROUX).
P22A **CELASTRACEAE**
e.g. *Mystroxylon* (AF,MA)

53c MA. Distal part of sympodial unit with subopposite phyllotaxy. Stipules foliaceous.
P8A **HAMAMELIDACEAE**
Dicoryphe (MA)

51* Branches with UE in monopodial series (F6 a,b,g).
54 Venation densely reticulate (F 18a1). Translucent dots. L. folding involute (F16 d). (RAUH, twigs supple). P16A **FLACOURTIACEAE**
e.g. *Euceraea* (AM)

54* Venation not densely reticulate (F 18a2).
55 Short twigs are leafy thorns: e.g *Flueggea*. (Twigs supple, leaves serrulate, no translucent dots).
P19 **EUPHORBIACEAE**
e.g. *Claoxylon* (Paleotrop.), *Acalypha*, Key W

55* Venation IV-V poorly visible. (Leaves glabrous).
P27A **AQUIFOLIACEAE**
Ilex

Key E

1. Petiole with a distal pulvinus (F15 m,p,q).
   2. Transversal cut of trunk (or not too small branch) with succession of concentric and radiate layers (F4 k). (Twigs rigid, elastic, self-supporting form of a liana?).
                                                                                     P5A **MENISPERMACEAE**
                                                                                     e.g. *Abuta* (AM), *Cocculus*
   2*. Transversal cut different.
      3. AS. Underside of lamina with pale pink, waxy indumentum. Branches erect, sympodial. Petiole faintly enlarged distally.   P8A **HAMAMELIDACEAE**
                                                                                                                          Rhodoleia (AS)
      3*. Indumentum different or absent.
         4. All part aromatic. Large petiolar scars. Sheathing petiole F15 a1 (Venation camptodromous).   P27A **ARALIACEAE**
                                                                                       e.g. *Dendropanax* (AM, AS)
         4*. Rhy/Per forming thin longitudinal slits, F2 e,f. (Leaves aromatic: *Croton*). (Leaves glandular, vestigial stipules?).   P19 **EUPHORBIACEAE**
                                                                                        e.g. *Croton*, see *key W*
1*. Petiole not pulvinate.
   5. Leaves or bark aromatic (or spicy smell).
      6. Petiolar base sheathing the stem (F15 a1). (Nodes enlarged, F11 h).
         7. Branches plagiotropic, modular (PETIT: *Piper*) or herb with monopodial branches and large leaves: *Pothomorphe*.   P4B **PIPERACEAE**
                                                                                        very numerous spp. of *Piper*
         7*. Some stems with distichous phyllotaxy. (TROLL)
                                                                     P5A **ARISTOLOCHIACEAE**
                                                                    e.g. *Aristolochia arborea* (AM)
      6*. No petiolar sheath.
         8. Periderm retaining a green shade on numerous successive internodes. All stems with spiral phyllotaxy. (Stems angular or grooved F11 a,b,h; some leaves opposite; KORIBA).   P2B **LAURACEAE**
                                                                      e.g. *Cinnamomum* (AS,AU)
         8*. Periderm different.
                Lamina with two basal glands or stellate-peltate hairs (F20 m,n,p). (Leaves aromatic; stipules?).   P19 **EUPHORBIACEAE**
                                                                     e.g. *Croton*, see *key W*
   5*. Leaves and bark with other smell.
      9. Some stems with distichous phyllotaxy (F9 h) or L. as in F10 h (*Anisophyllea*).
         10. Venation not scalariform.
            11. Paleotrop. Periderm retaining a green shade on numerous internodes. Trunk and branches monopodial. Leaves supratrinerved (F18 e). (Ramification immediate, F6 b; ROUX). (Occasional petiolar scar below or above the stem, F9 g).   P2B **LAURACEAE**
                                                                      e.g. *Cryptocarya* (mainly MA,AS,OC)

- **11*** Paleotrop. Trunk rhythmically ramified. (MASSART; distichous or pseudodistichous phyllotaxy with readily observable anisophylly, F10 h; leaves supratrinerved, F18 e). P27B **ANISOPHYLLEACEAE**
Anisophyllea (Paleotrop.)
- **10*** Venation scalariform (F17 j,k).
AS. Trunk rhythmically branched and leaves asymmetrical. Leaves trinerved (F18 d). Monopodial twigs. V. camptodromous, F17 c. (MASSART). P27B **ALANGIACEAE**
e.g. *Alangium* (AF,AS,AU)
- **9*** All stems with spiral phyllotaxy (F9 a,b,g).
  - **12** Bark with network of fibres (F3 a,b,c,d). (V. brochidodromous).
    - **13** Plant unbranched (CORNER). L. palmately lobate (F14 a). P17B **CARICACEAE**
Carica (AM, cult.)
    - **13*** Characters different. See 14a, 14b, and 14c:
      - **14a** AS. Branches erect, monopodial. Bark with no smell of bitter almond. (Leaves with peltate hairs (F20 p): *Octomeles*). (Stem with decurrent wings at the bases of petioles, F11 k). **TETRAMELACEAE**
Tetrameles, Octomeles (AS)
      - **14b** AS. Branches plagiotropic by apposition (F6 e). L. serrulate, cordate, petiole without pulvinus. No peltate hairs. Bark smelling of bitter almond. (AUBREVILLE). P17B **FLACOURTIACEAE**
Pangium (AS)
      - **14c** Branches decumbent. (Leaves with hairs on the upper side, at least at the base of the blade). P29A **BORAGINACEAE**
e.g. *Cordia*
  - **12*** Bark different.
    - **15** Base of lamina glandular (F19 n,p).
Lamina with 2 glands. (Rhy/Per forming longitudinal slits, F2 e,f; stellate hairs; leaves aromatic: *Croton*). P19 **EUPHORBIACEAE**
e.g. *Croton*, see key W
    - **15*** Glands absent or otherwise disposed.
      - **16** AS, AU. Periderm retaining a green shade on numerous internodes. Stems angular or grooved (F11 a,b). Leaves supratrinerved (F18 e). (KORIBA; young leaves reddish). P2B **LAURACEAE**
e.g. *Cinnamomum* (AS,AU)
      - **16*** Periderm or stem different. Leaves trinerved (F18 d).
        - **17** Upper side of lamina with hairs erect, oblique, hooked or raised on small protuberances (F20 s). (Use a hand lens). (PREVOST, CHAMPAGNAT). (V II ending in a tooth f. s. spp.). P29A **BORAGINACEAE**
e.g. *Cordia*
        - **17*** Indumentum different or plant glabrous.
          - **18** Rhythmic growth with series of scale-leaves delimiting the units of extension.

19 Decurrent wings or ridges at the bases of the petioles (F11 j,k).
  Petiolar scars circular (F15 e). (Leaves entire, trilobate or peltate, F14 c; RAUH).
  P2A HERNANDIACEAE
  Gyrocarpus, Hernandia

19* No decurrent wings or ridges.
  AS. Apical buds with several scale-leaves.
  P8A HAMAMELIDACEAE
  Altingia (AS)

18* Unit of extension different.
  20 Venation not very visible or epiphyte or prostrate plant). **P12A ERICACEAE**
  e.g. *Satyria* (AM)
  20* Characters different.
   21 Rhy/Per forming thin longitudinal slits (F2 e,f); this character being transient.
   P19 EUPHORBIACEAE
   e.g. *Strophioblachia* (AS), see *key W*

   21* Eye marks on trunk (F2 h). Vertical serial buds (F12 d). Stellate hairs on L.
   P16A FLACOURTIACEAE
   e.g. *Kiggelaria* (AF)

Key F

1 Bark with network of fibres (F3 a,b,c,d). (Branches need to be old enough to show this feature).
   2 Bark aromatic.
       In the developing UE, the young L. are conduplicate and drooping, F16 g. (ROUX, PETIT, MANGENOT, TROLL; stellate or peltate hairs: *Duguetia*, *Annona* spp.).
                                         P3B **ANNONACEAE**
                                      Almost all the non lianescent species
  2* Bark not aromatic.
     3 Leaves or buds with appressed hairs (F20 e).
       4 AS, OC. Branches disposed in tiers (MASSART). Branches not easy to break (fibrous bark!). Leaves with translucent dots: e.g. *Gonystylus*. Lamina with an intramarginal (F18 m) or a fimbrial vein (F17 q): *Dicranolepis*, *Aquilaria*.
                                       P25A **THYMELAEACEAE**
              e.g. *Aquilaria* (AS), *Gonystylus* (AS,OC), *Dicranolepis* (AF)

       4* Branches not disposed in tiers. TROLL, ROUX.   P3B ANNONACEAE

    3* Leaves and buds without appressed hairs.
       Venation densely reticulate. (Bark can be pull off in fibrous strips, faintly aromatic f. s. spp., branches not easy to break). Base of lamina glandular (F19 n), MASSART: *Napoleonaea*. TROLL: *Eschweilera*, *Lecythis*. MASSART, short shoots: *Couratari*.   P13A **LECYTHIDACEAE**
          e.g. *Couratari*, *Eschweilera*, *Lecythis* (AM), *Napoleonaea* (AF)

1* Bark without noticeable network of fibres (but the bark can be fibrous and difficult to break). Wait to see if the bark turns reddish, with an astringent exudate. (Examine trunks old enough, with a diameter of more than 2 cm).
   6 Inner bark exsudate turning reddish, astringent taste. Branches disposed in tiers. (MASSART, apices with short, not appressed hairs). P4A   **MYRISTICACEAE**
                                              e.g. *Iryanthera* (AM)
  6* Bark different.
     7 Venation camptodromous, the V III and V IV are more or less oriented in the same direction. (Ramification of trunk faintly rhythmic; MASSART?).
                                         P28A **ICACINACEAE**
                                               Emmotum (AM)
     7* Venation different.
       8 Indumentum of stellate hairs (F20 k,m) or glandular trichomes below lamina, or on buds or twigs.
         9 Branches or twigs with sloughing off of the rhytidome in straw-like scales (F2 b) or stellate-peltate hairs on the underside of lamina (F20 m,n,p,q).
           10 AS. Rhytidome peeling into straw-like scales. (MASSART?).
                                                 STYRACACEAE
                                                     Styrax (AS)

        10*   AM. Rhytidome different. Hairs stellate.   P28A **ICACINACEAE**
Dendrobangia (AM)

9*   Rhytidome different or stellate hairs with free arms (F20 k).
    11   Underside of L. with glandular trichomes.   P11B **MYRSINACEAE**
e.g. *Ardisia* (AM)
    11*   Leaves without glandular trichomes.
        12   Lamina glandular along its margin. (Rhy/Per forming thin longitudinal slits, F2 e,f).   P19 **EUPHORBIACEAE**
Key W
        12*   Lamina not glandular.
AF. Petiole slightly enlarged distally (F15 m). (ROUX).
P7B **OLACACEAE**
Octoknema (AF)

8*   Leaves and apices without stellate hairs or glandular trichomes.
    13   Leaves glandular (glands can be very small in *Diospyros*).
        14   Black layer under rhytidome or indumentum of appressed (even very small) hairs on the buds (F20 e).
            15   (MASSART; trunk bearing tiers of plagiotropic branches or wood very hard; f. s. spp).   P12B **EBENACEAE**
Diospyros

            15*   AF. Base of lamina glandular (F19 n). Branches not easy to break. (MASSART).   P13A **LECYTHIDACEAE**
Napoleonaea (AF)

        14*   Rhytidome and indument different. See 16a, 16b, and 16c:
            16a   Glands scattered on the underside of blade.
                17   Glands in small grooves located on the veins (F19 v). Inner bark yellowis-brown to orange-brown. (Vertical serial buds, F12 a; leaves turning yellow when drying).
P26A **XANTHOPHYLLACEAE**
Xanthophyllum (AS)

                17*   L. with scattered concave glands (F19 v). Bark different.   P23B **SIMAROUBACEAE**
Samadera indica (MA, AS)

            16b   Glands on petiole or base of lamina (F19 h,m,q).Venation embossed (F18 s). (Leaves or bark with taste of biter almond).   P 21A **ROSACEAE**
e.g. *Prunus*

            16c   Glands in stipular position (F19 a). (Self-supporting form of a liana).   P26A **POLYGALACEAE**
e.g. *Securidaca* (AM,AF)
13*   Leaves not glandular.
    18   Nodes enlarged (F11 h). Petiolar base more or less sheathing.

- 19 Sympodial branches composed of short modules, each with only one scale-leaf and one developed leaf. L. with spicy, aromatic smell. (Shrub or herb; PETIT). **P4B PIPERACEAE**
Piper

- 19* Branches different. Petiolar scars annular. Leaves aromatic. (TROLL). **P5A ARISTOLOCHIACEAE**
e.g. *Aristolochia* (AM)

- 18* Nodes not enlarged or petiole not sheathing.
  - 20 Petiole distally pulvinate (F15 m,p,q).
  Petiole with an abscission joint (F15 q). If V I protruding (F 18 r), see LEGUMINOSAE. **P21B CONNARACEAE**
  e.g. *Manotes* (AF), *Ellipanthus* (AF,AS)

  - 20* Petiole not distally pulvinate. (See **21**).
- 21 Venation IV-V not very visible.
  - 22 Bark and leaves aromatic.
    MA. Venation almost invisible. (MANGENOT). **P2A CANELLACEAE**
    e.g. *Cinnamosma* (MA)
  - 22* Bark and leaves not aromatic.
    - 23 Branches monopodial.
      - 24 Leaves entire. Young internodes angular (F11 c). (Periderm retaining a green shade on numerous internodes). **P7B OLACACEAE**
      e.g. *Heisteria* (AM,AF), *Strombosia* (AF,AS)

      - 24* Leaves coriaceous, entire. V III not visible. Existence of UE in monopodial series (F6 a). Growth distinctly rhythmic. **P11A THEACEAE**
      e.g. *Cleyera* (Himalaya, Japan, C. AM)

    - 23* Branches sympodial or leaves not at all coriaceous. See 25a, 25b, and 25c:
      - 25a Leaves coriaceous. Rhytidome becoming scaly. Branches sympodial. Petiole short. (Basal V II close to one another, F18 h; phyllotaxy, not distinctly distichous). **P12A ERICACEAE**
      e.g. *Cavendishia* (AM)

      - 25b MA. L. coriaceous, entire glabrous. Some leaves opposite to a scale-leaf (i.e. anisophylly). Rhytidome not scaly. **KALIPHORACEAE**
      Kaliphora (MA)

      - 25c AS. L. not at all coriaceous, entire, not acuminate, with a short petiole and its blade cuneate at base. Young internodes angular (F11 j). **P7B OPILIACEAE**
      e.g. *Lepionurus* (AS), *Opilia* (Paleotrop.)

- 21* Venation IV-V readily visible.
  - 26 Buds and young leaves hairy (observe the buds carefully with a hand lens).
    - 27 Bark aromatic.
      - 28 Branches monopodial. (V. camptodromous, F17c; scaly rhytidome: *Eusideroxylon*; L. supratrinerved, F18 e, glaucous beneath: *Cryptocarya*). **P2B LAURACEAE**
      e.g. *Cryptocaraya*, *Eusideroxylon* (AS)

28* Branches sympodial. (Leaf folding conduplicate, F16 b; V. brochidodromous). **P3B ANNONACEAE**
27* Bark not aromatic.
29 Branches with UE forming monopodial series.
30 Paleotrop. TROLL's model. Leaves not coriaceous. Appressed hairs. (Rhythmic growth; domatia, F20 t). **P23B MELIACEAE**
Turraea (AF,MA,AS,AU)
30* Neotrop. or characters different.
31 AM, AS. Some branches monopodial, (ROUX, TROLL, appressed hairs). (Numerous lenticels f. s. spp.). (Petiole canaliculate, F15 b: *Freziera*). **P11A THEACEAE**
e.g. *Freziera* (AM), *Adinandra* (AS)

31* Venation scalariform. Very small hairs minute on buds.
P7B OLACACEAE
e.g. *Ochanostachys* (AS)
29* Branches distinctly sympodial.
32 Venation brochidodromous (F17 a). Very small hairs minute on buds. **P28A ICACINACEAE**
e.g. *Gomphandra* (AS,OC), *Gonocaryum* (AS)

32* Venation camptodromous (F17 c). **P12B EBENACEAE**
e.g. *Maba* (AM)
26* Buds and young leaves entirely glabrous.
33 Leaf folding involute (F16 d) or petiole very short.
34 Bark that can be pulled off in fibrous strips (a little sticky f. s. spp.). (Twigs difficult to break; TROLL: *Eschweilera*, *Lecythis*; MASSART: *Couratari*). **P13A LECYTHIDACEAE**
e.g. *Couratari*, *Eschweilera*, *Lecythis* (AM)
34* AM, AF. Bark different.
35 Twigs brittle. (MANGENOT, TROLL; base of petiole enlarged, F15 n; leaves subsessile or shortly petiolate). **P22B HUMIRIACEAE**
e.g. *Sacoglottis* (AM, West AF), *Vantanea* (AM)

35* AF. Leaves with minute teeth.: *Brazzeia*; leaves entire: *Rhaptopetalum*, *Scytopetalum*. Young internodes angular. **LECYTHIDACEAE**
e.g. *Brazzeia*, *Scytopetalum* (AF)

33* Leaf folding different or this character not observable, petiole developped.
36 AS. Leaves serrulate. V. brochidodromous (F17 a). Teeth glandular. Branches plagiotropic. (ROUX). **P5B MELIOSMACEAE**
Meliosma (AS)
36* Leaves entire.
37 MA. Apex of leaves acute, mucronate (F14 c). Leaves glabrous and shiny. Periderm retaining a green shade. **PHYSENACEAE**
Physena (MA)
37* Characters different. (Domatia, F20 t, f. s. spp.).
38 AM. Translucent dots. V. scalariform. **P15A CLUSIACEAE**
Caraipa (AM)

38* No translucent dots.
  39 Leaves more or less coriaceous.
    49 V. IV-V readily visible. (Leaves shiny and coriaceous, PETIT: e.g. *Leptaulus*, AF, MA; inflorescences oppositefoliate f. s. spp.).   **P28A ICACINACEAE**
e.g. *Lasianthera* (AF), *Citronella* (AS)

    40* Twigs lenticellate and supple. Periderm siny. (External wood furroed, F3 f,g).
  P21A CHRYSOBALANACEAE
e.g. *Licania* (AM)
  39* Leaves not at all coriaceous.
Young twigs angular. Branches sympodial. Corky outgrowths on twigs where inflorescences appeared. V. IV-V not readily visible.   **P7B OLACACEAE**
e.g. *Cathedra* (AM)

Key G

1 Branches with lateral short shoots (F11 p), in some cases modified into spines. (CHAMPAGNAT, MASSART, RAUH).
    2 Plant without spines.
        3 Bark aromatic or young internodes angular (F11 a) or grooved (F11 b). (MASSART).   P2B **LAURACEAE**
e.g. *Aniba* (AM)
        3* Bark and twigs different.
            4 AM (or cultivated). Leaves pseudowhorled on woody expansions of the branches. (Leaves notched, F14 c; CHAMPAGNAT).
P30A **BIGNONIACEAE**
Crescentia (AM, cult.)

            4* Leaves grouped on well developed short shoots. Indumentum of erect, oblique or hooked hairs, or hairs on small expansions of the epidermis (F20 s), (use a hand lens). (L. with rigid hairs f. s. spp.).
P29A **BORAGINACEAE**
e.g. *Bourreria* (AM), *Cordia* (e.g. AS)
P29B **SOLANACEAE**
e.g. *Brunfelsia* (AM, ornam.)
    2* Plant spiny.
        3 Spines trifid (spines are modified leaves).   **BERBERIDACEAE**
Berberis (AM, ornam.)
        3* Spines not trifid.
            5 AM, in savannahs. Spines in pairs subtended by the leaves.   **CACTACEAE**
Pereskia (AM)

            5* Spines woody, solitary. V. not very visible.   P7B **OLACACEAE**
Ximenia
1* Short shoots absent or terminal (F6 e)
    6 Branches plagiotropic by apposition (F6e), (AUBREVILLE, FAGERLIND).
        7 Crushed bark or leaves giving out an aromatic, or resinous, or special spicy smell.
            8 Petiole distally pulvinate (F15 m,p).   P24B **RUTACEAE**
e.g. *Esenbeckia* (AM)
            8* Petiole not distally pulvinate.
                9 Young internodes angular (F11 a) or grooved (F11 b). Periderm retaining a green shade on numerous internodes, if suberising soonly: e.g. *Aniba*. (Ramification immediate, F6 b; AUBREVILLE; sclerenchymatous orange inclusions in bark, F4 t; f. s. spp.).   P2B **LAURACEAE**
e.g. *Aniba* (AM), *Ocotea*

                9* V II at first rectilinear then abruptly curved near the margin (F18 p). (Leaves auriculate, F14 b; spirodistichous phyllotaxy, F9 f; f. s. spp.).
P24A **ANACARDIACEAE**
e.g. *Campnosperma* (e.g. AS)
        7* Bark and leaves with different smell.
            10 AM, AS. Hairs thick, appressed on the young stems or leaves (F20 r). (AUBREVILLE). (Hairs arranged in crowns for a few spp).

10* Indument different or absent.
    11 Leaves with glandular trichomes or translucent dots. (Oval insertions of the twigs, F11 r; AUBREVILLE).

P11A **ACTINIDIACEAE**
e.g. *Saurauia* (AM,AS)

P11B **MYRSINACEAE**
e.g. *Ardisia* (AS)

11* Leaves without glandular trichomes or translucent dots.
    12 Young internodes angular (F11 a) or grooved (F11 b). Buds covered with appressed hairs (F20 e).
        Periderm suberising soonly (AUBREVILLE).

P2B **LAURACEAE**
e.g. *Aniba* (AM)

12* Internodes or indument different.
    13 Veins of higher order (IV-V) readily visible.
        14 Indumentum of erect or oblique hairs, or periderm of the upper side of the lamina with small protuberances, (use a hand lens). (KORIBA, FAGERLIND).

P29A **BORAGINACEAE**
e.g. *Cordia* (AM)

14* Indumentum different or absent.
    15 Leaves subsessile or shortly petiolate. (Base of lamina glandular or translucent dots; AUBREVILLE).

P13B **COMBRETACEAE**
e.g. *Terminalia*

15* AS, OC. Petiole well developed. Twigs lenticellate. V. camptodromous (F 17c).

P23B **MELIACEAE**
Vavaea (AS,OC)

13* Venation of order IV-V not very visible.
    16 Leaves auriculate (F14 b). (L. sessile or subsessile; AUBREVILLE).
        Leaves coriaceous. (In poor, siliceous soils).

P11A **BONNETIACEAE**
e.g. *Archytaea* (AM), *Ploiarium* (AS)

16* Leaves not auriculate. See 17a, 17b, and 17c:
    17a Leaf shortly petiolate or subsessile. (L. folding involute, F16 d).

P11A **THEACEAE**
e.g. *Gordonia* (AS, AM)

    17b Seashores. L. shortly petiolate or subsessile. Young leaf folding revolute (F16 e).

GOODENIACEAE
Scaevola (mainly AU,OC)

    17c Venation IV-V not very visible.

CORYNOCARPACEAE
Corynocarpus (AS, AU, OC)

6* Branches not plagiotropic by apposition. (Unbranched treelet f. s. spp.).

G 43

18   Leaves with pellucid dots or glandular trichomes (shiny or opaque) giving a waxy aspect of the underside of leaves.
   19   Leaves with pellucid dots (see *F19*). (Leaves or bark aromatic, or with strange or peculiar smell).
      20   Petiole articulate with the lamina (F15 q) or distally pulvinate (F15 p). CORNER: e.g. *Erythrochiton*. (Petiole winged, F15 d, f. s. spp.).
                                          P24B RUTACEAE
                          e.g. *Esenbeckia, Pilocarpus* (AM), *Toddalia* (AS)

      20*   Petiole not articulate and not distally pulvinate. See 21a, 21b, and 21c:
         21a   AS, AU, OC. (Leaves aromatic, terminal bud scaly).
                                          P13B MYRTACEAE
                                            e.g. *Tristania* (AS,AU,OC)

         21b   AM. Terminal bud not scaly.   **P2A CANELLACEAE**
                                                  Cinnamodendron (AM)

         21c   AM. L. not aromatic. Terminal bud not scaly. (Branch shedding leaving an oval scar, F11 r).   **P2A MYRSINACEAE**
                                                       Ardisia (e.g AS)

   19*   Leaves not pellucid dotted.
      22   Branches monopodial (RAUH) or Unbranched treelet (CORNER).
         23   Leaves and bark not aromatic. Trichomes glandular and shiny.
                                            P11B MYRSINACEAE
                        e.g. *Rapanea* (mainly AM), *Tapeinosperma (NG,AU,OC)*

         23*   Bark spicy. Underside of leaves whitish due to the presence of very numerous trichomes. (RAUH).   **WINTERACEAE**
                    e.g. *Drimys* (AM,AS,OC), *Zygogynum* (NG,OC)

      22*   Branches sympodial.
         23   AS. Indument waxy beneath the leaves. Petiole long, slightly enlarged distally. Leaves not coriaceous.
                                          P8A HAMAMELIDACEAE
                                                    Rhodoleia (AS)
         23*   Characters different. (Leaves aromatic).
            25   AM, AS. Modules with pseudowhorled leaves (F9 c).
                                               P2A ILLICIACEAE
                                                     Illicium (AM,AS)
            25*   AM, AF? Leaves not pseudowhorled.
                                             P2A CANELLACEAE
                                                     e.g. *Canella* (AM)

18*   Leaves without pellucid dots, glandular trichomes and without waxy indumentum. (Indument can be white because of presence of appressed hairs).
   26   Venation readily observable or hidden by the indumentum.
      27   Apices of twigs with leaves distinctly disposed in rosettes (F9 c) or bark that comes off in fibrous strips. (See also further: BORAGINACEAE, some *Cordia* having a fibrous bark).
      28   Buds and apices with indumentum of appressed hairs (F20 e).

29 Dense indumentum of thick hairs. Bark fibrous. Twigs not easy to break. **P25A THYMELAEACEAE**
e.g. *Funifera* (AM), *Lasiosiphon* (AF,MA,AS)

29* Characters different. (CORNER's model f. s. spp.).
30 Young leaf folding convolute (F16 c).
AF, MA, AS. Short internodes between most of the leaves. **P13A LECYTHIDACEAE**
Petersianthus (AF,AS), *Foetidia* (AF,MA,AS)

30* Young leaf not distinctly convolute.
31 Large, elongate, short petiolate leaves or bark that can be pulled off in fibrous strips. (LEEUWEN., KORIBA). **P13A LECYTHIDACEAE**
e.g. *Gustavia*, *Grias* (AM), *Barringtonia* (mainly AS,OC)

31* Bark and leaves different.
32 AM. Trunk unbranched (CORNER) with large, elongated L. (*Clavija*, *Theophrasta*) or modular architecture (KORIBA) with entire, subulate L. (*Jacquinia*).
P11B THEOPHRASTACEAE
Clavija, *Jacquinia*, *Theophrasta* (AM)

32* AF, AS, AU, OC Modular architecture (F8 f,g). Leaf blade with cuneate base. (KORIBA, bark faintly aromatic-spicy).
P27A PITTOSPORACEAE
Pittosporum (Paleotrop.)

27* Leaves not distinctly disposed in rosettes. Bark different. (See 33).

33 Leaves glandular (F19 j,m).
(Hairs stellate or peltate, aromatic leaves f. s. spp.). **P19 EUPHORBIACEAE**
e.g. *Croton*, Key W

33* Leaves not glandular.
34 Branches rhytidome that comes out in scales or strips (F2 b,c,d).
Growth not very ryhythmic. (L. with embossed venation, F18 s).
P12A ERICACEAE
e.g. *Gaylussacia* (AM), *Rhododendron* (North hemisphere)

34* Rhytidome different.
35 AS. Leaves profoundly incised. Petiole sheathing the stem (F15 a1). (CHAMBERLAIN, LEEUWENBERG). **ARALIDIACEAE**
Aralidium (AS)

35* Characters different.
36 Bark with networked structure (F3 b,j).
37 Petiole distally enlarged. **P16B CAPPARIDACEAE**
e.g. *Morisonia* (AM)

37* Petiole not distally enlarged.
38 Branches modular, sympodial by substitution (F6 f, F8 h, PREVOST). Petiole quite long. Hairs erect, oblique or hooked, on the upper side of lamina. **P29A BORAGINACEAE**

e.g. *Cordia* (e.g. AM)

    38* External wood furrowed (F3 j). (Internodes of unequal length; leaves coriaceous or, f.s.spp., deeply lobate).
        P6A **PROTEACEAE**
        e.g. *Panopsis, Roupala* (AM), *Macadamia* (AS,AU)

36* Bark different.
    39 Not very wood (shrub), sympodial architecture shoots displaced upwards away from their subtending leaf and some leaves apparently without axillary buds (F12 p) or L. subtending two unequal-sized leafy prophylls (F12 m) or L. alternating with smaller ones.
        40 Leaf-margin not serrulate. (Herb or shrub; periderm retaining a green shade on numerous internodes or rhytidome grey, shiny on the young twigs; hairs also on the upper side of the leaves. (Disagreeable smell; f. s. spp.). P29B **SOLANACEAE**
            e.g. *Dunalia* (AM), *Solanum*

        40* Leaves serrulate, hairy.   **HYDROPHYLLACEAE**
            e.g. *Wigandia* (AM)

39* Characters different
    41 Petiole well developed with thickened base.
        AM. Branches erect. Young twigs angular (F3 a,b).
        P5B **MELIOSMACEAE**
        Meliosma (AM)

    41* Petiole different.
        42 Branches sympodial, erect. See 43 a, b, c, and d:
            43a Underside of leaves shiny. V. densely reticulate.
                P22A **CELASTRACEAE**
                e.g. *Schaefferia* (AM), *Denhamia* (AU)

            43b Underside of leaves not shiny. Leaf blade entire with cuneate base (KORIBA; Gum in bark; petiole short; twigs lenticellate; f. s. spp.).
                P27A **PITTOSPORACEAE**
                Pittosporum (Paleotrop.)

            43c Underside of leaves not shiny. Leaves entire. Twigs lenticellate.   P27A **EBENACEAE**
                Diospyros (AS, cult.)

            43d Underside of leaves not shiny. Leaves entire. Twigs lenticellate.   P27B **CORNACEAE**
                Mastixia (AS)

    42* Branches monopodial or not erect. See 44 a, b, and c:
        44a Twigs lenticellate. (V. not very visible).
            P28A **ICACINACEAE**
            e.g. *Apodytes* (AS,AU,OC)

        44b Branches drooping. Stems and buds pubescent.
            P26A **POLYGALACEAE**

e.g. *Bredemeyera* (AM)

    44b Twigs lenticelate. Venation peculais (P 16A f)
P16A FLACOURTIACEAE
e.g. *Ludia* (AF,MA)

26* V. not very visible. Indumentum absent or not conspicuous.
  45 AM. In a mangrove habitat.
Leaves sessile, separated by short internodes. Young leaves convolute (F16 c). **P11A PELLICIERACEAE**
Pelliciera (AM)

  45* Plant not specifically in a mangrove habitat.
    46 AM. Leaves sessile, glabrous, its apices acute and spiny. Branches sympodial, each module consisting of one unit of extension. Most of the L. pseudowhorled at the ends of twigs. (KORIBA). **P11B THEOPHRASTACEAE**
e.g. *Jacquinia* (AM)

    46* Leaves not subulate or not sessile.
      47 Leaves more or less coriaceous.
        48 V III visible. See 49 a, b, c, and d:
          49a AF, MA. Leaves with a thick hyaline margin. Some veins secant (F17 n). Branches sympodial, not modular. **BREXIACEAE**
Brexia (AF,MA,Seychelles)

          49b L. short petiolate. Rhytidome that comes out in fibrous strips (F2 b,d).
P12A ERICACEAE
e.g. *Bejaria* (AM)

          49c MA. L. long petiolate, cuneate at base. Petiole canalicuate (F 15b).
MELANOPHYLLACEAE
Melanophyllum (MA)

          49d MA. Twigs lenticellate. V I protruding on upper side (F 18r). **DIDYMELACEAE**
Didymeles (MA)

        48* V III not visible.
          50 Marked rhythmic growth (scale-leaves).
P11A THEACEAE
e.g. *Ternstroemia* (AM)

          50* Modular architecture. Marked rhythmic growth (scale-leaves). Blade cuneate.
P27A PITTOSPORACEAE
e.g. *Pittosporum* (Paleotrop.)

      47* Leaves smooth. Branches sympodial, not modular.
P7B OPILIACEAE
e.g. *Agonandra* (AM)

Key H

1   Leaves with peltate (F20 p), stellate (F20 k,m,n) hairs, or with glandular trichomes, or with translucent dots. (Use a hand lens).
    2   Bark with network of fibres (F3 a,b,c).
       3   Branches hard to break owing to the presence of fibres.
          4   Indumentum of appressed hairs.   P25A **THYMELAEACEAE**
                                                                  e.g. *Daphnopsis* (AM)

          4*   AM, AS. Hairs peltate (F20 p).   **ELAEAGNACEAE**
                                                              e.g. *Elaeagnus* (North AM,AS)

      3*   Branches different (brittle, leaving oval scars after shedding, F11 g). (Elongated translucent pouches f. s. spp.).   P11B **MYRSINACEAE**
                                      e.g. *Rapanea* (AM), *Oncostemum* (MA), *Ardisia*
    2*   Bark without network of fibres.
       5   Underside of leaves with peltate or stellate hairs (giving in some cases a shiny, metallic aspect to the lamina).
          6   Rhy/Per with thin longitudinal slits (F2 e,f). Peltate hairs.
                                                                    P19 **EUPHORBIACEAE**
                                                                           Key W

        67*   Rhy/Per different.
          7   Leaves entire. Well developed midrib ending in a mucro and/or an indentation (F14 c). Midrib distinctly broader than V II-V III. (Petiole distally pulvinate: *Capparis*, *Morisonia* or not pulvinate: *Capparis*).
                                                                P16B **CAPPARIDACEAE**
                                                                       e.g. *Morisonia* (AM)

          7*   Midrib different. Petiole not pulvinate. (V. embossed, F 18s).
                                                                              STYRACACEAE
                                                                            e.g. *Styrax* (AM)
    5*   Indument different or none.
       8   Petiole distally enlarged or pulvinate (F15 m,p).
          9   Translucent dots. (Petiole winged or with abscission joint, F15 q).
                                                                     P24B **RUTACEAE**
                                                                   e.g. *Erythrochiton* (AM)
         9*   No translucent dots.
            10   Branches plagiotropic. (Sclerenchymatous, orange inclusions in bark, F4 t, f. s. spp.).   P16A **FLACOURTIACEAE**
                                                                   e.g. *Ryparosa* (AS)
            10*   Branches orthotropic. See 11a, 11b, and 11c:
                  11a   AS, AU, OC. Buds and apices with appressed hairs. (Stipules really absent?).   P18A **ELAEOCARPACEAE**
                          Elaeocarpus (AS,AU,OC), *Sloanea* (AM)

11b No appressed hairs. (Twigs flexible, Rhy/Per forming thin longitudinal slits, F2 e,f). **P19 EUPHORBIACEAE**
e.g. *Codiaeum* (AS, ornam.), *key W*

11c Leaves entire. Well developed midrib ending in a mucro and/or an indentation (F14 c). Midrib distinctly broader than V II-V III. **P16B CAPPARIDACEAE**
e.g. *Morisonia* (AM)

8* Petiole different (but slightly enlarged for *Rhodoleia* (AS), see *key E*.
12 Glandular trichomes. See 13a, 13b, and 13c:
13a (At altitude). Leaves inserted on small expansions of the twigs. (V II secant, F17 p). **MYRICACEAE**
e.g. *Myrica* (AM,AS,OC)

13b Twigs different. (MASSART, RAUH). Mangrove habitat, SCARRONE: *Aegiceras*. CHAMPAGNAT: *Maesa*. (Branches leaving oval scars on the trunk after shedding, F11 r; elongated pellucid pouches; f. s. spp.). **P11B MYRSINACEAE**
e.g. *Rapanea* (AM), *Oncostemum* (MA), *Aegiceras* (Paleotrop.)
Ardisia (AS,AU), *Maesa* (Paleotrop.)

13c Ramification almost continuous in trunk and main branches. Young internodes angular (F11a). L. shiny. **SAPINDACEAE**
Dodonaea

12* Leaves without glandular trichomes (possibly small translucent dots: LAURACEAE, MYRSINACEAE). (See also old branches: possibly fibrous bark in MYRSINACEAE).
14 (Mostly at altitude and AM). Shiny trichomes on the underside of blade. External wood with long, prallel furrows (F3 e). **P28B ASTERACEAE**
e.g. *Baccharis* (AM)

14* Characters different. Translucent dots.
15 Leaves entire.
16 Rhytidome sloughing off in scales or thin strips (F2 b,c). Translucent dots. ATTIMS: *Eucalyptus*; CHAM-PAGNAT: *Leucadendron*, (Young leaf folding plane-curved, F16 a). **P13B MYRTACEAE**
e.g. *Eucalyptus* (AS,OC,AU), *Leucadendron* (AU)

16* AM. Leaves elongated, cuneate, smooth. **RHABDODENDRACEAE**
Rhabdodendron (AM)

15* Leaves serrulate. AM. **MYOPORACEAE**
Myoporum (AM)

1* Leaves different.

17 Leaves glandular (i.e. with extrafloral nectaries). (Domatia in nerve axils for some LAURACEAE). See 18a, 18b, and 18c:
   18a Base and margin of lamina glandular (F19 q). (On shores or in rear-mangrove habitat: *Conocarpus*). **P13B COMBRETACEAE**
Strephonema (AF), *Conocarpus* (AM,AF)

   18b Shrub with serrulate leaves. **TURNERACEAE**
Turnera (AM)

   18c Glands located differently. Rhy/Per with thin longitudinal slits (F2 e,f), (this character is transient). **P19 EUPHORBIACEAE**
key W

17* Leaves not glandular.
  19 V II or V III secant (F17 n,p)
    20 Leaves entire. (Resinous smell?). **P24A ANACARDIACEAE**
e.g. *Ozoroa* (AF)

    20* Leaves serrulate. See 21a, 21b, and 21c:
     21a AM, AS. Thick, brown, rust-coloured hairs (F20 r). Petiole not canaliculate. (Hairs arranged in crowns f. s. spp.).
P11A ACTINIDIACEAE
Saurauia (AM,AS)

     21b Hairs not thick. Petiole canaliculate (F15 b). (Rhytidome scaly, stems scabrous). **P6B DILLENIACEAE**
e.g. *Curatella* (AM), *Dillenia* (AS,AU)

     21c Rhytidome scaly. Branches sympodial (F6 g). Hairs not thick (stellate f. s. spp.). Petiole not canaliculate. **CLETHRACEAE**
Clethra (AM,AS, Madeira)

  19* Venation not secant.
    22 Strong rhythmic growth noticeable by several scale-leaves and short internodes at the base of the UE.
     23 Crushed leaves or bark giving out a resinous smell.
Petiole not pulvinate. V II quite rectilinear then abruptly curved near the margin (F18 p). (SCARRONE, RAUH).
P24A ANACARDIACEAE
key Y

    24* Smell not resinous.
     25 AM (at altitude or on poor soils). Leaves shortly petiolate or sessile, coriaceous or somewhat fleshy. See 26a, 26b, and 26c:
      26a Leaf subsessile.L. folding involute
P11A BONNETIACEAE
Bonnetia (AM)

      26b L. shortly petiolate with a hyaline margin.
P12A CYRILLACEAE
e.g. *Cyrilla* (AM)

      26c V. not very visible. Leaves shortly petiolate.
P11A THEACEAE
e.g. *Ternstroemia* (mainly AM,AS)

    25* Petiole well developed.
        27 AS, AU. Cuneate leaves with long and red petiole. Well developed, circular petiolar scars (F15 e).
                          DAPHNIPHYLLACEAE
                              Daphniphyllum (AS,AU)

        27* Characters different. See 28a, 28b, 28c, and 28d:
            28a Young stems angular. (Apical buds with several large scale-leaves; MASSART, RAUH).
                            P2B **LAURACEAE**
                      e.g. *Licaria* (AM), *Litsea* (mainly AS,AU)

            28b AS. Apical buds with several small imbricate scale-leaves. (Minute stipules on the petiole?).
                          P8A **HAMAMELIDACEAE**
                                Altingia (AS)

            28c AS, AU, OC. (Suberisation of the periderm forming scales; no stipules? *Ixonanthes*, AS).
                          P22B **IXONANTHACEAE**

            28d AM. V. camptodromous (F17k) more or less scalariform (*Mahurea*, AM).
                          P15A **CLUSIACEAE**

  22* Growth different, not strongly rhythmic, but internodes between devlopped leaves can be short.
        29 Spine subtended by a leaf (F12 e).   P7A **NYCTAGINACEAE**
                            e.g. *Bougainvillea* (AM, ornam.)

        29* Spines absent or disposed otherwise. (See 30):

30 V. brochidodromous (F 17a). V II irregularly curved (F17 a2) or V. not very visible. Plant distinctly woody. Bark and leaves not aromatic.
  31 External wood furrowed (F3 e,f,g,h,j).
    32 Bark with networked structure (F3 j).   P6A **PROTEACEAE**
                        e.g. *Roupala* (AM), *Helicia* (AS,AU,OC)
    32* Bark without networked structure.
  31* External wood not furrowed.   P28B ASTERACEAE
    33 Venation conspicuously reticulate.
      34 Branches drooping.
        (Ramification continuous; leaf folding conduplicate, F16 b1; apices with ochre indumentum).   P26A **POLYGALACEAE**
                            e.g. *Diclidanthera* (AM)
      34* Branches erect.
        35 Petiole distally pulvinate (F15 p).
          36 Bark or leaves aromatic. Phyllotaxy spiral to subopposite. (RAUH).   P24A **BURSERACEAE**
                          e.g. *Protium* (e.g. AM)

36*   Not aromatic. Submarginal V. (F 18k). V II numerous. Stems with short internodes and leaves distinctly grouped
P23B **SIMAROUBACEAE**
e.g. *Soulamea* (AS,OC)

35*   Petiole different.
    37 Ramification continuous on trunk. (ATTIMS).
P28A **ICACINACEAE**
e.g. *Anisomallon* (New Caledonia)

    37* Rhythmic growth (RAUH).
      38 AM, AF. Leaf folding involute (F16 d). (Base of petiole enlarged, F15 n; leaves cordate; f. s. spp.).
P22B **HUMIRIACEAE**
e.g. *Humiria* (AM)

      38* L. lobed or serrulate. Heterophylly (F10 a): the leaves are compound when plant get older.
P23A SAPINDACEAE

33*   Veins IV-V not very visible. (Some APOCYNACEAE might not produce latex during the dry season: see *key B*). (Young leaf folding not involute).
    39 Leaves modified into phyllodes (flattened structures).
P20 **LEGUMINOSAE**
Acacia (AU, ornam.)

    39* Leaves not modified into phyllodes.
      40 Plant glabrous.
        41 Young leaf folding convolute (F16 c). Branches sympodial. (Leaves asymmetrical, serrulate). P11A **THEACEAE**
Laplacea (AM)

        41* Young leaf folding not convolute. (Leaves serrulate or spiny).
          42 Leaves rhomboidal, acuminate.   **SANTALACEAE**
Jodina, AM

          42* Leaves not rhomboidal. Leaves serrulate: *Ilex*; leaves spiny: *Villaresia*.   P27A **AQUIFOLIACEAE**
Ilex
P28A **ICACINACEAE**
Villaresia, AM

      40* Young stems and underside of L. hairy. See 43a, 43b, and 43c:

        43a Leaves entire or serrulate, more or less petiolate. (RAUH; L. turning yellow when drying; f. s. spp.).
SYMPLOCACAEAE
Symplocos (AM,AS,AU)

        43b Littoral shrub. Lenticels distinctly marked. Existence of short twigs. Leaves small and elongate. **SURIANACEAE**
Suriana

       43c   Appressed hairs (F20 e). L. serrulate, subsessile.
P11A **THEACEAE**
Gordonia (AM)

30*   Venation different, more or less camptodromous (F17 c,d) or, if brochidodromous, the V II forming regularly curved loops (F17 a1), or plant slightly woody, or bark aromatic.
44   Bark with distinctly parallel fibres.
       Periderm suberising early or sheathing leaves with canaliculate petiole, F15 a,b). (Rhytidome scaly; leaves serrulate; self supporting form of a liana).
P6B **DILLENIACEAE**
e.g. *Davilla* (AM)

44*   Bark different.
    45   Shrub with drooping branches. Periderm retaining a green shade on numerous internodes. (Venation slightly embossed on upper side of blade, F18 s; plant slightly woody; with glandular hairs).   P29B **SOLANACEAE**
e.g. *Cestrum* (AM, ornam.)

    45*   Architecture or periderm different.
       46   Angular, or cannelate young internodes (F11 a,b). Periderm retaining a green shade on numerous internodes.
       Ramification immediate, F6 b. (Appressed hairs; bark or leaves aromatic, or with disagreeable smell; domatia; f. s. spp).
P2B **LAURACEAE**
e.g. *Ocotea*

       46*   Periderm and young stems different.
         47   Young internodes angular or cannelate (F 11 a,b,j). See 48a, 48b, and 48c:
           48a   Young internodes angular or cannelate. Periderm suberising soonly.   P2B **LAURACEAE**
e.g. *Aniba, Licaria* (AM)

           48b   Shrub or small tree. Young internodes with decurrent ridge (F11 j).   P28B **ASTERACEAE**
e.g. *Gochnatia* (AM)

           48c   Shrub. Rhytidome sloughing off in strips (F2b,d).
ONAGRACEAE
e.g. *Ludwigia*

        47*   Young internodes (sub)cylindrical.
          49   Not in AF. Leaves more or less petiolate, not aromatic. (RAUH; leaves coriaceous or turning yellow when drying; f. s. spp.).   **SYMPLOCACEAE**
Symplocos (AM,AS,AU)

          49*   Not in AM. Mangrove. Continuous ramification. (ATTIMS).   P13B **COMBRETACEAE**
Lumnitzera (Paleotrop.)

Key J

1 Leaves pinnate (simply or bi/tripinnate) with more than three leaflets. (Stipules adnate to the petiole: ARALIACEAE, ROSACEAE *s.l.*; a basal pair of leaflets can mimic stipules (F14 n) for some species of the SAPINDALES).
    2 AF, MA, AS. Herb, shrub or small tree. Stipules appressed (F13 h). Teeth rosoid, F20 c. Venation more or less scalariform (F17 k). (CHAMBERLAIN).
        P10B **LEEACEAE**
        Leea (AF,MA,AS)
  2* Characters different.
    3 Bark or leaves aromatic (bizarre smell for some species).
Leaves with pellucid dots. (Aromatic or bizarre smell, spines modified into false stipules; f. s. spp.).   P24B **RUTACEAE**
        Fagara
    3* Plant not aromatic.
      4 Petiolules with an abscission joint (F15 h). Leaflets entire. Petiole pulvinate at its base (F15 l). (Lenticels elongating transversely, F2 k; pellucid dots; f. s. spp.; leaflets capable of moving autonomously for most species).
        P20 **LEGUMINOSAE**
        key Z

      4* Petiolules without abscission joint (F15 j) or leaflets sessile, not capable of moving autonomously. Leaflets entire or serrulate.
        5 Teeth rosoid (F20 c).
          6 AF. or temperate zones. (Lenticels elongating transversely, F2 k).
        P8B **ROSACEAE**
        e.g. *Hagenia* (East AF mountains), *Sorbus* (North AM,AS)

          6* AM. L. serrulate with thin petiolules.   P8B **STAPHYLEACEAE**
        Huertea (AM)

        5* Teeth or geographical distribution different. See 7a, 7b, and 7c:
          7a Young internodes grooved or angular (F11 a,b). (L. bipinnate; self-supporting form of a liana; f. s. spp.).   P23A **SAPINDACEAE**
        e.g. *Paullinia* (AM)

          7b South AF. Internodes not angular-grooved. Teeth not rosoid, F20a. (Intrapetiolar stipules: *Melianthus*).   P8B **MELIANTHACEAE**
        Melianthus, *Bersama* (AF)

          7c AM. Veins II ending in a tooth. Bases of the upper three L. fused. Rhythmic growth marked by scale-leaves.   P15B **OCHNACEAE**
        Godoya splendida (AM)

1* Leaves palmate or trifoliolate. (The "palmate" L. of some MORACEAE-Cecropioideae are in fact simple but deeply divided L. (see *key D*), and in this case a stipular hood is present).
    8 Bark with network of fibres (F3 a,b,c,d), (trunk and branches have to be observed).

9 Petiole not distally geniculate or orange aqueous exudate in bark, petioles or young stems. **P17B BIXACEAE**
Cochlospermum

9* Petiole distally pulvinate (F15 p). Bark without orange exudate (with broad coloured strips of fibres in some BOMBACACEAE). P18B **MALVACEAE** *s.l.*
See *key X*

8* Bark without network of fibres.
  10 Latex. (Leaves trifoliolate; RAUH). **P19 EUPHORBIACEAE**
e.g. *Hevea* (AM, cult.), *Key W*

  10* No latex.
    11 Leaflets entire.
      12 Petiole pulvinate at the base (F15 l).
Leaves trifoliolate. Lenticels elongating transversely (F2 k), (trunk and branches have to be observed). (Red sap in the bark f. s. spp.). (Plant spiny, e.g. *Erythrina*). **P20 LEGUMINOSAE**
key Z

      12* Base of petiole not pulvinate.
        13 Bark or leaves aromatic.
Wood soft. Stipule adnate to the petiole (F13 r). (Small tree with modular architecture; fimbrial V., F17 q, e.g. *Schefflera*, AM). **P27A ARALIACEAE**
Numerous genera

        13* Plant not aromatic.
          14 Midrib on underside of blade, ending in a mucro and/or an indentation (F14 c). Leaves trifoliolate. **P16B CAPPARIDACEAE**
e.g. *Euadenia* (AF), *Crateva*
TOVARIACEAE
Tovaria (AM)

          14* Midrib different (RAUH; twigs supple). **P19 EUPHORBIACEAE**
e.g. *Piranhea* (AM), *key W*

    11* Leaflets serrulate or crenate. See 15a, 15b, and 15c:
      15a AM. Leaflets more or less crenate or with smooth teeth. (Bark resinous, translucent dots). **CARYOCARACEAE**
Anthodiscus (AM)

      15b Leaflets serrulate. (Scaly rhytidome: *Ricinodendron*). **P19 EUPHORBIACEAE**
e.g. *Ricinodendron* (AF), *key W*

      15c AS. Reddish exudate in bark. **BISCHOFIACEAE**
Bischofia (AS)

Key K

(Leaves truely compound?, or are the "leaves" phyllomorphic branches: e.g. *Phyllanthus*, bearing simple and stipulate leaves? See *key C*).

1 Leaves with more than three leaflets or leaves bifoliolate.
    2 Sheathing petiole (F15 a1).
        3 Wood soft. Plant aromatic. Branches erect or monocaulous plant. (Minute stipules adnate to the petiole, F13 r).   P27A **ARALIACEAE**
        e.g. *Sciadodendron* (AM)

        3* Wood yellow. Plant not aromatic.   P5B **BERBERIDACEAE**
        e.g. *Mahonia* (AM, temp.)

    2* Characters different. See 4a, 4b and 4c:
        4a Subtropical AF, AS. Leaves bifoliolate. Short twigs modified into spines. (Petiolules without an abscission joint; minute stipules?, CHAMPAGNAT).
        BALANITACEAE
        Balanites (AF,AS)

        4b AF, AS. Leaves bi- or tripinnate. Bark with curious smell (myrosine). (Bark with network of fibres; rachis with prominent glands).   **MORINGACEAE**
        Moringa (AF,AS, ornam.)

        4c Plant not as in (3a) or (3b).
            5 Petiolules with an abscission joint or an annular furrow (F15 h).
                6 Base of petiole pulvinate (F15 l).
                      7 Embossed venation (F18 s).
                          8 (Tree or self-supporting form of a liana; young internodes not angular; basal V II ascending, F17 h; lamina inserted above the petiolule, F14 d; lenticels circular, F2 j).   P21B **CONNARACEAE**
                      e.g. *Agelaea* (Paleotrop.), *Connarus*

                      8* AS. TROLL, KORIBA. Shrub or small tree. (Leaflets opposite).
                      P21B **OXALIDACEAE**
                      Averrhoa (AS, cult.)

                7* Venation not embossed.
                    Large tree. (Leaflets alternate or opposite, lenticels transversely elongated, F2 k, translucent dots: e.g. *Myroxylon*).
                    P20 **LEGUMINOSAE**
                    e.g. *Alexa*, *Dipteryx* (AM)

            6* Base of petiole not pulvinate.
                9 Leaves with translucent dots.
                    Bark aromatic. (Trunk spiny f. s. spp.).   P24B **RUTACEAE**
                    e.g. *Fagara* (incl. *Zanthoxylum*)

                9* No translucent dots.
                    Bitter bark. Leaves hairy.   P23B **SIMAROUBACEAE**
                    Picramnia (AM)

    5* Petiolules without abscission joint (F15 j). (A thin groove may exist but without forming a true abscission joint).
      10 Bark with network of fibres (F3 j) and external wood furrowed (F3 j).
          P6A **PROTEACEAE**
          e.g. *Roupala* (AM), *Grevillea* (AS,NG,AU,OC)
      10* Bark or wood different.
        11 Leaflets entire or serrulate but teeth not glandular. See 12a, 12b, and 12c:
          12a Leaves without peltate hairs (in a few cases, stellate hairs: *Aglaia, Campnosperma*).   P23-24 **SAPINDALES**
              key Y
          12b Leaves with very small appressed peltate hairs (F20 p).
              JUGLANDACEAE
            e.g. *Oreomunnea* (AM), *Engelhardia* (AS)

          12b Shrub not very woody. Branches sympodial. Leaves with some hairs.   P29B **SOLANACEAE**
            e.g. *Cyphomandra* (AM)

        11* AS. Leaflets with glandular teeth. Thickened petiolar base; cut bark turning orange: *Meliosma*).   P5B **MELIOSMACEAE**
          Meliosma (AS), *Ophiocaryon* (AM)
1* Leaves palmate or trifoliolate.
    13 Sheathing petiole (F15 a1) or wood soft and bark aromatic (smell of carrot). Branches erect or monocaulous plant. (Minute stipules adnate to the petiole, F13 r).
          P27A **ARALIACEAE**
          e.g. *Cussonia* (AF,MA, ornam.), *Schefflera*
    13* Characters different.
      14 Leaves with pellucid dots. (Plant aromatic).   P24B **RUTACEAE**
        e.g. *Aegle* (AS), *Cusparia* (AM)

    14* Leaves without pellucid dots. Plant not aromatic.
      15 AM. Short shoots (F11 p) with subwhorled L. Rachis winged (F14 p). Trifoliolate leaves accompanied by smaller, simple leaves.
          P30A **BIGNONIACEAE**
          Crescentia (AM, cult.)
      15* Characters different.
        16 Some twigs with distichous phyllotaxy (F9 a).
          AS. Leaves trifoliolate. Basal V II ascending (F18 j).
          P21B **OXALIDACEAE**
          Sarcotheca (AS)
        16* All twigs with spiral phyllotaxy.
          17 Base of petiolule with an abscission joint (F15 h). Leaves not pellucid dotted. See 18a, 18b, and 18c:
            18a Tree or as yet self-supporting young liana. Leaves trifoliolate. Embossed venation, F18 s. (Numerous protruding lenticels, F2 j; heterophylly, F10 a; f. s. spp.).

              P21B **CONNARACEAE**
              Agelaea (Paleotrop.)

18b AM. Twigs not lenticellate. UE forming monopodial series. V. not embossed. Leaves trifoliolate. (Minute stipules?). **PICRODENDRACEAE**
Picrodendron (AM, Cuba)

18c Shrub. Rhythmic growth with short shoots. Leaves trifoliolate. **OXALIDACEAE**
Oxalis (e.g. AM)

17* Base of petiolule without an abscission joint (F15 j).
P23-24 **SAPINDALES**
key Y

Key L

1 Leaves simple.
  2 Alternate phyllotaxy (F9 b) on trunk and branches.
    3 Leaves needle-like, grouped in 2, 3 or 5. (At altitude). **PINACEAE**
Pinus (Temperate AS, AM)
    3* Characters different.
      4 Paleotropical. Phyllotaxy somewhat sophisticated (three foliar spirals, F9 f2). Leaves large, elongated. (Adventitious roots on the trunk; SCARRONE, STONE). **PANDANACEAE**
Pandanus (AF,MA,AS,OC)
      4* Phyllotaxy simply spiral (F9 b), or distichous.
        5 All branches and twigs with spiral phyllotaxy or plant monocaulous.
          6 Leaves elongated, sheathing the stem (F15 a2) or tubular leaf sheath.
            7 Plant weakly ramified. Lamina flat. Leaf base almost amplexicaul (*Dracaena*). **AGAVACEAE**
e.g. *Yucca* (AM), *Dracaena* (Paleotrop.), *Cordyline* (AM,AS,AU)

            7* Plant monocaulous. Corrugate leaves.
              8 Trunk erect, not ramified. (L. bifidous). Tubular leaf sheath. Pl **ARECACEAE**
e.g. *Geonoma* (AM), *Johannesteijsmannia* (Sumatra) young sapling (e.g. *Cocos*)

              8* AM. Plant weakly woody. Leaf sheath as in F15 a2). Trunk creeping or climbing plant at a young stage, CHAMBERLAIN, L. bifidous: e.g. *Asplundia*; CORNER, leaves simple, e.g. *Ludovia*. **CYCLANTHACEAE**
e.g. *Asplundia, Evodianthus, Ludovia* AM)

          6* Leaves decurrent to the stem, not sheathing (F15 a3). Plant highly ramified.
            9 Not in AF. Ramification of trunk distinctly rhythmic, forming tiers of branches (F7 e). Leaves very coriaceous and acuminate. (RAUH, MASSART). **ARAUCARIACEAE**
Araucaria (AM,NG,AU,OC)

            9* Mostly in mountains of the Southern hemisphere. Ramification rhythmic (F7 d). Leaves not coriaceous. **PODOCARPACEAE**
e.g. *Podocarpus*
        5* Some branches or twigs with distichous phyllotaxy.
          10 Growth of some twigs rhythmic with large leaves and reduced leaves. Growth becoming abruptly plagiotropic (MANGENOT). **AGAVACEAE**
Cordyline (AS)

10* Aerial stems without rhythmic growth. Stems enlarged at the nodes. (Trunk generally remaining erect but, f. s. spp., growing sideways at a certain stage). **BAMBUSACEAE**
   e.g. *Dendrocalamus, Gigantochloa* (AS), *Bambusa*

 2* Branches with opposite phyllotaxy.
   Not in AF. Opposite phyllotaxy. (MASSART). **ARAUCARIACEAE**
   Agathis (AS,NG,OC)

1* Leaves corrugate and compound (if not corrugate: ZAMIACEAE).
   11 Leaves pinnate. Closed tubular leaf sheath. (Heterophylly: juvenile leaf of simpler form than adult leaf). In most cases unbranched tree or small plant, but trees sprouting at the base also frequent (e.g. *Bactris, Chrysalidocarpus*).Pl **ARECACEAE**
   e.g. *Bactris, Euterpe, Roystonea* (AM)
   Chrysalidocarpus (MA), *Elaeis* (AF, cult.), *Veitchia* (OC)
   11* Leaves palmate.
     12 Closed tubular leaf sheath. Tree monocaulous or sprouting at its base (aerial branching, e.g. *Hyphaene*, subtropical, old world). Pl **ARECACEAE**
     e.g. *Mauritia* (AM), *Borassus* (AF,AS)
     Phoenix (EU, AF,AS), *Livistona* (AF,AS,AU)

     12* AM. Leaf sheath as in (F15 a2). **CYCLANTHACEAE**
     e.g. *Carludovica* (AM)

N.B. The genera submentioned in this key are often cultivated as ornamentals. In America, the ZAMIACEAE (e.g. *Zamia*) and the CYCLANTHACEAE (e.g. *Ludovia*: leaves simple, *Asplundia*: leaves bilobate, *Carludovica*: leaves palmate) are very like small palm trees.

Key M

1 Stipules (very small in CLUSIACEAE and VOCHYSIACEAE).
  2 Latex (resinous or not resinous).
    3 Stipules or their scars conspicuous. Venation densely reticulate (F 18a1).
          P9B **MORACEAE**
          e.g. *Ficus* (AS), *Bagassa* (AM)
    3* Stipules small.
      4 Stipules lateral, small or minute (sometimes reduced to very small black dots or hair-like).
        5 Venation brochidodromous.
          6 Yellow or white latex in bark or leaves. Existence of UE in monopodial series, (RAUH, MASSART). (Young internodes angular).
            P15A **CLUSIACEAE**
            e.g. *Symphonia* (AM,MA)
          6* Latex white. Venation IV-V not very visible, underside of L. shiny. Intramarginal vein (F18 m). (L. 3-whorled f. s. spp., false stipules?).
            P31B **APOCYNACEAE**
            e.g. *Allamanda, Couma* (AM)
        5* Venation camptodromous.
          7 Tree. White latex. Venation densely reticulate. (F 18a1).
            P12B **SAPOTACEAE**
            Sarcosperma (AS)
          7* Shrub. Latex white. Venation not densely reticulate (F 18a2). Twigs sympodial. P19 **EUPHORBIACEAE**
            Chamaesyce (~*Euphorbia*), see *key W*
    4* Intrapetiolar stipules (F13 b).
      8 Buds with appressed hairs. Leaves glandular. P25B **MALPIGHIACEAE**
        e.g. *Spachea* spp. (AM)
      8* Leaves glabrous. P31B **APOCYNACEAE**
        e.g. *Kopsia* (AS)
  2* Aqueous exudate or gum.
    9 Gum in the bark. Interpetiolar stipules. P32 **RUBIACEAE**
      e.g. *Cinchona* (AM, cult.)
    9* AM. Wood of old individual may be resinous. P14A VOCHYSIACEAE
1* No stipules.
  10 Pairs of young leaves appressed for a time during their development (F16 h). Gum-resin yellow or orange. (Leaves with black dots or stellate hairs, latex orange, e.g. *Vismia, Harungana*). (Lamina with undulate resiniferous ducts, F17 r, e.g. *Garcinia*). (Bark with network of fibres f. s. spp.). P15A **CLUSIACEAE**
    e.g. *Vismia* (AM,AF), *Harungana* (AF,MA), *Garcinia*

10* Young leaves not observable, or very small at that moment.
    11 Hemiepiphyte with aerial roots growing towards the soil. P15A **CLUSIACEAE**
                                                                                           e.g. *Clusia* (AM)
  11* Characters different.
    12 Branches plagiotropic, regularly spaced on the trunk (F7 a).
      13 Internodes subcylindrical. (Latex white). (ROUX: e.g. *Picralima*, MANGENOT: e.g. *Funtumia*).   P31B **APOCYNACEAE**
                                                                                      e.g. *Funtumia, Picralima* (AF)

      13* Young internodes angular. (Latex more or less yellow; swollen petiolar base, F13 p; f. s. spp.).   P15A **CLUSIACEAE**
                                                                      Symphonia (AM,AF), *Garcinia*

    12* Architecture somewhat different or not identifiable.
      14 Latex.
        15 Twigs lenticellate.
            Leaves opposite or in whorls of 3 or 4. Latex white, pale yellow or opalescent. (Underside of L. pale-green, with V III-IV not very visible; LEEUWENBERG, PREVOST; apical bud embedded between the petiolar bases, F13 p; interpetiolar ridge; f. s. spp.).   P31B **APOCYNACEAE**
                                              e.g. *Alstonia, Tabernaemontana, Rauvolfia*

        15* Lenticels absent or very inconspicous.
          16 Latex pale yellow, yellow, orange or becoming brownish. (ROUX: e.g. *Garcinia*, MASSART: e.g. *Platonia, Symphonia*, RAUH: e.g. *Mammea*, KORIBA: e.g. *Cratoxylum*; swollen petiolar base, F13 p; bark with network of fibres, F3 a; numerous parallel V II; f. s. spp.).
                                                                    P15A **CLUSIACEAE**
             Mammea, Symphonia (AM), *Cratoxylum*(AS), *Garcinia*

          16* Leaves arranged in small groups at the end of the twigs. (AUBREVILLE; latex white).   P12B **SAPOTACEAE**
                                                                   e.g. *Pradosia* (AM)
  14* Resin or coloured exudate other than latex.
      17 Venation not parallelodromous.
        18 AS V II abruptly curved near the margin (F18 n). Growth and ramification rhythmic. (Bark a little resinous; crushed leaves giving off a resinous smell).
                                                                  P24A **ANACARDIACEAE**
                                                                        Bouea (AS)

        18* AU. Sticky exudate. Leaves opposite or in whorls of four.   P13B **MYRTACEAE**
                                                                e.g. *Syncarpia* (AU)

    17* AS. V. parallelodromous (F17 k). Resin in trunk. Petiole decurrent to the stem (F 15a3). P27B **ARAUCARIACEAE**
                                                                      Agathis (AS,NG,OC)

Key N

1   Venation pinnate, base of lamina not trinerved (F18 e,f,g) or venation not readily observable (e.g. in the case of the presence of a dense indumentum. Possibly scale-leaves subtending green stems).
   2   Stipules forming a ring, at least at their base (F13 c) or sheathing petiolar bases (F13 f).
      3   Leaves serrulate. Aromatic. (ATTIMS; nodes swollen).
                                                4B **CHLORANTHACEAE**
                              e.g. *Hedyosmum* (AM,AS), *Ascarina* (MA,AS,OC)

      3* Leaves entire or plant not aromatic. See 4a, 4b, and 4c:
         4a LEEUWENBERG, ROUX, SCARRONE, or other architectural model. (Helicoidal anisoclady f.s. spp., F6 d2).   P32 **RUBIACEAE**
                                                            e.g. *Gaertnera* (Paleotrop.)

         4b AU, OC. FAGERLIND. Ramification rhythmic. Branches plagiotropic by apposition (F6 e).   P31A **LOGANIACEAE**
                                                            e.g. *Logania* (AU,OC)

         4c Young stems grooved. Stipules larges, bilobed. Venation camptodromous. Appressed hairs (P20 e).   **DIALYPETALANTHACEAE**
                                                            Dialypetalanthus (AM)

   2* No stipular ring.
      5   Stipules entirely **intra**petiolar (F13 b,p).
         6   Leaves entire.
            7   The stipule is an expansion of the petiolar base (F13 p). Buds without appressed hairs.   P31A **LOGANIACEAE**
                                                          e.g. *Anthocleista* (AF,MA), *Fagraea* (AS,AU,OC)

            7* Stipules free from petiole, more or less triangular and appressed to the stem (F13 b). Hairs appressed, F20 e. (FAGERLIND; L. glandulate f. s. spp.).   P25B **MALPIGHIACEAE**
                                                e.g. *Burdachia*, *Byrsonima*, *Spachea* (AM)

         6   Leaves serrulate.
            8   AU, OC. Large intrapetiolar stipules.   P26B **CUNONIACEAE**
                                                      e.g. *Geissois* (AU,OC)

            8* AM. Stipules small. Leaves trinerved. V. camptodromous, (F17 c).
                                                           P9A **ULMACEAE**
                                                            Lozanella (AM)

   5* Stipule not entirely intrapetiolar (F13 a,c,d,e,g).
      9   **Inter**petiolar stipules or reduced annular stipules (F13 d,e).
         10 Leaves entire (exceptionally with large teeth for a few RUBIACEAE; stipule-like expansions at the nodes, leaves 3-whorled, peeling rhytidome: ONAGRACEAE, *Fuchsia*, AM).
            11 Peeling rhytidome as in (F2 a,b,d) or young stems without lenticels or venation camptodromous (F17 c). (Many different architectural models).   P32 **RUBIACEAE**
                                                              Numerous genera

P31A **LOGANIACEAE**
e.g. *Buddleja, Norrisia* (AS)

11* Characters different. (Twigs distinctly lenticellate). See a, b, and c:
   12a AM. Young liana? Underside of lamina with dense indumentum. Venation brochidodromous (F17 a). Lenticels numerous.
TRIGONIACEAE
Trigonia (AM)

12b AF, AS. Strong rhythmic growth. Modular erect branches. Plant glabrous. Venation brochidodromous (F17 a).
CTENOLOPHONACEAE
Ctenolophon (W AF, Malaisia)

12c Plant different, leaves without dense indumentum.
   13 Stipules more or less elongated (F13 d). (Minute teeth?). In mangrove: *Bruguiera, Rhizophora*. (ATTIMS, AUBREVILLE). ROUX: *Blepharistemma, Cassipourea*.
P26B **RHIZOPHORACEAE**
e.g. *Bruguiera* (Paleotrop.), *Rhizophora*
Blepharistemma(India), *Cassipourea* (AM,MA,Ceylon)

13* AS, OC. Large interpetiolar stipules sheathing the apical buds (F13 e). (Leaves coriaceous).
P26B CUNONIACEAE

10* Leaves serrulate. (Peeling rhytidome not scaly, numerous lenticels).
   14 AF, MA. Venation scalariform (F17 k). (ROUX).
P10A **RHAMNACEAE**
Lasiodiscus (AF,MA)

14* Venation not scalariform. (Minute or glandular teeth). See 15a, 15b, and 15c)
   15a Phyllotaxy opposite or whorled. (MASART, ROUX). (Lenticels numerous, F2 j; external wood furrowed; f. s. spp.). Hollow stems: *Gynotroches*.
P26B **RHIZOPHORACEAE**
e.g. *Gynotroches* (AS,OC), *Pellacalyx* (AS), *Cassipourea*

   15b Not in AF (mostly in AU, OC). Large interpetiolar stipules sheathing the apical buds (F13 e). (Leaves coriaceous; RAUH). P26B **CUNONIACEAE**
e.g. *Lamanonia* (AM), *Schizomeria* (AS,NG,AU)

   15c Phyllotaxy opposite. (RAUH). (No lenticels).
P26B **LOGANIACEAE**
e.g. *Buddleja, Norrisia* (AS)

9* Stipules not really interpetiolar (at least a part of the stem is free from stipule insertion between te leaves).
   16 AS. Stipule adnate to the petiole (F13 g). (FAGERLIND).
P31A **LOGANIACEAE**
Neuburgia

16* Stipule not adnate to the petiole.
    17 Trunk with a more or less spiral phyllotaxy.
        19 Leaves serrulate.
            AF. Venation scalariform (F17 k). (ROUX).
                      P22A CELASTRACEAE
                      Catha (East AF, cult.)

        19* Leaves entire.
            20 (TROLL). Stipules vestigial. L. not glandular. Pinnate, embossed (F18 s), and camptodromous venation
                      P14A LYTHRACEAE
                      Lagerstroemia (AS, ornam.)

            20* (MASSART). Stipules well developed. L. trinerved, glandular (F19 m) with embossed venation.
                      P10A RHAMNACEAE
                      e.g. *Colubrina*, *Krugiodendron* (AM)

    17* Opposite or whorled phyllotaxy in trunk. See **21**.

21 Leaves entire. (Scale-leaves subtending gree stems: RHAMNACEAE, *Colletia*, AM).
    22 Leaves supratrinerved (F18 e) or apressed hairs (F20 e).
        23 L. supratrinerved. (MANGENOT).   **P31A LOGANIACEAE**
                                                    Strychnos
        23* Appressed hairs. Twigs lenticellate. (Tree, shrub or young liana).
            (L. glandular f. s. spp.).   **P25B MALPIGHIACEAE**
                                                  Numerous genera
    22* Venation different or no appressed hairs. (If stipules very thin and translucent dots: MYRTACEAE: e.g. *Myrcia*, *Psidium*, AM).
        24 Branches or small twigs sympodial. See 25a, 25b, and 25c:
            25a Underside of L. pale green and shiny. (Leaves opposite or subopposite, the two L. of each pair slightly unequal in size).   **P17A VIOLACEAE**
                                                  e.g. *Rinorea* (AM)

            25b Leaves opposite or subopposite, pairs of leaves with equal sized L. Stipules minute. (Laticiferous threads, F4 u; orange layer under the rhytidome; f. s. spp.).   **P22A CELASTRACEAE**
                                                  (incl. Hippocrateoideae)

            25c Twigs supple. Rhy/Per with longitudinal slits (F2 e,f).
                                              P19 EUPHORBIACEAE
                                          e.g. *Mischodon* (AS), *Key W*
        24* AM. AF. Branches monopodial.
            26 Branches erect: *Vochysia*, plagiotropic: *Qualea*. V. pinnate (F18 f,g); Leaf folding conduplicate (F16 b1. Cauline gland accompanying the stipule, F19 b,: *Qualea*).   **P14A VOCHYSIACEAE**
                      e.g. *Qualea*, *Vochysia* (AM), *Erismadelphus* (West AF)

            26* Short shoots modified into spines.   **P10A RHAMNACEAE**
                                                          e.g. *Scutia*

21* Leaves serrulate (vestigial teeth for some QUIINACEAE).
    27 Leaves serrulate with V II secant (F17 p).
        28 AM. Small stipules. (Indumentum of abundant hairs f. s. spp).
                                                  P26B BRUNELLIACEAE
                                                             Brunellia (AM)

28* AM. Four linear lanceolate stipules at each node. Leaves glabrous. Sympodial trunk. (CHAMBERLAIN, KORIBA). P15B **QUIINACEAE**
e.g. *Quina* (AM)

27* Venation different.
29 Paleotrop. Underside of lamina, buds or young stems with appressed hairs (F20 e). (Indument shiny: *Sericolea*). P18A **ELAEOCARPACEAE**
e.g. *Aceratium* (AS,AU,OC), *Sericolea* (NG)

29 Neotrop. or appresed hairs.
30 Leaves glandular (F19 q).
Branches modular, each module with a few (or only one) pairs of leaves. (Latex scanty). P19 **EUPHORBIACEAE**
e.g. *Tetrorchidium* (AF), key W

30* Leaves not glandular.
31 Leaves opposite or subopposite, the two L. of each pair slightly unequal in size. Branches sympodial. (Underside of L. pale green and a little shiny). P17A **VIOLACEAE**
e.g. *Rinorea* (AM)

31* Leaves opposite or subopposite, pairs of leaves with equal sized L. Stipules minute. (Branches sympodial, F6 f). (Laticiferous threads, F4 u; orange layer, i.e. phelloderm, under the rhytidome; f. s. spp.). P22A **CELASTRACEAE**
e.g. *Zinowiewia* (AM), *Lophopetalum* (AS,AU), *Salacia*

1* Venation readily observable. Venation palmate (F18 c) or leaves trinerved (F18 d).
32 Petiole not distally pulvinate.
33 Leaves not glandular.
34 Bark fibrous (meshes of the network of fibres much elongated, F3 c). (Some herbs anisophyllous, F10 g). P9A **URTICACEAE**
e.g. *Boehmeria*

34* Bark not fibrous.
35 AM. Venation densely reticulate (F18 a1). P9A **ULMACEAE**
Lozanella (AM)

35* Subtrop. South AM, AU. Leaves finely serrulate. P18A **ELAEOCARPACEAE**
Aristotelia (AM,AU,OC)

33* Leaves glandular (margin of lamina glandular, F19 q). P10A **RHAMNACEAE**
e.g. *Colubrina* (mainly AM)

32* Petiole with a distal pulvinus (F15 p).
Rhy/Per with thin longitudinal slits (F2 e,f), this character is transient. (Anisophylly, F10 g, f. s. spp.). P19 **EUPHORBIACEAE**
e.g. *Mallotus* (key W)

Key O

1 Leaves distinctly 3- or 5-nerved (F17 j,m) or supratrinerved (F18 e).
   2 Venation typical, scalariform with extending secondary veins (F17 j). (Peeling rhytidome scaly, F2 b,d).   P14B **MELASTOMATACEAE**
Numerous genera
   2* Venation different. See 3a, 3b, and 3c:
      3a Branches plagiotropic. Rhytidome not scaly. Lamina supratrinerved (F18 e). (Liana at a young stage; stem prickly; f. s. spp.).   P31A **LOGANIACEAE**
Strychnos

      3b Stems more or less quadrangular (F11 d).   P30A **VERBENACEAE**
e.g. *Gmelina* (Paleotrop.), *Citharexylum* (AM)

      3c Section of stem more or less circular or oval-shaped. If internodes grooved, than furrows on two opposite sides only of one internode (F11 e).
         4 Heterophylly (F10 a). Liana at a young stage. (Leaves becoming compound on an older plant).   P30A BIGNONIACEAE

         4* No such heterophylly. Young leaf involute (F16 d). (At altitude; petiole glandular; f. s. spp.).   **CAPRIFOLIACEAE**
e.g. *Viburnum* (mainly temperate AM,AS)

1* Venation pinnate (F18 f,g,h) or parallelodromous (F17 f) or almost invisible or leaves very small and scaly.
   5 Base of internodes swollen (F11 g).
      6 Stems angular. Leaves entire or serrulate. (Shrub or herb; sectorial anisoclady, F6 d$_\Lambda$). Axillary spines: e.g. *Barleria*.   P30B **ACANTHACEAE**
Numerous genera

      6* AM. Stems cylindrical. Leaves entire. (Herb).   P7A **NYCTAGINACEAE**
Mirabilis (AM, ornam.)

   5* Base of internodes not swollen.
      7 Leaves aromatic. Leaves serrulate.
         Rain forest. Twigs retaining a green shade.   P4B **CHLORANTHACEAE**
e.g. *Sarcandra* (AS)

         Dry forest. Periderm suberising soonly.   **LAMIACEAE**
e.g. *Dendrohyptis* (AM)

      7* Plant different.
         8 AM. Plant monocaulous.   P31A **LOGANIACEAE**
Potalia (AM)

         8* Plant ramified.
            9 Paleotrop. Leaves very small and scaly, in whorls of four, eight or more. (External wood furrowed, F3 g).   **CASUARINACEAE**
Casuarina (mainly AS,AU,OC)

            9* Plant different.

10   Some twigs plagiotropic with opposite phyllotaxy and twisting of the internodes (F9 k).
   11   Veins of orders IV-V faintly visible. (Small trees)
      12   Trunk with rhythmic branching. (MASSART, KORIBA, MANGENOT, rhytidome remaining thin in old trees).
         P14B **MELASTOMATACEAE**
         e.g. *Mouriri* (AM), *Memecylon* (AF,MA)

      12* Trunk with continuous branching. (ROUX).
         GENTIANACEAE
         e.g. *Tachia* (AM)
   11* Venation distinctly reticulate.
      13   Bark with network of fibres.
         Leaves with translucent dots or lines. (Young leaves appressed, F16 h; MANGENOT).
         P15A **CLUSIACEAE**
         Marila (AM)
      13* Bark different.
         AM. Small tree. Helicoidal anisoclady (F6 d2). Petiolar bases of leaf pairs almost united. (ATTIMS, ROUX).   P31A **LOGANIACEAE**
         Antonia (AM)
10* All stems different.
   14   Venation scalariform (F17 k).
      Mainly in temperate zones or in tropical AS. Periderm of the young twigs smooth and shiny.   P27B **CORNACEAE**
         e.g. *Cornus* (AS)
   14* Venation different.
      15   Petiolar scars more or less united in pairs (F15 g).
         16   (Base of petiole with stipuliform expansions, if not: e.g. *Coutoubea*; LEEUWENBERG, SCARRONE, KORIBA, FAGERLIND).
            P31A **LOGANIACEAE-GENTIANACEAE**
            e.g. *Anthocleista* (AF,MA), *Coutoubea* (AM)
            Fagraea (AS,AU,OC)

         16* (RAUH). No stipular expansions. (Venation camptodromous).   **GARRYACEAE**
            Garrya (North AM)
      15* Petiolar scars different.
         17   Underside of lamina with black dots or lines, or resiniferous ducts (F17 r), or leaves glandular. (If coloured exudate, see *key M*).
            18a   Black dots, (ROUX), e.g. *Vismia* or resin ducts, (FAGERLIND), e.g. *Tovomita*.
               P15A **CLUSIACEAE**
               e.g. *Tovomita, Vismia* (AM)

            18b No appressed hairs. (Base of lamina lamina with 1-2 cupuliform glands: *Citharexylum*).
               P30A **VERBENACEAE**
               e.g. *Citharexylum* (AM)

18c Buds with appressed hairs (F20 e). (base of bladeglandular). P21A **MALPIGHIACEAE**

17* Leaves neither dotted nor glandular. See **19**.

19 External wood with elongated parallel furrows (F3 e). (Small plants).
P28B **ASTERACEAE**
e.g. Heliantheae (e.g. *Calea, Verbesina*, AM)

19* Wood different.
    20 Rhytidome peeling to become scaly on young branches. See 21a, 21b, and 21c:
        21a All stems erect. Ramification delayed (F6 a): *Buddleja* or immediate (F6 b): *Nuxia*. (Leaves serrulate or faintly crenate). P31A **LOGANIACEAE**
Nuxia (AF,MA), *Buddleja*

        21b Rhytidome that comes out in fibrous strips (F2 b). Some stems plagiotropic. Leaves entire. Ramification delayed. P14A **LYTHRACEAE**
e.g. *Woodfordia* (Paleotrop.)

        21c Rhytidome sloughing off in strips. (F 2 d). L. serrulate.
HYDRANGEACEAE
Hydrangea (AS, ornam.)

20* Rhytidome of branches different.
    22 Anisophylly (F9 d, F10 g), often not very conspicuous. Ramification delayed (F6 a). (L. 3-whorled; sectorial anisoclady, F6 d1; f. s. spp.).
P30B **ACANTHACEAE**
e.g. *Bravaisia* (AM)

    22* Plant different.
        23 Young leaf folding involute (F16 d). (scaly or stellate hairs f.s.spp).
CAPRIFOLIACEAE
e.g. *Viburnum* (AM,AS)

        23* Leaf folding different.
            24 Leaves 3-whorled or, if opposite, young stems more or less quadrangular (F11 d). (Mangrove: AVICENNIACEAE: *Avicennia*).
P30A **VERBENACEAE**
e.g. *Citharexylum* (AM), *Tectona* (AS, cult.)

            24* Plant different. (Leaves entire).
                25 MA, AS. Venation IV-V (almost) not visible. (If translucent dots: MYRTACEAE). P14B **MELASTOMATACEAE**
Memecylon (MA,AS), *Mouriri* (AM)

                25* Venation readily visible.
                    26 Not in AM. In a mangrove or rear-mangrove habitat. L. oval. (Vertical aerial roots; stems glandular, F19 a). P14A **LYTHRACEAE**
Sonneratia (Paleotrop.)

                    26* AM or not associated with mangrove.
                        27 AS. Branches sympodial. (Peeling rhytidome).
CRYPTERONIACEAE
Crypteronia (AS)

27* AM. Petiolar base swollen (F13 p). (FAGERLIND). P15A **CLUSIACEAE** e.g. *Tovomita* (AM)

Key P

1 Base of lamina distinctly 3- or 5-nerved (F17 j,m) or supratrinerved (F18 e).
   2 Leaves not aromatic. Venation typical (leaves trinerved and venation scalariform, F17 j). (Peeling rhytidome scaly, F2 b,d).   **P14B MELASTOMATACEAE**
e.g. *Miconia* (AM)
   2* Leaves aromatic or venation different.
      3 Crushed leaves aromatic. No translucent dots
Periderm retaining a green shade on numerous internodes. Leaves supratrinerved (F18 e). (Venation scalariform f. s. spp., F17 k).
P2B LAURACEAE
Cinnamomum (AS,AU,OC)
      3* Leaves not aromatic or leaves with translucent dots.
         4 Venation polygonal, without blind veinlets (F18 b). L. entire. Base of internodes swollen (F11 g). (Laticiferous threads in the leaves, F4 u; ROUX).
GNETACEAE
Gnetum
         4* Plant different.
            5 Weak rhythmic growth.
               6 Stems more or less quadrangular (F11 d). Base and upperside of blade glandulate: e.g. *Gmelina*.   **P30A VERBENACEAE**
e.g. *Gmelina, Premna* (AF,AS)

               6* Stems sub cylindrical. Small translucent dots.   **P13B MYRTACEAE**
e.g. *Rhodamnia* (AS,OC)
            5* Strong rhythmic growth.
               7 Suberization forming scales (F2 a). Stems quadrangular. Petioles short.   **BUXACEAE**
e.g. *Buxus*

               7* Mostly boreal. Twigs lenticellate. Petioles long.   **SAPINDACEAE**
Acer (some spp. in tropical AS)

1* Venation pinnate (F18 f,g,h) or parallelodromous (F17 f) or invisible or leaves very small and scaly.
   8 Leaves with translucent or black dots, or leaves glandular or with small grooved spots on underside of blade, or glandular trichomes (use a hand lens).
      9 Petiole or lamina with typical glands (see F19).
         10 Buds, underside of lamina or young stems with appressed hairs (F20 e). Venation brochidodromous (F17 a).   **P25B MALPIGHIACEAE**

         10* Leaves, buds and stems without appressed hairs. (Hairs oblique f. s. spp.) or venation camptodromous (F17 c,d).
            11 AS, AU, OC. Leaves 3-whorled. Base and upper side of lamina with 1-2 cupuliform glands. (V I ramified at its end, F18 q).
P30A BIGNONIACEAE
Deplanchea (AS,AU,OC)

11* Leaves subopposite or whorled.
 12 Internodes angular. Base of lamina glandular.
               P 30A **VERBENACEAE**
               e.g. *Citharexylum* (AM)

 12* Internodes not angular. (MASSART: *Terminalia*; ATTIMS, in mangrove, glanduar trichomes: *Laguncularia*). (V. camptodromous, petiole or base of lamina glandular; peeling rhytidome scaly, F2 b,c,d; f. s. spp.). P13B **COMBRETACEAE**
    e.g. *Laguncularia* (AM,AF), *Terminalia* (AS,OC), *Combretum*

9* No typical glands (glandular trichomes, translucent or black dots, etc.).
 13 Petiole distally enlarged (F15 m,p,q) or with sheathing stipuliform base (F13 p).
   (Petiole with an abscission joint, F15 q). P24B **RUTACEAE**
         e.g. *Metrodorea* (AM), *Evodia* (AF,AS,AU,OC)
 13* Petiole different.
  14 Rhytidome sloughing off in scales or strips (F2 b,c,d). (Branches and trunks have to be observed carefully).
   15 Shrub with decumbent, monopodial branches. Bark not aromatic (disagreeable smell; leaves with translucent pouches or glandular trichomes; f. s. spp.). P13B **COMBRETACEAE**
                e.g. *Combretum*

   15* Branches not decumbent or branches sympodial. Pellucid dots. (Leaves aromatic; intramarginal vein, F18 k,n; if branches plagiotropic than opposite-pseudodistichous phyllotaxy by twisting of internodes, F9 k).
    RAUH: e.g. *Pimenta*, *Eugenia malaccensis*; TROLL: e.g. *Psidium*; KORIBA: e.g. *Eugenia*. ATTIMS, the adult tree with spiral phyllotaxy: *Eucalyptus*. P13B **MYRTACEAE**
    e.g. *Pimenta*, *Psidium* (AM), *Eucalyptus* (AS,AU,OC), *Eugenia*

  14* Peeeling rhytidome different.
   16 Architecture sympodial, some modules becoming erect to form the trunk, other modules becoming plagiotropic to form the branches (F8 e). (Branches consisting of short modules for *Calyptranthes*; venation as in F18 k,m). P13B **MYRTACEAE**
            e.g. *Calyptranthes* (AM), *Eugenia*

   16* Architecture different or characters not observable (tree too high). See 17a, 17b, and 17c:
    17a Venation brochidodromous with one, or in some cases, two intramarginal veins (F18 k). (Architecture: F8 e).
               P13B **MYRTACEAE**
           e.g. *Myrceugenia* (AM), *Eugenia*

    17b Venation different. Periderm retaining a green shade on numerous internodes. (Laticiferous threads in the leaves, F4 u). P3A **MONIMIACEAE**
              e.g. *Xymalos* (AF)

17c Shrub. External wood furrowed (F3 e). Large translucent marginal pouches on lamina. **P28B ASTERACEAE**
e.g. *Porophyllum* (AM)

8* Leaves without translucent / black dots or glandular trichomes. Leaves opposite or subwhorled, (trunk with alternate phyllotaxy f. s. spp.).
  18 Peeling rhytidome of trunk or branches in strips or scaly (F2 b,c,d). Venation not parallelodromous.
    19 Some twigs with opposite phyllotaxy and twisting of the internodes (F9 k) or pairs of large L. alternating with pairs of small L. (F10 j).
      20 Veins IV-V not observable. See 21a, 21b, and 21c:
        21a (Young leaves revolute, F16 e; MASSART, MANGENOT). **P14B MELASTOMATACEAE**
e.g. *Memecylon* (AF), *Mouriri* (AM)

        21b ROUX-ATTIMS. Leaves subsessile. **P14A LYTHRACEAE**
Pemphis (Paleotrop. coasts)

        21c Young internodes not angular. (MASSART; leaves glandular; old leaves turning orange-red; f. s. spp.). **P13B COMBRETACEAE**
e.g. *Terminalia* (AS), *Combretum*

      20* Veins IV-V visible.
Young internodes angular (F11 h). **P14A LYTHRACEAE**
e.g. *Lagerstroemia* (AS,AU, ornam.)

    19* All stems with opposite-decussate or opposite phyllotaxy accompanied by bending or twisting of the petioles (F9 j). (Whorled phyllotaxy f. s. spp.).
      22 Venation camptodromous, without intramarginal veins (F17 c). **P14A LYTHRACEAE**
Lagerstroemia (AS,AU)

      22* Venation different, brochidodromous (F17 a).
        23 Young leaf folding involute (F16 d). (Mainly in temperate zones). **CAPRIFOLIACEAE**
e.g. *Viburnum* (AM,AS)

        23* Leaf folding different. See 24a, 24b, and 24c:
          24a Branches modular (F8 f). Intramarginal veins (F18 k). **P13B MYRTACEAE**
e.g. *Calyptranthes* (AM)

          24b Venation different. Shrubs or weakly woody. Peeling rhytidome scaly (F2 d). (Branches erect, sympodial: e.g. *Lafoensia*; in moutains: e.g *Fuchsia*, or in swamps: e.g. *Ludwigia*; whorled phyllotaxy; f. s. spp.). **ONAGRACEAE**
e.g. *Lafoensia, Fuchsia* (AM), *Ludwigia*

          24c Short shoots modified into axillary spines. **P7A NYCTAGINACEAE**
e.g. *Pisonia*

  18* Rhytidome different or venation parallelodromous (F17 f).

25 Intramarginal or fimbrial veins (F18 k,m,n; F17 q) or parallelodromous venation.
   26 Intramarginal veins. (Really no latex? see also CLUSIACEAE, *key L*)
      27 Branches erect.
         28 AF. Young internodes with two decurent ridges starting from petiole insertion (F11 j). **OLINIACEAE**
Olinia (AF)

         28* AM. Leaf folding conduplicate (F16 b1). (Stipules minute; MANGENOT). P14A **VOCHYSIACEAE**
Ruizterania (AM)

      27* Branches plagiotropic. See 29a, 29b, and 29c:
         29a Trunk with opposite phyllotaxy. Intramarginal veins (F18 k,n). (See F8 e). P13B **MYRTACEAE**
e.g. *Eugenia*

         29b AS. Trunk with spiral phyllotaxy. (MASSART). P14A **LYTHRACEAE**
Duabanga (AS)

         29c AS. Numerous parallel V II. Branches sympodial (KORIBA). Underside of leaves waxy. P14A **CLUSIACEAE**
Mesua (AS)

   26* AS, NG, OC. Venation parallelodromous. Petiole decurrent to the stem (F15 a3). **ARAUCARIACEAE**
Agathis (AS,NG,OC)

25* Venation different or not very visible.
   30 Petiole distally enlarged, (F15 p,q), (abscission joint f. s. spp.).
      31 AS. Stems quadrangular (F11 d) or lenticels not protruding. P30A **VERBENACEAE**
e.g. *Teijsmanniodendron* (AS)

      31* AM. Stems not quadrangular. Lenticels protruding (F2 n). (Petiole with an abscission joint, F15 q). P30A **BIGNONIACEAE**
e.g. *Tabebuia* (AM)

   30* Petiole not distally enlarged.
      32 Leaves serrulate, crenate or lobate, or bark with network of fibres (F3 a,b,c), (observe trunk and branches).
         33 Leaves serrulate, crenate or lobate.
            34 Periderm retaining a green shade on numerous internodes.
               35 Young stems not quadrangular. Periderm retaining a green shade on numerous internodes or bark with network of fibres. (Anisophylly, F10 g, much pronounced in *Glossocalyx*; teeth glandular, F20 b; laticiferous threads in leaves, F4 u; f. s. spp.). P3A **MONIMIACEAE**
e.g. *Glossocalyx* (AF), *Mollinedia* (AM)

               35* AS. Bark not fibrous. Leaves serrulate.

AUCUBACEAE
e.g. *Aucuba* (Himalaya to Japan)
34* Periderm different.
36 Young stem more or less quadrangular (F11 d). Bark without network of fibres. (Young rhytidome thin and shiny; vertical serial buds, F12 a). P30A **VERBENACEAE**
e.g. *Premna* (AF,AS)

36* Lenticels protruding (F2 n). Bark not fibrous. P30A **OLEACEAE**
e.g. *Tessarandra* (AM), *Forsythia* (AS, ornam.)

33* Leaves entire (or teeth very inconspicuous: MONIMIA-CEAE).
37 Twigs not easily breakable, (bark with network of fibres, F3 a). P25A **THYMELAEACEAE**
e.g. *Lophostoma* (AM), *Craterosiphon* (AF), *Phaleria* (AS,OC)

37* Twigs easily breakable.
38 External wood furrowed (F3 j). (Bark with network of fibres. F3 j). P6A **PROTEACEAE**
e.g. *Panopsis* (AM)

38* Wood different. (Apices of young internodes flattened; periderm retaining a green shade on numerous internodes (in twigs or in young trunk); branches sympodial). P3A **MONIMIACEAE**
e.g. *Siparuna* (AM)

32* Leaves entire. Bark without network of fibres. See **39**.

39 Bark or leaves with an aromatic, resinous, or special spicy smell.
40 AS. V II abruptly bent near the margin (F18 p). Branches erect, ramification delayed (F6 a). UE with several scale-leaves (F10 f). P24A **ANACARDIACEAE**
Bouea (AS)
40* Venation different.
41 Branches sympodial, firstly erect than becoming oblique (MANGENOT; periderm retaining a green shade on numerous internodes, special smell in bark, small stellate hairs). P3A **MONIMIACEAE**
e.g. *Siparuna* (AM)

41* Branches monopodial. Young internodes angular. P2B **LAURACEAE**
e.g. *Beilschmiedia* (AM), *Caryodaphnopsis* (AS)
39* Bark and leaves not aromatic.
42 Leaves coriaceous. Two parallel longitudinal ribs extending below petiole insertion (F1 j). (Marginal or submarginal vein, appressed hairs). **BUXACEAE**
e.g. *Buxus* (AM, AS), *Notobuxus* (AF)
42* Characters different.
43 Branches modular, bifurcate (F8 d), each module consisting of only one UE (F6 f). See **44a, 44b,** and **44c**:

44a Two unequal modules at a same node, one of these can be absent, i.e. anisoclady (F8 g).
    45 Mostly in AM. (External wood furrowed, F3 g; anisophylly; rust-coloured indument on the apices.). **P7A NYCTAGINACEAE**
e.g. *Neea* (AM), *Pisonia*

    45* Paleotrop. Twigs pseudomonopodial (only one leader at node).
P28A ICACINACEAE
Cassinopsis (AF,MA)

44b AF, MA. Two unequal modules. Leaves glabrous, shiny.
MONTINIACEAE
e.g. *Grevea* (AF,MA)

44c AS. V. scalariform (F17 k). **P27B CORNACEAE**
e.g. *Cornus* (AS)

43* Branches different, not bifurcate (monopodial or sympodial).
    46 AS. V. scalariform (F17 k). (Temperate zones). **P27B CORNACEAE**
e.g. *Cornus* (AS)

    46* Venation not scalariform.
        47 Some leaves 3-whorled.
            48 Young stems quadrangular (F11 d). (Also *Olea europea*).
P30A VERBENACEAE
e.g. *Clerodendron*

            48* Young stems different. **P30A OLEACEAE**
Jasminum

        47* Leaves never 3-whorled.
            49 Venation IV not distinctly visible.
                50 Underside of leaves with small cavities (= glandular trichomes). **P30A OLEACEAE**
e.g. *Osmanthus* (AS), *Fontanesia* (AS, China)
Comoranthus (MA), *Linociera*

                50* Leaves different.
                    51 Not in AM. (Leaves somewhat fleshy-coriaceous; nodes enlarging after leaves have fallen).
P14B MELASTOMATACEAE
e.g. *Memecylon* (MA,AS)

                    51* Young stems grooved on two opposite sides (F11 e). **P13B MYRTACEAE**
e.g. *Campomanesia* (M)

            49* Venation distinctly reticulate.
                52 Young internodes quadrangular (F11 d). (Vertical serial buds, F12 a; interpetiolar ridge inconspicuous; f. s. spp.). **P30A VERBENACEAE**
e.g. *Callicarpa* (AS)

                52* Young internodes subcylindrical.
                    53 Lenticels protruding, F2 n; Indumentum of very short erect or spherical hairs.
P30A OLEACEAE
e.g. *Noronhia* (MA), *Linociera*

53* Lenticels not protruding, or absent. See 54a, 54b, and 54c:

   54a AM. Axillary buds with several scale-leaves (F10 f). (Indumentum abundant; if minute stipules present, see *key M*).
P14A VOCHYSIACEAE
Callisthene (South Brazil)

   54b AS. ROUX. Petiolar bases covering the apical bud. V. densely reticulate.
P14A CLUSIACEAE
Poeciloneuron (India)

   54c AS, AU, OC. Branches sympodial (Internodes flattened distally).
SANTALACEAE
e.g. Santalum (AS,AU,OC)

Key Q

1   Stipules present.
    2   Leaflets serrulate or V II secant (F17 n,p) or fimbrial vein (F17 q).
        3   Leaflets serrulate with V II ending in a tooth. (F17 p).
            4   AM, AS. Stipules fused, interpetiolar.
                5   Young leaf folding involute (F16 d).   **P8B STAPHYLEACEAE**
e.g. *Turpinia* (AM,AS)

                5*   AU. Leaves hairy.   **P26B CUNONIACEAE**
Vesselowskya (AU)

            4*   Characters different.
                5   AM. (RAUH). Stipules small. L. hairy.   **P26B BRUNELLIACEAE**
Brunellia (AM)

                5*   AM. Stipules narrow and elongated. Leaves glabrous. (Sympodial trunk).
P15B QUIINACEAE
Froesia, Touroulia (AM)

        3*   Leaflets with different venation.
            6   Interpetiolar stipules (F13 d,e). (L. with more than three leaflets).
P26B CUNONIACEAE
e.g. *Weinmannia* (AM,MA,AS)

            6*   AM. Stipules not interpetiolar. Leaves with three crenate leaflets, teeth not acute.   **CARYOCARACEAE**
Caryocar (AM)

    2*   Leaflets entire. Venation different.
        7   Stipules quite large.
            Stipules elongated (interpetiolar, F13 d,e), leaving large scars. L. entire or serrulate (often coriaceous). Leaves pinnate (L. palmate, intrapetiolar stipules: *Geissois*).   **P26B CUNONIACEAE**
e.g. *Geissois* (AU,OC)

        7*   Stipules small.
            8   AM. In savannahs or deciduous forests. Sympodial branches consisting of short modules.   **P8B ZYGOPHYLLACEAE**
e.g. *Guaiacum, Bulnesia* (AM)

            8*   AM. Architecture different. (Lenticels becoming transversely elongated, F2 k).   **P20 LEGUMINOSAE**
e.g. *Taralea, Platymiscium* (AM), key Z

1*   No stipules.
    9   Leaves pinnate with more than three leaflets.
        10   Bark or leaves aromatic, or leaves with translucent dots.   **P24B RUTACEAE**
e.g. *Flindersia* (AS,AU,OC), see *key Y*

        10*   Neither translucent dots, nor aromatic smell.
            11   AS. Young stems quadrangular (F11 d). Rachis winged (F14 p).
P30A VERBENACEAE
Peronema (AS)

11* Stems not quadrangular or rachis not winged. (L. bipinnate f. s. spp.).
    12 Petiolule without an abscission joint (F15 j) or young leaflets involute (F16 d).
        13 N. or S. temperate. Young leaflets involute. Leaflets with small teeth. (Stipuliform glands f. s. spp.). **CAPRIFOLIACEAE**
Sambucus (AM,AS)

        13* Young leaf folding different. A few species of the Sapindales have oppositate leaves. e.g. *Dysoxylum*, **MELIACEAE**
e.g. *Acer, Paranephelium*, **SAPINDACEAE**

    12* Petiolule with an abscission joint (F15 h), (not well marked f. s. spp.). Rachis swollen at nodes (F14 m). Young leaflets folding not involute.
        14 Lenticels protruding, F2 n; leaves simply pinnate: *Spathodea* (AF), *Tecoma* (AM), or L. bipinnate: *Jacaranda* (AM), *Millingtonia* (AS); CORNER: *Colea* (MA).
P30A **BIGNONIACEAE**

        14* AM, AS. (In temperate zones, large axillary buds: *Fraxinus*).
P30A **OLEACEAE**
Fraxinus (North AM,AS, ornam.), *Schrebera* (AF,AS)

9* Leaves palmate or trifoliolate.
    15 Leaves with translucent dots or aromatic. P24B **RUTACEAE**
e.g. *Myllanthus* (AM)

15* Neither translucent dots nor aromatic smell.
    16 Petiole distally enlarged or pulvinate (F15 m,p,q).
        AM. Branches erect, sympodial. (Protruding lenticels, F2 n, f. s. spp.).
P30A **BIGNONIACEAE**
e.g. *Tabebuia* (AM, ornam.)

    16* Petiole different.
        17 Lenticels protruding (F2 n).
        Existence of short shoots (F11 p) or branches decumbent: *Parmentiera*. Branches erect: e.g. *Godmania*.
P30A **BIGNONIACEAE**
e.g. *Parmentiera, Godmania* (AM)

        17* Lenticels not protruding.
            18 Petiolule with an abscission joint (F15 h).
                19 Petiolar base suberized or young internodes more or less quadrangular (F11 d) or, if leaves are 3-whorled, section of internodes triangular. P30A **VERBENACEAE**
e.g. *Teijsmanniodendron* (AS), *Vitex*

                19* AF. Petiole and internodes different. (Rhy/Per with longitudinal slits?, F2 e,f). P19 **EUPHORBIACEAE**
Oldfieldia (AF)

            18* AM, temperate EU, AS. Petiolule without abscission joint (F15 j). P23A **SAPINDACEAE**
e.g. *Aesculus* (AM,EU,AS, ornam.), *Billia* (AM)

Key R

1 Some twining stems.
    2 Leaves opposite or whorled. See 3a, 3b, and 3c:
        3a Base of internodes swollen (F11 g). Laticiferous threads in the leaves, F4 u. (Latex yellowish; venation, see F18 b). **GNETACEAE**
e.g. *Gnetum urens* (AM)

        3b Latex white or pale yellow. (Most of the ASCLEPIADACEAE are herbs with glandular leaves, their woody species can produce a very corky rhytidome).
P31B **APOCYNACEAE**
e.g. *Forsteronia* (AM), *Landolphia* (AM, AF,MA)
Parsonsia (AS,AU,OC)

        3c Latex red.
L. glandular. Appressed hairs (F20 e). P25B **MALPIGHIACEAE**

L. not glandular. Hairs erect.
P22A **CELASTRACEAE**
e.g. *Prionostema* (AM), *Salacia* (e.g.AF)

  2* Distichous or spiral phyllotaxy.
    4 Stipules.
        5 Stipule hood-like (F13 m). P9B **MORACEAE**
e.g. *Ficus* (e.g. AS)

        5* Stipule different.
            6 Leaves trinerved (F18 d) or palmately nerved (F18 c).
Latex white or coloured. (Stipules; lamina with basal glands; Rhy/Per with thin longitudinal slits, F2 e,f).P19 **EUPHORBIACEAE**
e.g. *Manniophyton* (AF)

            6* Leaves or leaflets with pinnate venation.
Petiolules with an abscission joint (F15 h). Leaves trifoliolate or pinnate. Red exudate in bark. (Lenticels becoming transversely elongated, F2 k). P20 **LEGUMINOSAE**
e.g. *Machaerium* (AM), *Mucuna*

    4* No stipules.
        7 Latex white, milky (F3, see text). Venation pinnate or palmate. Hairs appressed (F20 e) or hairs simple, stellate (F20 m), erect or spherical (use a hand lens). (Young stems soft, compressible; woody or herbaceous liana).
P29A **CONVOLVULACEAE**
e.g. *Ipomoea*

        7* Red exudate. Modules of trunk distally twining (F5 b).
P21B **CONNARACEAE**
e.g. *Manotes* (AF), *Connarus*

1* No twining stems.
    8 Tendrils.
        9 Margin of leaves or leaflets not serrulate. Leaves entire, bilobate or bifoliolate. Tendrils circinnate (F5 n) or bifid. (Bark producing a red exudate).
P20 **LEGUMINOSAE**
e.g. *Bauhinia*

9* Leaflets serrulate. Tendrils circinnate.   P23A **SAPINDACEAE**
    e.g. *Paullinia*
8* No tendrils.
  10 Stipular hood or sheathing stipule, leaving an annular scar (F13 m). Climbing plant or (hemi)epiphyte. Thigmonastic twigs (see F5) for *Ficus* sect. *Sycidium*. Latex white (Moroideae) or blackish (Cecropioideae).
    P9B **MORACEAE**
    Cecropioideae, e.g. *Coussapoa* (AM)
    Moroideae, e.g. *Ficus*
  10* No stipular hood. Stipules small or non-existent.
    11 Leaves simple.
      12 Opposite phyllotaxy.
        Hemiepiphyte (*Clusia*), with aerial roots, or epiphyte (*Clusia, Garcinia*) or weakly prostrate (*Vismia*).   P15A **CLUSIACEAE**
        e.g. *Clusia* (AM), *Vismia* (AF), *Garcinia* (AU)

      12* Spiral or distichous phyllotaxy.
        13 AF. Weakly prostrate plant. Venation pinnate (F18 f,g,h). Apices with very short hairs. (ROUX).   P12B **SAPOTACEAE**
        Very few spp. of *Chrysophyllum* (AF)

        13* Paleotrop. Venation densely reticulate (F18 a1).
        P9B **MORACEAE**
        e.g. *Streblus* spp., *Trophis* spp. (AS?)
    11* Leaves compound. (Spiral phyllotaxy).
      Plant weakly prostrate? (F5 e). Young internodes angular or cannelate (F11 a,b). Latex white, milky.   P23A **SAPINDACEAE**
      e.g. *Paullinia* (AM)

Key S

1 Some stems twining (F5 a,b).
   2 Stipules.
      3 MA, AS. Bark with network of fibres (F3 a,b,c,d). Petiole distally enlarged (F15 p) or very short. (Trunk angular).   **P18B STERCULIACEAE**
Byttneria (MA,AS)

      3* Bark without network of fibres. Rhy/Per with thin longitudinal slits (F2 e,f), this character being transient. (Base of lamina glandular, F19 m).
P19 EUPHORBIACEAE
e.g. *Omphalea*, see *key W*

   2* No stipules.
      4 Base of petiole decurrent to the stem (F11 m).
         5 Leaves or bark aromatic. (Herb; nodes enlarged, F11 h; prophyll appressed to the stem, F12 h; f. s. spp.).   **P5A ARISTOLOCHIACEAE**
e.g. *Aristolochia* (mainly AM)

         5* Venation campylodromous (F17 e).   **DIOSCOREACEAE**
Dioscorea

      4* Petiolar base not decurrent (F11 n).
         6 Petiole enlarged at both ends. Young twigs soon becoming woody. (Section of trunk with radiate concentric zones, F4 k). (Leaves peltate, F14 e; e.g. *Cissampelos*).   **P5A MENISPERMACEAE**
Numerous genera

         6* Plant different.
            7 Long (prehensile?) petiole able to bend to reorientate the lamina (F15 m).
P28A ICACINACEAE
e.g. *Pyrenacantha* (AF,AS), *Phytocrene* (AS)

            7* Petiole different.
               8 Leaves entire or lobate but margin not serrulate. Underside of lamina with appressed, stellate or spherical hairs (use a hand lens). (Woody plant with compressible young stem or herb). (Latex?).
P29A CONVOLVULACEAE
e.g. *Ipomoea*

               8* Leaves serrulate. (Hairs simple).   **P28A ICACINACEAE**
e.g. *Pyrenacantha* (AF), *Natsiatum* (AS)

1* No twining stems.
   9 Twigs prehensile (F5 c,d) or presence of tendrils (F5 n,p,q) or of hooks (F5 f).
      10 Twigs prehensile, becoming woody.
         11 Stipules. Twigs ending in a hook (F5 f): *Bandereia*; twig-tendril: *Bauhinia*.
P20 LEGUMINOSAE
e.g. *Bandereia* (AF)

         11* No stipules. Twigs hook-like. Bark aromatic.   **P2A HERNANDIACEAE**
Sparattanthelium (AM)

10* No prehensile woody twigs.
  12 Tendrils.
    13 Axillary tendrils (F5 n,p) inserted in a plane formed by the stem and the petiole.
      14 Stipules.
        15 Petiole glandular (F19 f) or underside of lamina glandular.
          P16B PASSIFLORACEAE
          e.g. *Passiflora* (AM,AS,AU), *Adenia* (Paleotrop.)

        15* Leaves not glandular. (Tendrils simple or bifid; lenticels becoming transversely enlarged, F2 k).
          P20 LEGUMINOSAE
          Bauhinia
      14* AS. No stipules. Petiole not glandular. Leaves cordate.
        P7B OLACACEAE
        Erythropalum (AS)
    13* Tendrils inserted otherwise.
      16 Tendrils oppositifoliate (F5 q). Stipules. (Rays of phloem, F4 k; L. serrulate, teeth glandular, F20 b). **P10B VITACEAE**
        e.g. *Tetrastigma* (AS,AU), *Cissus*
      16* Tendrils not oppositifoliate.
        Two tendrils inserted on the petiole (F5 u), (reduced into stipules, or leaves spiny; f. s. spp.). **SMILACACEAE**
        Smilax

        Herbaceous vine. Tendril (or short twig bearing several tendrils) forming a right angle with the stem and the petiole. (Numerous genera). **CUCURBITACEAE**
        Numerous genera
  12* Plant without tendrils (plant bearing hooks).
    17 Stipules. (Peduncles of inflorescences modified into hooks, F5 f; venation scalariform; f. s. spp.). **P10A RHAMNACEAE**
      e.g. *Gouania* (mainly AM)

    17* AM. No stipules. Leafless short shoots modified into hooks.
      P2A HERNANDIACEAE
      Sparattanthelium (AM)

9* No prehensile twigs, tendrils, or hooks. (Plant weakly prostrate or epiphyte).
  18 Stipule or sheathing petiolar base (F15 a1).
    19 Leaves glandular.
      Underside of lamina with glands disposed in two symmetrical arches: *Manniophyton*; 2 glands at the base of the lamina (F19 j): e.g. *Croton*.
      P19 EUPHORBIACEAE
      e.g. *Manniophyton* (AF), *Croton, key W*
    19* Leaves not glandular.
      20 Petiolar base sheathing the stem.
        Plant climbing by means of adventitious roots. (Bark or leaves aromatic). (Stipule adnate to the petiole f. s. spp.).
        P27A ARALIACEAE
        e.g. *Hedera* (EU,AS,AU)

20* Petiolar base not sheathing. See 21a, 21b, and 21c:
 21a Petiole not distally pulvinate, quite long. Lamina trinerved (F18 d). (L. serrulate; stinging hairs; thigmonastic twigs or rigid hairs fastening the stem to the support f. s. spp.).
   P9A URTICACEAE
   e.g. *Urera* (AM,AF,MA,OC)

 21b Short twigs modified into spines. Petiole not distally pulvinate.
   P9A ULMACEAE
   e.g. *Celtis* (e.g. AM)

 21c Stems not spiny. Petiole distally pulvinate (F15 p).
   P18B TILIACEAE
   e.g. *Grewia* (AF,AS,AU)

18* No stipules. Petiolar base not sheathing. (Leaves entire).
22 AS. Base of lamina asymmetrical, distinctly trinerved.
   P27B ALANGIACEAE
   Alangium

22* Lamina symmetrical. Petiole quite long, distally pulvinate (F15 p).
   P5A MENISPERMACEAE
   e.g. *Tiliacora* (AF,AS,AU)

Key T

1 Plant with green stems and very reduced leaves.
   2 AM, AF, MA. Leaves reduced into hairs or spines forming areoles (F13 z). Epiphyte with flattened or cylindrical green stems. **CACTACEAE**
e.g. *Epiphyllum*, *Rhipsalis* (mainly AM, ornam.)

   2* Paleotrop. Twining plant. Leaves reduced to scales subtending leaf-like linear cladodes. **ASPARAGACEAE**
Asparagus (e.g. AU)

1* Plants with well developed leaves.
   3 Bark with network of fibres (F3 a,b,c). No stipules.
Bark aromatic (with pleasant smell). Twining plant or climbing by means of hooks (F5 h, *Artabotrys*) or with prehensile twigs (F5 j, *Rauwenhoffia*, *Uvaria*). P3B **ANNONACEAE**
e.g. *Artabotrys* (AF,AS), *Rauwenhoffia* (AS,AU)
Uvaria (Paleotrop.)

   3* Bark different or leaves stipulate.
     4 Some twining stems (F5 a,b).
       5 No stipules, possibly stipule-like petiolar scars. Rhytidome scaly. Tangential section of bark disclosing long parallel ridges. Petiole canaliculate, F15 b.
Scaly hairs: *Tetracera*; tick hairs: *Davilla*, *Doliocarpus* (petioles and stems scabrous); thin hairs: *Pinzona*. P6B **DILLENIACEAE**
e.g. *Hibbertia* (mainly AU,OC), *Tetracera*
Davilla, *Doliocarpus*, *Pinzona* (AM)

     5* Plant different. (Stipules present or absent).
       6 Ochrea (F13 s,t) present, its annular scar readily observable. (Short twigs, F11 p; young leaf revolute, F16 e; xylem lobate, F4 d). P6B **POLYGONACEAE**
e.g. *Coccoloba* (AM)

      6* Stems without ochrea or annular petiolar scar.
       7 Veinlets III and IV easily observed or indumentum very hairy.
        8 Stipules.
         9 Leaves serrulate.
          10 AS, MA. Teeth small or minute. Leaves not glandular. (Laticiferous threads in bark, F4 u). P22A **CELASTRACEAE**
e.g. *Celastrus* (MA,AS)

          10* AM. Teeth small or minute. L. not glandular. (Leaf underside shiny). P17A **VIOLACEAE**
Anchietea, *Corynostylis* (AM)

         9* Leaves entire.
Apices with oblique or appressed (F20 e) hairs. (Twining character sometimes not evident to observe). (Underside of lamina glandular, F19 w). P25B **DICHAPETALACEAE**
Dichapetalum (e.g. AM, AF)

8* No stipules.
- 11 Leaves or bark aromatic.
  AS. Vertical serial buds (F12 a). (Node with one large appressed prophyll, F12 h; extremities of stems with young leaves in a zigzag line). **SCHISANDRACEAE**
  e.g. *Schisandra, Kadsura* (AS)
- 11* Leaves and bark not aromatic.
  - 12 Leaves glandular.
    AF, MA. Base of lamina glandular (F19 m).
    P25B **MALPIGHIACEAE**
    Acridocarpus (AF,MA)
  - 12* Leaves not glandular. See 13a, 13b, and 13c:
    - 13a Basal V II grouped (F18 h). Hairs appressed, stellate or spherical (hand lens!). (Young stems compressible; phloem protruding in the wood, F4 n). P29A **CONVOLVULACEAE**
      e.g. *Dicranostyles* (AM), *Neuropeltis* (AF)
    - 13b Leaves coriaceous. (Petiole faintly enlarged distally, F15 m; Venation embossed, F18 s).
      ICACINACEAE
      e.g. *Leretia* (AM), *Desmostachys* (AF,MA)
      Neostachyanthus, Pyrenacantha (AF), *Sarcostigma* (AS)
    - 13c AS. Leaves glabrous, smooth. (Old petioles suberized). **SABIACEAE**
      Sabia (AS)

7* Veinlets III-IV almost invisible. Leaves glabrous or not very hairy. No stipules. Small glandular grooves, here and there, on the underside of the lamina (F19 v) or base of lamina glandular.
P26A **POLYGALACEAE**
e.g. *Moutabea* (AM)

4* No twining twigs.
- 14 Some twigs or branches prehensile or vegetative axes bearing hooks or tendrils. (Inflorescence ending in a tendril: *Antigonon*, POLYGONACEAE).
  - 15 Twigs prehensile by thigmonastism (F5 j).
    - 16 Leaves simple, entire. Venation pinnate. (Glands in stipular position, F19 a, f. s, spp.). P26A **POLYGALACEAE**
      e.g. *Bredemeyera* (AM), *Securidaca* (mainly AF)
    - 16* Venation embossed on upper side of blade, camptodromous, scalariform. No glands. P0A **RHAMNACEAE**
      e.g. *Ventilago* (AS)
  - 15* Plant different.
    - 17 Twigs modified into tendrils.
      - 18 Leaves glandular. (Tendril at the axil of a leaf, F5 n).
        P16B **PASSIFLORACEAE**
        e.g. *Passiflora* (AM)
      - 18* Leaves not glandular.

19 AS. Tendril at the base of a leafy twig, (tendril more or less perpendicular to the plane defined by the trunk and the twig, F5 p). **LOPHOPYXIDACEAE**
Lophopyxis (AS)

19* AF. Short twig ending in a tendril. Young leaves revolute (F16 e). Ochrea minute (F13 t).
P6B **POLYGONACEAE**
Afrobrunnichia (AF)

17* Twigs different (hooks or prehensile petioles f. s. spp.).
  20 Some leaves with hooks or with tendrils (F5 u,v) or prehensile petioles.
    21 Tendrils or hooks.
      22 AF. Bifid hook at the apex of the leaf (F5 v). Heterophylly well marked (F10 a).
DIONCOPHYLLACEAE
e.g. *Dioncophyllum* (AF)

      22* Petiole bearing tendrils (F5 u).
SMILACACEAE
Smilax

    21* No tendrils and no hooks. (Petiole prehensile).
      23 Plant weakly woody. P29B **SOLANACEAE**
e.g. *Solanum* (e.g. AM)

      23* Plant woody. (Laticiferous threads in the leaves, F4 u). P28A **ICACINACEAE**
e.g. *Desmostachys* (AF,MA), *Pyrenacantha* (AF)

20* Leaves different but plant bearing hooks (F5 f,g).
  24 AF, AS. Short shoots (F11 p) bearing the leaves, and long shoots bearing terminal hooks (F5 h). (Minute stipules?). **ANCISTROCLADACEAE**
Ancistrocladus (AF,AS)

  24* Plant different.
    25 AM. Underside of lamina lepidote (F20 p). Hooks (or thick woody tendril) subtended by the leaves. P16B **PASSIFLORACEAE**
Ancistrothyrsus (AM)

    25* Plant different.
      26 Paleotrop. Hooks opposite or subopposite. (Stipules). P22B **LINACEAE**
Hugonia (mainly AF,AS)

      26* Hooks alternate or twigs ending in a hook. See 27 a, 27b, and 27c:

        27a V. scalariform and camptodromous (F17 c,k). Stipules.
P10A **RHAMNACEAE**

e.g. *Ventilago* (mainly AS)

27b Paleotrop. Venation different. L. serrulate. Young L. folding involute (F16 d). (Stipules).
P22B **LINACEAE**
Indorouchera (AS)

27c Paleotrop. Hook inserted at the base of a twig. No stipules.
P7B **OLACACEAE**
Anacolosa (mainly AS)

14* No prehensile twigs. Neither hooks nor tendrils. (Plant supporting itself by its branches or by its spines, petiolar bases or adventitious roots or epiphyte).

28 Stipules (hairy or fringed f. s. spp.) or the two stipule-like prophylls modified into spines (F12 n).

29 Ochrea (very short f. s. spp.) leaving an annular scar (F13 s,t). Young leaf folding revolute (F16 e). (Xylem lobate, F4 d).
P6B **POLYGONACEAE**
e.g. *Coccoloba* (AM)

29* Plant different.

30 Young trunks with lenticels in longitudinal rows. Apices with appressed hairs (F20 e). (Young internodes angular).
P25B **DICHAPETALACEAE**
Dichapetalum

30* Lenticels different or non-existent or no appressed hairs. See 31a, 31b, and 31c:

31a Underside of the lamina with well marked midrib ending in a small mucro or an indentation (F14 c). Some twigs with distichous phyllotaxy (F9 h). (stipules modified into spines f. s. spp.).
P16B **CAPPARIDACEAE**
e.g. *Capparis*

31b Underside of lamina different. (Lenticels becoming trasversely elongated, F2 k). P20 **LEGUMINOSAE**
e.g. *Dalbergia*

31c AM, in deciduous forests. Phyllotaxy spiral, but becoming distichous at the end of the twigs. (CHAMPAGNAT, MANGENOT; Powerful odour of garlic, stipules modified into spines; f. s. spp.).
P7A **PHYTOLACCACEAE**
e.g. *Seguieria* (AM)

28* No stipules (pssibly axillary spines).

32 AM. Spines paired near leaf axils. Leaves supratrinerved. (CHAMPAGNAT). P28B **ASTERACEAE**
e.g. *Dasyphyllum* (AM)

32* Characters different.

33 AF. L. with a intramarginal vein (F18 m) or a fimbrial vein (F17 q). **P25A THYMELAEACEAE**
Dicranolepis (AF)

33* Leaves different.

34 AF. Liana bearing zigzagging plagiotropic branches (MANGENOT). **P28A ICACINACEAE**
Rhaphiostylis (AF)

34* Plant different.

35 Nodes enlarged (F11 h). Sympodial branches. Plant aromatic. **P4B PIPERACEAE**
Piper

35* Plant different.

36 Hairs scaly (F20 p). (Bark fibrous; Twigs oriented backwards). **ELAEAGNACEAE**
e.g. *Elaeagnus* (e.g. AS)

36* Hairs different or none. See 37.

37 AS, AU, OC. Venation parallelodromous (F17 f). (Plant climbing by means of adventitious roots). **PANDANACEAE**
Freycinetia (AS,AU,OC)

37* Venation different.

38 Leaves coriaceous. Venation not well marked. See 39a, 39b, and 39c:

39a Rhytidome scaly. Basal V II grouped (F18 h). **P12A ERICACEAE**
e.g. *Satyria* (AM), *Agapetes* (AS,AU,OC)

39b AM. Petiole short. (Heterophylly well marked between understorey-leaves and canopy-leaves; lenticels in longitudinal rows; underside of lamina with glandular striae). **P11A MARCGRAVIACEAE**
e.g. *Marcgravia, Norantea* (AM)

39c MA, AS. **P21B OXALIDACEAE**
Dapania (MA,AS)

38* Leaves not coriaceous or venation distinctly reticulate.

40 External wood furrowed (long parallel furrows, F3 e). Plant weakly prostrate. (CHAMPAGNAT). **P28B ASTERACEAE**
e.g. *Piptocarpha* (AM)

40* External wood not furrowed.

41 Adventicious clamp-roots. Main shoot with very long internodes. Annular prophyll (F12 j). Petiolar base sheathing the stem (F15 a1). (Roots modified into tendrils, F5 w; f. s. spp.). **ARACEAE**
e.g. *Heteropteris* (AM)

41* No such roots.

42 Leaves thick. (Hemiepiphyte). L. grouped: *Markea*. **P29B SOLANACEAE**
e.g. *Solandra, Markea* (AM)

42* Leaves not thick. See 43a, 43b, 43c, and 43d:

43a Branches zigzaguing, sympodial. **P29B SOLANACEAE**
e.g. *Lycianthes* (AM)

43b Branches not zigzaguing. Both sides of the blade hairy.
P29A **BORAGINACEAE**
e.g. *Tournefortia* (AM)

43c Upper side of the blade glabrous. Branches plagiotropic, monopodial (ROUX). P11B **MYRSINACEAE**
e.g. *Embelia* (AS)

43d AM. Hemiparasitic, epiphytic. Distichous pyllotaxy. Venation poorly visible. **EREMOLEPIDACEAE**
e.g. *Antidaphne* (AM)

Key U

1   AF, MA, AS. Trunk sympodial, each module ending in a tendril (F5 m). (Xylem in blocks, F4 g).   **P28A ICACINACEAE**
Iodes (AF,MA,AS)

1*   Plant different.
  2   Trunk twining (F5 a,b), (but twigs might be not).
    3   Internodes swollen at the base (F11 g).
      4   Leaves glabrous. Venation "polygonal", without blind veinlets (F18 b). (Laticiferous threads in the leaves, F4 u; woody liana).   **GNETACEAE**
Gnetum

      4*   Venation different. L. without threads. Weakly woody.
        5   AM. Venation pinnate (F18 f,g,h).   **P30B ACANTHACEAE**
Mendoncia (AM,MA)

        5*   AF, MA, AS. Stems not quadrangular. Leaves trinerved (F18 d), (entire, serrulate or lobate).   **P30B ACANTHACEAE**
e.g. *Thunbergia* (AF,AS)

    3*   Base of internodes not swollen.
      6   Interpetiolar ridge (F13 y).
        7   Stipules.
          8   Stipules interpetiolar (F13 d).   **P32 RUBIACEAE**
e.g. *Manettia* (AM), *Atractogyne* (AF), *Paederia*

          8*   Stipules free (F13 a), or if fused no extending from one petiolar to the other.
            9   AM. Indumentum of erect or oblique hairs. (Numerous lenticels).   **TRIGONIACEAE**
Trigonia (AM)

            9*   Hairs appressed (F20 e). (Numerous lenticels; stipules small; L. glandulate: e.g. *Banisteriopsis*, *Stigmaphyllon*, or not, e.g. *Tetrapteris*, interpetiolar stipules f. s. spp.).   **P25B MALPIGHIACEAE**
e.g. *Hiptage* (MA,AS,OC)

        7*   No stipules.
          10   External wood with long parallel furrows (F3 e). Leaves not glandular.   **P28B ASTERACEAE**
e.g. *Mikania* (AM)

          10*   External wood not furrowed or leaves glandular.
            11   Rhytidome peeling in fibrous strips (F2 d).
              12   Temperate zones; young leaf folding involute (F16 d).   **CAPRIFOLIACEAE**
e.g. *Lonicera* (EU, North AM,AS)

              12*   Venation brochidodromous and scalariform (F17 j).   **P14B MELASTOMATACEAE**
e.g. *Leandra*, *Topobea* (AM)

11* Rhytidome different.
    13 Stems, leaves or buds hairy.
        14 Buds with erect hair. **P30A OLEACEAE**
e.g. *Jasminum bifarium* (AS)
        14* Appressed hairs (F20 e).
P25B MALPIGHIACEAE
e.g. *Heteropteris* (AM)
    13* Plant glabrous.
        15 AS. Stems distinctly quadrangular (F11 d). Leaves entire, trinerved (F18 d). **P30A OLEACEAE**
Myxopyrum (AS)

        15* AS. Young internodes angular. Leaves not trinerved. **P31A LOGANIACEAE**
e.g. *Gardneria, Gelsenium* (AS)

6* No interpetiolar ridge.
    16 Venation campylodromous (F17 e) or supratrinerved, but without blind veins. (Leaves lobate; prickly stems; f. s. spp.). **DIOSCOREACEAE**
Ripogonum (NG,AU,OC), *Dioscorea*

    16* Venation different.
        17 Leaves supratrinerved (F18 e), trinerved (F18 d) or venation palmate (F18 c).
            L. trinerved (F18 d) or venation palmate. L. opposite or 3-whorled. (Stems quadrangular). **P30A VERBENACEAE**
e.g. *Clerodendron*

        17* Venation pinnate (F18 f,g,h).
            18 Rhytidome becoming scaly (F2 b).
(Wood with strands of phloem, F4 m; bark with disagreeable smell; underside of L. with coloured spots; leaves glandular; f. s. spp.). **P13B COMBRETACEAE**
e.g. *Combretum*

            18* Rhytidome different.
                19 Petiole not distally enlarged.
                    20 Minute stipules or hairs appressed (F20 e).
P25B MALPIGHIACEAE
e.g. *Banisteriopsis* (AM)

                    20* No stipules. No appressed hairs.
                        21 Venation distinctly reticulate, brochidodromous (F17 a). (L. 3-whorled f. s. spp.; petiole modified into a spine: *Clerodendron*).

P30A VERBENACEAE
e.g. *Aegiphila, Petrea* (AM)

                        21* Phyllotaxy opposite, subopposite or occasionally alternate. (V. camptodromous, F17 c; blade with basal glands in the American *Combretum*; bark with disagreeable smell f. s. spp.).
P13B COMBRETACEAE
e.g. *Combretum*

  19* Petiole distally enlarged (F15 m). (Leaves 3-whorled f. s. spp.). **P30A OLEACEAE**
e.g. *Jasminum* (AF,AS)

2* Trunk not twining.
 22 Existence of twining or prehensile (thigmonastic) twigs (F5 c).
  23 Stipules small or minute (young stems have to be observed).
   24 Apices with appressed hairs (F20 e). (Leaves glandular). Some twining twigs. (Interpetiolar ridge, F13 y). **P25B MALPIGHIACEAE**
e.g. *Banisteriopsis* (AM)

   24* No appressed hairs. Leaves without glands. (Thigmonastic twigs; xylem lobate: F4 n, in blocks: F4 p or in rings: F4 q). (Laticiferous threads in the leaves, F4 u; leaves subopposite; orange-coloured layer under the rhytidome; f. s. spp.). **P22A CELASTRACEAE**
incl. Hippocrateoideae, e.g. *Salacia*

  23* No stipules. Small twining twigs or twigs forming hooks. (If there are small leaflets at the base of a tendril, the plant may be a BIGNONIACEAE, see *key V*). (Branches not easy to break owing to the presence of a fibrous bark). **P25A THYMELAEACEAE**
e.g. *Craterosiphon* (AF), *Enkleia* (AS)

 22* No twining or prehensile twigs.
  25 Plant climbing by means of hooks (F5 g,k). See 26a, 26b, and 26c:
   26a Hook subtended by a leaf or a scale-leaf (F5 g). Leaves supratrinerved (F18 e). (Wood with strands of phloem, F4 m). **P31A LOGANIACEAE**
Strychnos

   26b Hook subtended by a laminar leaf. Venation pinnate (F18 f,g). **P32 RUBIACEAE**
e.g. *Canthium* (Paleotrop.), *Uncaria*

   26c Old petiolar base hook-like (F5 k). Stems quadrangular. L. trinerved (F18 d) or venation palmate (F18 c). (L. 3-whorled f. s. spp.). **P30A VERBENACEAE**
e.g. *Clerodendron*

  25* Plant without hooks.
   27 Stipules or stipular expansions on petioles. See 28a, 28b, and 28c:
    28a Interpetiolar stipules (F13 c,d,f). Epiphyte, e.g. *Hillia* (AM). (If appressed hairs, L. glandular, and stipules small: MALPIGHIA-CEAE). **P32 RUBIACEAE**
e.g. *Sabicea* (AM), *Canthium* (Paleotrop.)

    28b Intrapetiolar stipules (F13 b,p).
     29 Epiphyte. Stipular expansions (F13 p). **P31A LOGANIACEAE**
e.g. *Fagraea* (AS,AU,OC)

     29* Stipule free from petiole (F13 b). L. trinerved (serrulate). (Anisophyllic herb f. s. spp.). **P9A URTICACEAE**
e.g. *Pilea*

28c Stipules small, normally inserted (F13 a). Shoot short modified into spines. **P10A RHAMNACEAE**
Scutia (AM,MA)

27* No stipules, or stipular expansions.

30 Leaves supratrinerved. Basal secondary veins ascending with tertiary V. perpendicular to the midrib (F17 j). (Anisophylly, F10 g; clamp-roots). **P14B MELASTOMATACEAE**
e.g. *Adelobotrys* (AM)

30* Venation not scalariform.

31 Parasitic plant with sucker-roots. (Nodes enlarged). (Red or yellow inflorescences in the LORANTHACEAE). **P7B LORANTHACEAE**
e.g. *Phoradendron* (AM), *Dendrophthoe* (Paleotrop.)
P7B VISCACEAE
Dendrophthora (AM), *Viscum* (AF,MA,AS)

31* Plant not parasitic. (Plant weakly prostrate).

32 Rhytidome becoming scaly (F2 b). Glands or glandular trichomes. (Bark with disagreeable smell f. s. spp.). **P13B COMBRETACEAE**
Combretum

32* Rhytidome different. See **33**.

33 Base of internodes swollen (F11 f) or geniculate. **P30B ACANTHACEAE**
e.g. *Anisacanthus* (AM)

33* Internodes different.

34 AM. V. brochidodromous (F17 a). Rhy/Per more or less shiny. Stem almost cylindrical, not quadrangular. Lenticels protruding (F2 n). **P30A BIGNONIACEAE**
Schlegelia (AM)

34* Plant different.

35 Paleotrop. Stem not spiny. See 36a, 36b, and 36c:

36a Leaves (supra)trinerved (F18 d,e) or venation pinnate (F18 f,g). (Petiole with an articulation f. s. spp.). **P30A OLEACEAE**
Jasminum (AF,AS)

36b Venation pinnate (F18 f,g). (Plant sarmentous, L. with minute stellate hairs). **P3A MONIMIACEAE**
Palmeria (NG,AU)

36c Plant climbing by means of divaricate twigs. **P31A LOGANIACEAE**
Strychnos

35* Stems spiny.

37 L. supratrinerved (F18 e), somewhat coriaceous. Petiole not articulate. (Wood with strands of phloem, F4 m). **P31A LOGANIACEAE**
Strychnos

37* Venation pinnate. Leaves thin and soft. **P7A NYCTAGINACEAE**
e.g. *Pisonia aculeata* (AU)

Key V

1 Phyllotaxy opposite or whorled (or seeming "alternate" due to a pronounced anisophylly for a few spp.).
    2 Leaves with a terminal tendril (F5 r) or a scar at the apex of the rachis. Leaves modified into trifid hooks for *Macfadyena* (AM). Trunk twining: *Pandorea* (AS, AU). (Glands located on the stem in the proximity of the nodes; lenticels protruding, F2 n; stellate hairs; f. s. spp.).   **P30A BIGNONIACEAE**
Numerous genera: e.g. *Arrabidaea* (AM)
    2* Leaves without tendril or apical scar.
        3 Petiole prehensile (F5 s). In AF, AU, or in temperate zones.   **P5B RANUNCULACEAE**
Clematis (AF,AU)
        3* Petiole not prehensile. (L. trifoliolate articulate, F15 q).   **P30A OLEACEAE**
Jasminum (Paleotrop.)
1* Spiral or distichous phyllotaxy.
    4 Trunk or its extremity twining. (Trunk twisted but plant not twining for some LEGUMINOSAE).
        5 Leaves palmate.
            6 Section of trunk or branches with radiate concentric zones (F4 k).   **P5A MENISPERMACEAE**
e.g. *Disciphania* (AM), *Burasaia* (MA)
            6* Trunk and branches different.
                7 Paleotrop. Petioles modified into hooks (F5 t). Petiolar base decurrent (F11 m).   **P2A HERNANDIACEAE**
Illigera (mainly AS)
                7* Petioles different. Petiolar base not decurrent (F11 n). (Xylem lobate, F4 n; young stem with well-developed pith).   **P29A CONVOLVULACEAE**
e.g. *Merremia*
        5* Leaves (bi)pinnate or trifoliolate.
            8 Stipules. (Venation protruding, F18 r).   **P20 LEGUMINOSAE**
key Z
e.g. *Derris* (AM), *Dioclea* (mainly AM), *Mucuna*
            8* No stipules.
                9 Temperate South AM and AS. Leaves bipinnate.   **P5B LARDIZABALACEAE**
e.g. *Parvatia* (AS)
                9* Leaves pinnate. (If stipules minute: LEGUMINOSAE).
Trunk sympodial. (Basal V II ascending, F18 j; embossed venation, F18 s; xylem in rings, F4 c,h; f. s. spp.).   **21B CONNARACEAE**
e.g. *Agelaea* (Paleotrop.), *Connarus*
    4* Trunk not twining.
        10 Some twigs prehensile. (Xyleme in rings, F4 c,h, or simple, F4 a).   **P20 LEGUMINOSAE**
e.g. *Millettia* (Paleotrop.), key Z

10* No prehensile twigs.
  11 Tendrils axillary (F5 n) or oppositifoliate (F5 q).
    12 Tendrils circinnate (F5 n) or bifid.
      13 Leaf with more than 3 leaflets. (Leaflets serrulate).
        14 Petiole glandular (F19 f).   **P16B PASSIFLORACEAE**
e.g. *Passiflora*

        14* Petiole not glandular.   **P23A SAPINDACEAE**
e.g. *Paullinia, Serjania* (AM)

      13* Leaves bifoliolate or bilobate. (Young trunk flattened, F11 f).
P20 LEGUMINOSAE
e.g. *Bauhinia*

    12* Tendrils oppositifoliate (F5 q). (Rays of phloem, F4 k).
P10B VITACEAE
e.g. *Tetrastigma* (AS,AU), *Cissus*

  11* Plant without axillary or oppositifoliate tendrils.
    15 Existence of a different kind of tendril or presence of hooks (F5 t).
      16 Rachis of leaf ending in a tendril (F5 r).
        17 Petiolule pulvinate (F15 h). L. bipinnate.
P20 LEGUMINOSAE
Entada, key Z

        17* AM. Petiolule not pulvinate. L. pinnate.
POLEMONIACEAE
Cobaea (AM)

      16* No terminal tendrils.
        18 Plant climbing by the means of woody tendrils.
P20 LEGUMINOSAE
Mimosoideae: e.g. *Pseudoprosopis* (AF)

        18* Paleotrop. Petiole twining (F5 s) or modified into a hook (F5 t). Leaves trifoliolate. Petiolar base decurrent (F11 m).   **P2A HERNANDIACEAE**
Illigera (mainly AF,AS)

    15* Plant without tendrils or hooks. (Weakly prostrate, climbing by means of hairs, spines, roots, petiolar bases or plant leaning on its branches).
      19 Venation parallelodromous (F17 f).
        20 Leaves compound. Grapnel-climber. Support provided by a cirrus (extended leaf axis) or a flagellum (a modified inflorescence axis fused to the internode and leaf sheath).
P1 ARECACEAE
e.g. *Desmoncus* (AM), *Ancistrophyllum* (AF), *Calamus* (AS)

        20* Leaves simple, distinctly bilobate.
AM. Plant climbing by means of adventitious roots.
CYCLANTHACEAE
e.g. *Asplundia, Evodianthus* (AM)

      19* Venation different.

21a Sheathing petiolar base (F15 a1) or bark aromatic. (Clamp-roots or bristly hairs and L. pinnate or trifoliolate). P27A **ARALIACEAE**
e.g. *Aralia* (AS), *Cephalaralia* (AU)

21b Petiole not sheathing. Smell different. (Leaves compound (bifoliolate: *Bauhinia*), rarely simple (bilobate: *Bauhinia*); leaves or stems spiny f. s. spp.). P20 **LEGUMINOSAE**
e.g. *Machaerium* (AM), *Acacia*, *Bauhinia*
key Z

21c Axillary spines. Translucent dots or aromatic smell. Petiole not sheathing. P24B **RUTACEAE**
e.g. *Luvunga* (AS)

Key W

PHY ~ Phyllanthoideae, OLD ~ Oldfieldioideae, ACA ~ Acalyphoideae, CRO ~ Crotonoideae, EUP ~ Euphorbioideae. Classification according to Webster [42], except for his tribe Galearieae, and for the Bischofiaceae and Picrodendraceae which are families of their own.

1   No white or opalescent latex, and no coloured exsudate.
    2   Some twigs with distichous phyllotaxy.
       3   Branches with UE in monopodial series. Leaves not glandular.
          4   Venation pinnate.
             5   ROUX. V I ending in a minute apical mucro. Stipules minute. Underside of leaves glabrous, not glaucous beneath.   **PANDACEAE**
                e.g. *Galearia* (AS), *Microdesmis* (AF,AS), *Panda* (AF)

             5*   One of these characters different.
                6   Young internodes angular.
                   7   ROUX. TROLL. Growth of plagiotropic branches weakly rhythmic. (Underside of leaves glaucous f. s. spp.).
                                                        PHY - **Phyllantheae**
                  e.g. *Sauropus* (AS), *Glochidion* (mainly AS,OC), *Phyllanthus*

                 7*   Plagiotropic branches with rhythmic growth.   PHY - **Amanoeae**
                                                         e.g. *Amanoa* (AM)

                6*   Young internodes subcylindrical.
                    TROLL. V I not ending in an apical mucrounderside of leaves hairy f. s. spp.).   PHY - **Antidesmeae**
                        e.g. *Antidesma* (Paleotrop.), *Thecacoris* (mainly AF,MA)
          4*   Leaves trinerved.
              Leaves pubescent. TROLL.   PHY - **Antidesmeae**
                                     e.g. *Antidesma* (AS)
      3   Branches distinctly sympodial or leaves glandular.
          8   Venation pinnate. Leaves not glandular.
             9   Petiole distally pulvinate.
                Indumentum of simple hairs.   PHY - **Aporuseae**
                                     e.g. *Aporusa* (AS)
             9   Petiole not distally pulvinate.
                10   Leaves entire. See 11a, 11b, and 11c:
                    11a   Young internodes angular. L. glabrous, coriaceous f. s. spp. (Fimbrial vein: *Bridelia* spp., TROLL; ROUX: e.g. *Drypetes*).
                                                      PHY - **Wielandieae**
                                                  e.g. *Savia* (AM,AF,MA),
                                                      PHY - **Bridelieae**
                        e.g. *Bridelia*, *Cleistanthus* (Paleotrop.)
                                                  PUTRANJIVACEAE
                                                         e.g. *Drypetes*

        11b   Stipules large. No peltate hairs. (TROLL).
                                                  ACA - **Chaetocarpeae**
                                                       e.g. *Chaetocarpus*

        11b Small peltate hairs. (MASSART).   ACA - **Pereae**
                                                          e.g. *Pera* (AM)
    10* Leaves serrulate.
            Young internodes angular.   **PUTRANJIVACEAE**
                                                        e.g. *Putranjiva* (AS)
  8* Leaves trinerved or glandular. See 12a, 12b, and 12c:
      12a Two glands at the base and the upper side of the lamina. Petiole distally
            pulvinate. L. serrulate. (No latex?).   CRO - **Adenoclineae**
                                                      e.g. *Glycydendron* (AM)

      12b Glands on the underside of the lamina.   PHY - **Phylantheae**
                                                            e.g. *Phyllanthus*

      12c No glands. Peltate hairs. Foliaceous stipules.   CRO - **Crotoneae**
                                                    Crotonoides? (AM, Guianas)

2* All stems with spiral phyllotaxy or leaves opposite. (If young tree not branched, with large, elongate leaves: ACA - **Agrostistachydeae** - *Agrostistachys*, AS).
  13 Leaves simple. Venation pinnate.
      14 Trunk and branches with marked rhythmic growth (scale-leaves) and ramification, i.e. RAUH's, or AUBREVILLE's model for *Baccaurea*.
          15 No stellate or peltate hairs.
              16 Petiole distally pulvinate.
                    (RAUH, AUBREVILLE). Two glands at base and upperside of
                    the lamina: *Protomegabaria*).   PHY - **Aporuseae**
                    e.g. *Baccaurea* (AS,OC), *Maesobotrya*, *Protomegabaria* (AF)

              16* Petiole not distally pulvinate. (RAUH).   PHY - **Aporuseae**
                                                            e.g. *Richeria* (AM)
                                                             PHY - **Acalypheae**
                                                             e.g *Cleidion* (AS)
          15* Minute peltate or stellate hairs.
              17 Petiole distally pulvinate (Stilt-roots). Branches monopodial
                    (RAUH).   PHY - **Uapaceae**
                                                             e.g. *Uapaca* (AF,MA)

              17* Branches sympodial, but with UE forming monopodial series. L. glabrous: *Pycnocoma*; L. hairy, petiole short: *Ptychopyxis*.
                                                            PHY - **Pycnocomeae**
                            e.g. *Pycnocoma* (AF,MA), *Ptychopyxis* (AS)

      14* Branches monopodial (no marked rhythmic growth), or sympodial.
          18 Petiole distally pulvinate, long enough for this character to be observed.
              19 Leaves glandular. See 20a, 20b, and 20c:
                  20a RAUH. L. entire. Underside of L. glandular at the junction
                        between petiole and lamina.   PHY - **Spondiantheae**
                                                             Spondianthus (AF)

- **20b** Base of lamina with two glands at its upper side.
  - **21a** L. entire.   ACA - **Caryodendreae**
    e.g. *Caryodendron* (AM)
  - **21b** KORIBA. L. serrulate. Translucent dots.
    ACA - **Acalypheae**
    e.g. *Mareya* (AF)
  - **21c** RAUH. Red exsudate L. serrulate. 2-4 glands at the junction between blade and petiole.   CRO - **Codieae**
    e.g. *Pausandra* (AM)
- **20c** Base of lamina with two glands at its underside.
  - **22** L. serrulate.   ACA - **Alchorneae**
    e.g. *Alchornea*
  - **22*** L. entire. Shrub or small tree. Leaves entire or lobed (aromatic f.s.spp.). glands at the base of the lamina. Hairs simple, stellate or peltate.   CRO **Crotoneae**
    e.g. *Croton*

19* Leaves not glandular.
- **23** Branches with UE in monopodial series (F6 a).
  - **24** Leaves entire. Petiole grooved. V. camptodromous.
    ACA - **Pogonophoreae**
    e.g. *Pogonophora* (AM,AF)
  - **24*** Leaves serrulate.
    - **25** Underside of L. densely covered with stellate hairs.   ACA - **Chrozophoreae**
      e.g. *Sumbaviopsis* (AS)
    - **25*** Underside of L. different.   ACA - **Epiprineae**
      e.g. *Cephalomappa* (AS)
- **23*** Branches markedly sympodial.   ACA - **Cheiloseae**
  e.g. *Neoscortechinia* (AS)

18* Petiole not distally pulvinate or very short.
- **26** No peltate or stellate hairs.
  - **27a** Petiole long. (RAUH).   ACA- **Acalypheae**
    e.g. *Claoxylon* (Paleotrop.), *Acalypha*
  - **27b** Petiole short. L. entire. Thorny shrub. (CHAMPAGNAT).
    PHY - **Phyllantheae**
    e.g. *Flueggea* (Paleotrop.)
  - **27c** Opposite phyllotaxy. L. entire.   OLD - **Mischodontineae**
    Mischodon (AS)
- **26*** Peltate or stellate hairs.
  - **28** AM. Young internodes angular. Indumentum not orange coloured.
    - **29** L. entire. No glands.   ACA- **Pereae**
      Pera (AM)

  29*   L. serrulate. Glands on the underside.
                ACA **- Alchorneae**
                  e.g. *Alchornea*

  28*   AS. Young internodes subcylindrical. Orange peltae hairs.
                PHY **- Hymenocardieae**
                  Hymenocardia (AS)

13* Venation palmate, or leaves (faintly) trinerved or compound.

 30   Twining lianas (glabrous, pubescent or hirsute).   ACA- **Plukenetieae**
      e.g. *Dalechampia, Plukenetia* (AM), *Cnesmone* (AS), *Tragia*
 30* Plant not twining.
  31 Petiole distally pulvinate.
   32 Leaves glandular.
    33 Glands at the base and the upper side of the lamina.
     34 Stellate hairs.   ACA - **Caryodendreae**
                  e.g. *Discoglypremna* (AF)

     34* No stellate hairs. (KORIBA).   ACA - **Alchorneae**
                  e.g. *Conceveiba* (AM)

    33* Glands at the base and the underside of the lamina (glands
      disposed between the veins).   ACA - **Alchorneae**
                  e.g. *Alchornea*
   32* Leaves not glandular.
    35 Phyllotaxy opposite. Minute stellate-peltate hairs f. s. spp.
      (CHAMPAGNAT).   ACA - **Acalypheae**
                  e.g. *Mallotus* (Paleotrop.)

    35*   Phyllotaxy alternate.   PHY - **Antidesmeae**
                  e.g. *Hieronyma* (AM)
  31* Petiole not distally pulvinate.
   36 Leaves simple.
    37 glandular.
     38 L. serrulate or entire, more or less peltate in
       Macaranga.   ACA - **Acalypheae**
                e.g. *Macaranga* (Paleotrop.)

     38* L. entire. (Orange peltate hairs f. s. spp.; really no
       opalescent latex?).   CRO - **Crotonodeae**
                   e.g. *Croton*
    37* Leaves not glandular.
     39 L. entire. RAUH.   PHY - **Antidesmeae**
                  Hyeronima (AM)

     39* L. serrulate or entire, more or less peltate in
       Macaranga. Peltate hairs: e.g. *Mallotus*. Giant herb:
       Ricinus.   ACA - **Acalypheae**
             e.g. *Ricinus, Mallotus* (Paleotrop.)

   36*   Leaves trifoliolate. (RAUH).   OLD - **Hyaenancheae**
                  Piranhea (AM)

1*   Latex white or opalescent, or coloured aqueous or latex-like exsudate.
    40   Latex or exsudate other than white (possibly whitish-opalescent), or petiole distally pulvinate, or leaves tri- or palmately nerved.
        41 Leaves simple. Venation pinnate.
            42 Latex clear, uncoloured. Petiole not distally pulvinate.
                43   (No stipules f. s. spp.).   CRO - **Codieae**
                    e.g. *Blachia* (AS), *Codiaeum* (AS,AU,OC), *Ostodes* (AS)

                43*   Leaves serrulate, shortly petiolate. (Short shoots modified into spines f. s. spp.).   EUP - **Hippomaneae**
                    e.g. *Sebastiana* (AM)

            42*   Coloured exsudate or white latex, or petiole distally pulvinate.
                44   Leaves not glandular.
                    45   Latex white. Petiole distally pulvinate with two distal glands.
                        CRO - **Elateriospermeae**
                        e.g. *Elateriospermum* (AS)

                    45*   Latex red. KORIBA, LEEUWENB. Petiole distally pulvinate.
                        CRO - **Codieae**
                        e.g. *Sagotia* (AM), *Baloghia* (AS,OC
                44*   Leaves glandular.
                    46   Two glands on the upper side and the base of the lamina.
                      47   Exsudate redddish. KORIBA's model.
                          CRO - **Trigonostemoneae**
                          e.g. *Trigonostemon* (AS)

                      47*   Latex-like exsudate? KORIBA. (L. glabrous: *Grossera*; peltate hairs: *Crotonogyne*).   CRO - **Aleuritideae**
                        e.g. *Crotonogyne* (AF), *Grossera* (AF,MA)

                46*   Glands disposed otherwise.
                    48   Two or four glands at the junction between petiole and lamina. (RAUH's model).   CRO - **Codieae**
                      e.g. *Pausandra* (AM)

                    48*   Petiole distally pulvinate. One gland at the base of the main vein.   CRO - **Jatropheae**
                      Vaupesia (AM)
    41*   Leaves compound (e.g. *Hevea*) or palmate venation or leaves trinerved.
        49   Twining liana. Latex opalescent or red.
            50   Two glands at the base of the lamina or at the distal end of the petiole.   ACA - **Omphaleae**
                Omphalea

            50*   Glands disposed on the margin. Petiole not distally pulvinate? (L. lobed).   CRO - **Aleuritideae**
                Manniophyton (AF)
        49*   Shrub or tree.
            51   Leaves not glandular.

52 Tree with monopodial rhythmic growth, RAUH. Leaves palmately compound, its underside with stellate or peltate hairs.
          CRO - Ricinodendreae
          Ricinodendron (AF)

52* Characters different. (Sympodial growth).
 53 Branches orthotropic. (Petiole not distally pulvinate).
  54 Stipules modified into ramified hairs. (LEEUWEN-BERG).   **CRO - Jatropheaee**
          Jatropha (AM)

  54* Stipules different.
   55 L. more or less lobate, its magin entire. (LEEUWENBERG).   **CRO - Manihoteae**
          Manihot (AM)

   55* L. cordate, entire.
    56 Stellate-peltate hairs on both side of the lamina.   **CRO - Aleuritideae**
         e.g. *Neoboutonia* (AF)

    56* No stellate-peltate hairs.
          CRO - Codieae
         e.g. *Strophioblachia* (AS)

 53* Branches plagiotropic. (TROLL). Leaves glabrous.
          CRO - Aleuritideae
         e.g. *Domohinea* (MA)

51* Leaves glandular.
 57 Glands disposed on the upper side of the leaf or on its margin.
  58 Petiole distally pulvinate.
   (RAUH). Two glands near the junction between blade and petiole. Leaves trifoliololate: *Hevea*.
          CRO Micrandreae
        e.g. *Hevea, Micrandra* (AM)

  58* Petiole not distally pulvinate.
   59 Stipules modified into ramified hairs. Petiole with distal gland(s).   **CRO - Jatropheae**
         e.g. *Jatropha* (AM)

   59* Stipules different. See 60a, 60b, and 60c:
    60a Two glands on the base and the margin of the lamina.   **CRO - Manihoteae**
         e.g. *Endospermum* (AS,OC)

    60b Two glands at the base of the lamina. (Leaves lobed).   **CRO - Aleuritideae**
         e.g. *Aleurites* (AS,OC)

    60c Leaves palmatilobate with glandular teeth. Stellate hairs.   **CRO - Ricinodendreae**
         e.g. *Givotia* (MA)

 57* Glands disposed on the underside of the leaf.

	61	Hairs simple. Shrub or small tree of sympodial architecture. Leaves entire, round shaped, diversely lobed or peltate. CRO - **Jatropheae**

e.g. *Jatropha*

        61* Stellate or peltate hairs, also with simple hairs f. s. spp. Shrub or small tree. RAUH, SCARRONE, LEEUWENBERG. Orange or red exsudate. L. entire, serrulate, or lobed (aromatic f. s. spp.). Glands at the base of the lamina.   CRO - **Crotoneae**

e.g. *Croton*

40*   Latex white. Simple leaves with pinnate venation. Petiole not distally pulvinate (all EUPHORBIOIDEAE and a few CROTONOIDEAE).
    62 Venation camptodromous (but with an intramarginal vein f. s. spp.).
        63 Two glands on the upper side of the main vein: *Glycydendron*; or two glands on the distal end of the petiole: AM. *Tetrorchidium*, or two glands on the upper side and the base of the lamina: AF. *Tetrorchidium*.
            CRO - **Adenoclineae**

e.g. *Glycydendron* (AM), *Tetrorchidium* (AF)

        63* Two glands on the underside and the base of the lamina.
            EUP - **Stomatocalyceae**

e.g. *Pimeleodendron* (AS)

    62* Venation brochidodromous (F17 a).
        64 Leaves glandular.
            65 Leaves entire. See 66a, 66b, and 66c:
                66a Branches plagiotropic. TROLL. Two glands at the base and the underside of the lamina. Ramification not very rhythmic.
                    EUP - **Hippomaneae**

e.g. *Maprounea* (AM)

              66b Branches orthotropic. RAUH, two glands at the distal end of the petiole: *Sapium*. KORIBA, leaf round shaped, two glands at the base and the underside of the lamina (glands disposed between the veins): *Omalanthus*.   EUP - **Hippomaneae**

e.g. *Omalanthus* (AS,AU), *Sapium*

              66c Branches plagiotropic, sympodial. Two glands at the base and the underside of the lamina. Ramification distinctly rhythmic.
                    EUP - **Hippomaneae**

e.g. *Actinostemon* (AM)

        65* Leaves serrulate.
            67 Branches orthotropic. Two glands at the distal end of petiole.
              68 Latex clear. KORIBA (trunk spiny).
                    EUP - **Hippomaneae**

e.g. *Hura* (AM)

              68* Latex white. RAUH or KORIBA.   EUP - **Hureae**

e.g. *Sapium*

- **67*** KORIBA, one solitary gland on the upper side and the distal end of the petiole.   EUP - **Hippomaneae**
  e.g. *Hippomane* (AM)
- **64*** Leaves not glandular.
  - **69** Leaves entire. (L. crassulescent f. s. spp.).
    - **70** Some twigs with alternate, distichous phyllotaxy (or nearly so).
      - **71** NOZERAN: *Anthostema*. PREVOST?: *Dichostemma*.   EUP - **Euphorbieae**
        e.g. *Anthostema* (AF,MA), *Dichostemma* (AF)
      - **71*** MANGENOT. L. crassulescent.   EUP - **Euphorbieae**
        Pedilanthus (AM, ornam.)
    - **70*** All twigs with spiral or opposite phyllotaxy.
      - **72** (FAGERLIND).   EUP - **Hippomaneae**
        e.g. *Senefeldera* (AM)
      - **72*** RAUH, KORIBA, or other model. (Plant spiny or succulent; f. s. spp.). L. opposite, branches sympodial: *Chamaesyce*.   EUP - **Euphorbieae**
        Synadenium (AM,AF,MA), *Chamaesyce*, *Euphorbia*
  - **67*** Leaves serrulate.
    - **73** Branches orthotropic. Leaves spiny, glabrous. (RAUH).   EUP - **Pachystromateae**
      Pachystroma (AM, Brasil)
    - **73*** Branches plagiotropic. PREVOST: *Excoecaria*; NOZERAN: *Mabea*.   EUP - **Hippomaneae**
      Excoecaria (AF,AS), *Mabea* (AM)

Key X

Subfamilial delimitation of Malvaceae (ex-Malvales) according to Alverson *et al.* [46].

1   Leaves all simple.
    2   Some twigs with distichous phyllotaxy (F9 h).
       3   AS. Leaves with peltate hairs (F20 p).
          4   Peltate hairs only on the upper side. Branches monopodial (F6 a) plagiotropic. (ROUX). **Helicterioideae**
Durioneae, e.g. *Durio* (AS, cult.)

          4*   AM. (TROLL?). Peltate hairs on both sides of the lamina. **Tilioideae?**
e.g. *Mollia* (AM)

       3*   No peltate hairs (hairs simple or stellate).
          5   Leaves distinctly asymmetrical. (TROLL, ROUX).
             6   Leaves very hairy or scabrous. (Branches sympodial, domatia, F20 t, f. s. spp.). (TROLL: *Grewia, Colona, Luehea*; MASSART: *Desplatsia*). **Grewioideae**
e.g. *Luehea* (AM), *Desplatsia* (AF), *Colona* (AS) *Grewia* (Paleotrop.)

             6*   Leaves not hirsute or scabrous (but generally hairy).
                7   Leaves entire or with some large teeth. (TROLL). **Dombeyoideae**
e.g. *Scaphopetalum* (AF), *Pterospermum* (AS)

                7*   Leaves serrulate. Petiole distally pulvinate. (ROUX) The summit of trunk becoming oblique: erect form of a TROLL's model). **Byttnerioideae**
Guazuma (AM)

       5*   Leaves symmetrical or bases faintly asymmetrical.
          8   Branches monopodial (F6 a). (Axillary flowering).
             9a   Leaves entire MASSART's, NOZERAN's or TROLL's model. (Apex of lamina serrulate; short petiole; f. s. spp.). **Sterculioideae**

MASSART or NOZERAN   e.g. *Theobroma* (AM, cult.)
MASSART
                                        e.g. *Heritiera* (AF,MA,AS)
TROLL
                                          e.g. *Scaphopetalum* (AF)

             9b   Leaves serrulate or lobate with serrulate margin. (TROLL). **Helicterioideae**
e.g. *Helicteres* (AS)
Grewioideae
e.g. *Lueheopsis* (AM)

             9c   Leaves palmately lobate (F14 a). Stipule hood-like (F13 m). (Terminal flowering; MASSART). **Sterculioideae**
Triplochiton (AF, cult.)
       8   Branches sympodial. (Terminal flowering).
          10   Trunk rhythmically ramified. (MASSART, FAGERLIND).

    11 Leaves entire. **Malvoideae**
e.g. *Matisia*, *Quararibea* (AM)

    11* Inflorescence oppositifoliate. Petiole not distally pulvinate.
      L. palmatilobate. **Malvoideae**
Fremontodendron (AM, California)

   10* Branches regularly disposed on trunk (TROLL, ROUX, PETIT).
     12 TROLL's model. (Base of leaves with domatia, F20 t, f. s. spp.). **Grewioideae**
e.g. *Apeiba*, *Triumfetta* (AM), *Grewia* (Paleotrop.)

     12* PETIT's model. Leaves serrulate. **Helicterioideae**
e.g. *Helicteres* (AM)

2* All stems with spiral phyllotaxy (or monocaulous plant).
  13 Climbing plant (AM: weakly prostrate plant with prickly twigs, AS: twining plant). **Byttnerioideae**
Byttneria (AM,AS)

  13* Tree or shrub.
    14 Petiole not distally pulvinate (or faintly enlarged or twisted).
     15 Existence of UE in monopodial series (F6 g).
      16a RAUH's model. (V II ending in a tooth: *Dombeya*). **Malvoideae**
e.g. *Hibiscus* (mainly AS, ornam.)
Dombeyoideae
Dombeya (AF,MA)

      16b MASSART **Bombacoideae**
e.g. *Montezuma* (AM)

      16c CHAMPAGNAT. **Byttnerioideae?**
e.g. *Melochia* (mainly AM)

    15* Branches distinctly sympodial (F6 e,f).
     17a SCARRONE. **Sterculioideae**
e.g. *Hildegardia* (e.g. AF)
Grewioideae
e.g. *Heliocarpus* (AM)

     17b AUBREVILLE. **Sterculioideae?**
e.g. *Firmiana* (Paleotrop.)

     17c KORIBA. **Sterculioideae?**
e.g. *Firmiana* (Paleotrop.)
Malvoideae
e.g. *Ochroma* (AM)
Byttnerioideae
Kleinhovia (AS)

     17d PETIT. **Malvoideae**
      (Abaxial foliar glands). e.g. *Gossypium* (mainly AM, cult.)
Helicterioideae
e.g. *Ambroma* (AS,AU)

14* Petiole distally pulvinate (F15 p).
    18 Branches plagiotropic by apposition (F6 e). Axillary, not terminal, flowering. **Sterculioideae**
        e.g. *Firmiana* (AF,AS), *Pterocymbium* (AS,OC), *Sterculia*

    18* Branches not plagiotropic by apposition, or flowering terminal.
        19 Branches monopodial (F6 a).
            20 Branches erect, orthotropic growth (RAUH).
                21 L. hairy beneath, glaucous or ferrugineous aspect.
                    (L. entire or lobate; heteroblasty: the shape and width of L. variable in relation to their location on the UE, F10 e; f. s. spp.). **Sterculioideae**
                    e.g. *Sterculia*? (AM), *Cola* (AF)

                21* Leaves (sub)glabrous, shiny. **Bombacoideae**
                    e.g. *Catostemma* (AM)
            20* Branches plagiotropic. (MASSART).
                22 Leaves serrulate, V II ending in a tooth (F17 p).
                    Sterculioideae?
                    Mansonia (AF,AS)
                22* Leaves different.
                    23 Short shoots (F11 p) bearing cordate leaves (F14 a). **Brownlowioideae**
                      e.g. *Berrya* (AS)

                    23* Leaves not cordate, hairs lepidote (F20 p).
                      Sterculioideae?
                    e.g. *Heritiera* (AS)

        19* Branches sympodial (F6 f), (terminal flowering).
            24 (SCARRONE; leaves often broad, turning yellow before falling, L. entire: *Brownlowia*; L. serrulate: *Clappertonia*). **Brownlowioideae**
                e.g. *Clappertonia* (AF), *Brownlowia* (AS,OC)
            24* (KORIBA).
                25 AM. Leaves cordate (F14 a) or quite broad.
                    Bombacoideae?
                    Cavanillesia (AM)
                25* AS. Leaves oval with cuneate base.
                    Helicterioideae
                    Neesia (AS)

1* Leaves palmately compound (some simple leaves also present f. s. spp.).
    26 Branches plagiotropic by apposition (F6 e).
        27 (Buttressed tree). **Sterculioideae?**
            e.g. *Heritiera* (AS)

        27* (Trunk swollen, often spiny). **Bombacoideae**
            e.g. *Pseudobombax* (AM), *Bombax* (AF,AS, cult.)
    26* Branches not plagiotropic by apposition.
        28 Branches plagiotropic with spiral phyllotaxy. (Trunk spiny f. s. spp.).

Small shoots ending in short internodes. Ramification of trunk and branches rhythmic (MASSART). **Bombacoideae**
e.g. *Ceiba, Catostemma, Chorisia, Pachira* (AM)
Adansonia (AF,MA,AU)

28* Branches more or less erect, orthotropic growth. (Branches monopodial, i.e. RAUH's model if trunk is also monopodial and rhythmically ramified).
29 AM. **Bombacoideae**
e.g. *Eriotheca, Gyranthera, Pochota* (AM)

29* AF. (Trunk with flattened buttresses; leaves with VIII secant on a hyaline margin). **Sterculioideae**
e.g. *Cola* (AF)

Key Y

N.B. When tasting bark, caution!! Some Anacardiaceae are toxic or cause blisters. Families of the Sapindales are described in plates 23A, 23B, 24A, and 24B.

1 Leaves stipulate.
    2 Mainly AM. Leaves bipinnate. Self-supporting form of a liana. **SAPINDACEAE**
                                                                                 e.g. *Paullinia* (AM)
    2* Leaves simply pinnate. See a, b, c:
          a) Leaves imparipinnate (F14 h). Leaflets serrate-serrulate. (Intrapetiolar stipules). **MELIANTHACEAE**
                                                                               Melianthus, Bersama (AF)

          b) Bitter bark. (Caution with toxic bark!). Leaflets serrulate. Rhythmic growth. **SIMAROUBACEAE**
                                                                                   Picrasma (AM,AS,OC)

          c) No bitter bark. (Leaflets serrulate, leaflets stipellate (F14 h); f. s. spp). **BURSERACEAE**
                                                                                   e.g. *Garuga* (AS,AU,OC)

1* Leaves not stipulate (with small leaflets at the base of the rachis (F14 n) or with a pair of basal spines in stipular position; f. s. spp.).
    3 Leaves with translucent dots (use a hand lens and observe against the light). (Crumpled leaves have an aromatic or peculiar smell). Some MELIACEAE have also translucent dots but these are very small.
        4 Leaves compound.
            5 Spiral phyllotaxy. (Leaflet with an abscission joint: *Fagara*).
                Leaves paripinnate: e.g. *Fagara*; imparipinnate: e.g. *Pilocarpus* (AM).
                Leaves trifoliolate: e.g. *Moniera, Esenbeckia* (AM).; Leaves palmati-compound: e.g. *Vepris* (Paleotrop.) Leaflets alternate: e.g. *Murraya* (AS).
                Short twigs: e.g. *Aegle* (AS). **RUTACEAE**
                                                                                   Most of the species

            5* Opposite phyllotaxy. L. pinnate: e.g. *Evodia*; L. trifoliolate: e.g. *Balfourodendron*; L. palmate: e.g. *Myllanthus*. **RUTACEAE**
                                                      e.g. *Balfourodendron, Myllanthus* (AM)
                                                                         Evodia (Paleotrop.)

        4* Leaves simple.
            6 Alternate phyllotaxy.
                7 Bark or leaves aromatic.
                    Petiole winged (F15 d) or distally enlarged: e.g. *Esenbeckia, Toddalia*; petiole not winged: e.g. *Angostura*. **RUTACEAE**
                                        e.g. *Angostura, Esenbeckia* (AM), *Toddalia* (AF,MA,AS)
                7* No aromatic smell.
                    Bitter bark. Underside of blade glandulate. **SIMAROUBACEAE**
                                                  e.g. *Samadera indica* (AS)
            6 Opposite phyllotaxy.
                Petiole distally enlarged or pulvinate (F15 q) or with stipuliform sheathing base (F13 p). **RUTACEAE**
                                            e.g. *Metrodorea* (AM), *Tetractomia* (AS)

3* Leaves without translucent dots.
   8 AS, OC. Veins II abruptly curved near the margin (F18 p). (Leaves or bark aromatic or with a resinous smell). **ANACARDIACEAE**
e.g. *Astronium* (AM), *Pentaspadon* (AS,OC)

8* AS, NG, OC. Veins II not abruptly curved near the margin.
   9 Leaves simple. See 10a, 10b, and 10c:
      10a V II abruptly curved near the margin (F18 p). (Leaves or bark aromatic). (Mangifereae). **ANACARDIACEAE**
e.g. *Anacardium* (AM, cult.), *Gluta* (MA,AS), *Mangifera* (AS, cult.)

      10b V II or V III secant at the margin (F17 n): e.g. *Ozoroa*. Fimbrial vein (F17 p): e.g. *Holigarna*. **ANACARDIACEAE**
e.g. *Ozoroa* (AF), *Holigarna* (AS)

      10c AS, OC. Branches plagiotropic by apposition (F6 e). **MELIACEAE**
Vavaea (AS, OC)

9* Leaves compound.
   11 Petiolules distally pulvinate (F15 k), at least in terminal leaflet.
      12 Bark or leaves resinous or petiolar base canaliculate (F15 b). V. brochidodromous (F17 a). (Rhythmic growth: e.g. *Crepidospermum*; not rhythmic: e.g. *Protium*) **BURSERACEAE**
e.g. *Aucoumea* (AF, cult.), *Trattinickia* (AM)
Canarium (Paleotrop., ornam.), *Dacryodes*, *Protium*

      12* AS. Bark and leaves not resinous. V. camptodromous (F17c). Petiole different. **MELIACEAE**
Trichilia (e.g. AM), *Walsura* (AS)

   11* Petiolules not distally pulvinate.
      13 Petiolar base canaliculate (F15 b) or with a narrow raised edge.
         14 V. brochidodromous (F17 a). Rhythmic growth noticeable due to short internodes, or brittle monopodial branches, or resinous smell. Leaves imparipinnate (F14 h). (Leaflets stipellate f. s. spp.). **BURSERACEAE**
e.g. *Tetragastris* (AM)

         14* V. camptodromous (F17c). Growth not very rhythmic. No resinous smell, bark aromatic f. s. spp. (Trunk and branches sympodial, flexible, L. paripinnate (F14 g,k), e.g. *Guarea*; leaves with a terminal leaflet, e.g. *Trichilia*; L. bipinnate: *Melia*). **MELIACEAE**
e.g. *Guarea* (AM,AF), *Trichilia* (AM,AF)
Melia (Paleotrop., ornam.)

   13* Petiolar base different, i.e. flat or convex (F15 c).
      15 Cut bark producing an aromatic resin, cut bark or crumpled leaves smelling resin.
         16 V II abruptly curved near the margin (F18 a,p) or V. forming peripheral series of arches which almost touch the margin (marginal vein f. s. spp.). Rachis more or less winge: e.g. *Schinus*.

ANACARDIACEAE
e.g. *Schinus* (AM), *Sorindeia* (AF,MA)

16* Venation different (brochidodromous, F17 a)
    17 Branches sympodial. Petiolar base decurrent (F11 j), rachis not canaliculate.
ANACARDIACEAE
e.g. *Tapirira* (AM), *Sorindeia* (AF,MA)

    17* Branches monopodial with rhythmic growth.
BURSERACEAE
e.g. *Tetragastris* (AM)

15* No smell of resin, but smell could be weakly aromatic in MELIACEAE and a few SAPINDACEAE. (Latex whitish or coloured f. s. spp.).
    18 Latex (see text of F3) whitish or coloured (caustic f. s. spp.). Latex-like resin in a few BURSERACEAE. See 19a, 19b, and 19c:
        19a Latex sparse, milky white. Branches with growth weakly rhythmic and bearing more or less separated leaves. Young branches stout and flexible, not rigid and brittle. **MELIACEAE**
Trichilioideae: e.g. *Aglaia, Chisocheton* (AS,AU,OC)

        19b Latex white or coloured, even brownish or turning black (if necessary, wait to observe colour change) or venation as in F18 p. (Latex generally thick and sticky, toxic f. s. spp.). (Branches sympodial). Latex white-beige: e.g. *Astronium, Rhus, Trichoscypha,* or turning yellow: e.g. *Thyrsodium,* or black: e.g. *Gluta, Loxopterygium.*
ANACARDIACEAE
e.g. *Astronium, Loxopterygium, Thyrsodium* (AM)
Trichoscypha (AF), *Gluta* (MA, AS), *Rhus*

        19c Latex white, in bark or in leaves. Young branches not flexible, but brittle and rigid. **SAPINDACEAE**
e.g. *Harpullia* (AS,AU,OC)
    18* No latex. (See **20**).

20 Leaves ending in a terminal leaflet (F14 h). (Leaflets opposite: e.g. *Simaba, Quassia* (AM); alternate leaflets: e.g. *Simarouba* (AM).
    21 Bitter bark. (Axiliary buds somewhat displaced from leaf axil f. s. spp.).
        V. not very visible: e.g. *Simarouba* (AM), *Odyendea* (AF), *Eurycoma* (AS).
        Stipuliform spines: *Harrisonia* (Paleotrop., better included under RUTACEAE).
        Base of petiolule with an abscission joint (F15 h), leaves hairy: *Picramnia*, better included under PICRAMNIACEAE).
        Leaflets glandulate: e.g. *Perriera* (MA), *Ailanthus* (AS). **SIMAROUBACEAE**

    21* Bark not bitter.

22 Venation as in a), b) or c):
    a) V II abruptly curved near the margin (F18 a,p) or secant, F17 n, (at least at the apex of leaflets): e.g. *Astronium* (AM), *Rhus*.
    b) Intramarginal vein (F18 m): *Spondias* (AM,AS, cult.).
    c) Small intercostal V II oriented "backwards" (F18 p): e.g. *Sorindeia* (AF,MA), *Thyrsodium* (AM,AF).   **ANACARDIACEAE**
Most of the non-aromatic species

22* Venation different.
   23 Young branches brittle and rigid. Young internodes angular (F11 a) or cannelate (F11 b). If strong rhythmic growth and leaflets serrulate: *Azadirachta* (**MELIACEAE**).
      Leaflets alternate, e.g, *Cupania, Schleichera, Toechima, Toulicia*, opposite, e.g. *Cossinia*, or L. trifoliolate: *Allophylus*. (Rachis winged f. s. spp.).   **SAPINDACEAE**
      e.g. *Toulicia* (AM), *Cossinia* (MA, AST, OC), *Schleichera* (AS, cult)
      Lepisanthes (mainly AS), *Toechima* (NG,AST), *Allophylus*

   23* Young branches bending appreciably before breaking or young internodes cylindrical. (Pulvinate petiolule: PICRAMNIACEAE).
     24 Young branches flexible. (Hairs lepidote (F20 p): *Aglaia, Lepidotrichilia, Trichilia*). Leaflets alternate: e.g. *Trichilia* pp. Large leaves with stout petiole enlarged at base: *Aphanamixis*.
                                       MELIACEAE
      e.g. *Trichilia* (AM), *Aphanamixis* (AS), *Aglaia* (AS,AU,OC)

    24* Young branches brittle, breaking with a sharp cracking noise when bent. (Young internodes subcylindrical).
      25 Bark aromatic or rhythmic growth. Leaflets entire. Domatia (F20 t): e.g. *Trichilia*.   **MELIACEAE**
       Trichilia (e.g. AM), *Sandoricum* (AS, ornam.)

      25* Bark not aromatic. Growth weakly rhythmic. (L. crenate f. s. spp., trifoliolate: *Allophylus*).   **SAPINDACEAE**
       e.g. *Allophylus*

20* Leaves not ending in a terminal leaflet (with a terminal pair of opposite leaflets (F14 g), a terminal mucro (F14 g,j) or a bud of leaflets (F14 k)). (Leaflets opposite or alternate; L. bifoliolate: a few SAPINDACEAE).
  26 Foliar glands (F20 d).   **SIMAROUBACEAE**
       Ailanthus (AS, AU)
26* No foliar glands.
   27 Young branches brittle and rigid. See 28a, 28b, and 28c:
     28a Young internodes angular (F11 a) or cannelate (F11 b). Not aromatic. (Heterophylly: simple leaves in sapling or in trunk sprout; f. s. spp.). Rachis winged: e.g. *Harpullia* (AS, AU, OC); Domatia (F20 t): e.g. *Arytera, Mischocarpus, Toechima* (AS, AU, OC). L. bipinnate: e.g. *Dilodendron* (AM), *Macphersonia* (AF, MA).   **SAPINDACEAE**
       e.g. *Melicoccus* (AM), *Blighia* (AF, cult.), *Nephelium* (AS, cult.)

     28b Young internodes more or less cylindrical or resinous smell. (Venation brochidodromous (F17 a), branches monopodial).   **BURSERACEAE**
       e.g. *Dacryodes* (~*Pachylobus*) *edulis* (AF)

28c Bitter bark. Branches brittle. (Leaflets opposite). **SIMAROUBACEAE**
e.g. *Simaba* (AM)

27* Branches stout and flexible, bending appreciably before breaking.
   29 Bitter bark or underside of leaflets glaucous, with almost invisible venation. (Young twigs angular, leaflets alternate). **SIMAROUBACEAE**
e.g. *Simarouba* (AM)
   29* Bark and leaflets different. (Bark slightly aromatic).
      30 Apical buds or base of UE with several scale-leaves separated by short internodes. (Bark aromatic). Bitter bark or garlic smell: e.g. *Cedrela*. **MELIACEAE**
Swietenioideae, e.g. *Cedrela*, *Swietenia* (AM)
Entandophragma (AF), *Toona* (AS)

      30* Stems growing differently: the internodes are quite equal in length between the developed leaves. (Leaves with a terminal bud, F14 k: *Guarea, Chisocheton*). **MELIACEAE**
Trichilioideae, e.g. *Aglaia* (AS,AU,OC), *Trichilia* (e.g. AF)
Guarea (AM,AF), *Chisocheton* (AS,OC)

Key Z

CSL ~ Caesalpinioideae, MIM ~ Mimosoideae, PAP ~ Papilionoideae. Systematics according to Polhill & Raven [43], and to Herendeen & Brunneau [44].
Remark: " Serial buds " does not exclude the presence of some leaves with solitary buds.

1 Some twigs with distichous phyllotaxy.
  2 Leaves simple or simply pinnate.
    3 Leaves simple or with rachis ending in a terminal leaflet (F14 h).
      4 Leaflets opposite (at least the apical pair).
        5 AM. Buds solitary or in a weak zigzag line (F12 d). Growth weakly rhythmic. (Rachis winged, F14 p; leaves stipellate, red exsudate; f.s.spp.).
                                        PAP-**Swartziaeae**
                                             Swartzia (AM)

        5* Axillary buds solitary. Leaflets not stipellate. (Conspicuous scaly buds, translucent dots; f. s. spp.). Also *Dicorynia*, **Cassieae**. PAP-**Millettieae**
                                          e.g. *Lonchocarpus* (mainly AM)
      4* Leaflets alternate or leaves simple.
        6 Liana.
            Tendrils: *Bauhinia*; hooked twigs, leaves bilobate or trinerved (F18 d): *Bandereia*. Spiny stems: *Dalbergia, Machaerium* spp. CSL-**Cercideae**
                                          Dalbergia, Machaerium (AM)
                                          Bandereia (AF), *Bauhinia*
        6* Tree or shrub. See **7**.

    7 Rachis ending in a leaflet or leaves simple with a pulvinate petiole (F15 p,q).
      8 AM. Translucent dots.
        9 Vertical serial buds.
                                            PAP-**Sophoreae**
                                          e.g. *Myroxylon* (AM)

        9* AF, MA. Buds solitary.
                                          PAP-**Swartzieae**
      8* No translucent dots.   e.g. *Cordyla* (AF, MA)
        10 Leaves simple.   PAP-**several tribes**
            Red exudate in bark: **Swartzieae** (*Swartzia*, AM).
            Petiole pulvinate, e.g. **Dalbergieae** (*Dalbergia ecastaphyllum*, AM), **Millettieae** (*Poecilanthe*, AM), **Swartzieae** (*Bocoa*, AM) and **Sophoreae** (*Baphia*, Paleotrop.). L. serrulate, petiole not pulvinate, V. camptodromous (F17 c): e.g. **Phaseoleae**, *Flemingia* (AS).

        10* Leaves compound   PAP, except **Cassieae** (CSL)
            Red exudate in bark: **Dalbergieae** (*Machaerium*, AM, *Pterocarpus*).
            Well devloped buds (*Pterocarpus*).
            Fimbrial vein, F17q: **Dalbergieae** (*Machaerium*, AM).
            Spiny plant: **Dalbergieae** (*Dalbergia, Machaerium*).
            Short shoots: **Aeschynomeneae** (*Ormocarpopsis*, MA).
            Leaflets subopposite. (UE in monopodial series): **Cassieae**e (*Dicorynia*, AM).

    7* Leaves simple. Petiole not distally pulvinate.
        11 Leaves entire.   **PAP-Dalbergieae**
                                                                    e.g. *Inocarpus* (NG,OC)

        11* Leaves serrulate.   **PAP-Swartzieae**
                                                  e.g. *Lecointea, Zollernia* (AM)

3* Leaves compound, with rachis not ending in a terminal leaflet (F14 g,j).
    12 AM. Rachis with cupuliform glands at the nodes of the rachis (F19 d).
        (Buds forming a condensed branch system, F12 c).   **MIM-Ingeae**
                                                                    e.g. *Inga* (AM)

    12* Rachis without cupuliform glands (possibly with other kinds of glands or with glands disposed otherwise).
        13 Plant climbing by means of tendrils or spines (L. bilobate f. s. spp.).
                                                                  CSL-Cercideae
                                                                       Bauhinia

        13* Plant different.
            14 Stems with longitudinal ridges (F11 j). Buds solitary. (Underside of leaflets glaucous). Gland at the base of the petiole: *Chamaecrista* or on rachis: *Senna*).   **CSL-Cassieae**
                                              e.g. *Chamaecrista* (AM), *Senna, Cassia*
                                                                CSL-Caesalpinieae
                                                                e.g. *Poeppigia* (AM)

            14* Stems different (rarely serial buds).
                15 Leaflets not stipellate. (Intrapetiolar stipules, F13 b; twisted petiolules or L. glandular, F19 r; f. s. spp.).
                    Leaflets bifoliolate: e.g. *Cynometra, Macrolobium, Pelto-gyne*.
                    Leaflets alternate: e.g. *Crudia* (AM), *Kingiodendron* (AS).
                    Leaflets with translucent dots: *Hymenaea, Peltogyne*.
                    Serial buds in *Hymenaea*.   **CSL-Detarieae/Amherstieae**
                                        e.g. *Brownea, Hymenaea, Macrolobium* (AM)
                                        Anthonotha, Berlinia, Gilbertiodendron (AF)
                          Maniltoa (AS,OC), *Amherstia* (AS, ornam.), *Cynometra*

                15* Leaflets stipellate (F14 h).
                    16 TROLL. Rachis ending in a short segment bearing no leaflets distally. Axillary bud a short condensed branch system.
                                                          PAP-Robinieae
                                                  e.g. *Coursetia* (syn. *Humboldtiella*), (AM)

                    16* Shrub. Leaves paripinnate. Branches decumbent, but its extremity becoming erect.   **PAP-Robinieae**
                                                  e.g. *Sesbania* (AM, cult.)

2* Leaves bipinnate. (Serial buds existing in some axils, F12 a,c,d).
    17 Young internodes with well marked longitudinal ridges and grooves. (One gland at the base of the petiole and, f. s. spp., at the point of leaflet insertion).
                                                                MIM-Acacieae
                                                                Acacia

    17* Young internodes different.
        18 Existence of short shoots (F11 p).
            19 Mostly AM. TROLL. Rachis not glandular.   **MIM-Ingeae**
                                          e.g. *Calliandra* (AM,MA,AS, ornam.)

        19*   Mostly MA. Glands at the base of the petiole.   **MIM-Mimoseae**
                          e.g. *Dichrostachys* (MA, paleotrop.), *Gagnebina* (MA)
    18* No short shoots.
      20   Leaves glandular on the rachis or rachilla. (Leaflets very small).
        21   TROLL's model but the branches turn to be erect.
                                                    MIM-Mimoseae
                                                      e.g. *Newtonia* (AF)
        21* Branches plagiotropic. (TROLL).   **MIM-Ingeae**
                                                      e.g. *Enterolobium* (AM)
      20* Leaves not glandular. See 22a, 22b, and 22c:
        22a Leaflets not thigmonastic. Rachis ending in a terminal pinna.
                                                        CSL-Caesalpinieae
                                                      e.g. *Caesalpinia* (AM)
        22b Leaflets not thigmo. No terminal pinna.
                                                          MIM-Mimoseae
                                                      e.g. *Zapoteca* (AM)
        22c Leaflets thigmonastic.   **MIM-Mimoseae**
                                                      e.g. *Mimosa bimucronata* (AM)
1*   All stems with spiral phyllotaxy or leaves opposite (or whorled) or sympodial branches consisting of short modules (with only 1-2 well developed leaves).
     23   Leaves opposite or whorled. Leaflets alternate to opposite, V. camptodromous: *Tipuana*.   **PAP-Dalbergieae-Dipterygeae**
                                        e.g. *Platymiscium, Taralea, Tipuana* (AM)
     23* Alternate phyllotaxy.
        24   Leaves simple or simply pinnate.
           25   Leaves simple or existence of rachis with a terminal leaflet (F14 h). (Several leaves should be observed).
              26   Plant twining. See 27a, 27b, and 27c:
                27a Some leaves with at least five leaflets. (Short shoots: *Derris*).
                                                     PAP-Millettieae
                                                  e.g. *Millettia* (Paleotrop.), *Derris*

                27b Leaves with no more than three leaflets. Leaflets stipellate (F14 h).   **PAP-Phaseoleae**
                                                     e.g. *Dioclea* (mainly AM), *Mucuna*

                27c Leaves simple.   **PAP-Sophoreae**
                                                    e.g. *Bowringia* (AF,MA)
           26* Plant not twining.
              28   Buds solitary.
                29   AM, AF. Strong rhythmic growth noticeable due to the presence of short internodes. (Leaflets stipellate, F14 h).   **PAP-Dalbergieae**
                         e.g. *Andira* (AM,AF), *Centrolobium, Vaitarea* (AM)
                         Hymenolobium, Pterocarpus podocarpus (AM)

29* No strong rhythmic growth.
    30 Leaves trifoliolate, stipellate or glandular.
        (Plant spiny: *Erythrina*).   PAP-**Phaseoleae**
                    e.g. *Clitoria, Erythrina* (ornam.)

    30* Leaves neither trifoliolate, stipellate, or glandular.
        31 Existence of UE in monopodial series.
            32 Branches plagiotropic.
                33 Bark without spermatic odour.
                          PAP-**Millettieae**
                          e.g. *Lonchocarpus*
                33* Spermatic odour in bark.
                          PAP-**Sophoreae**
                          e.g. *Ateleia* (AM)

            32* Branches erect.
                34 Br. decumbent.   PAP-**Robinieae**
                          e.g. *Gliricida* (AM)

                34* Branches erect. PAP-**Sophoreae**
                    e.g. *Acosmium, Alexa* (AM), *Sophora*

        31* Branches distinctly sympodial.
            35 Leaflets alternate.   PAP-**Sophoreae**
                      e.g. *Bowdichia* (AM)

                Phyllotaxy distichous in the branches
                of the young individuals. CSL-**Cassieae**
                      e.g. *Dicorynia* (AM)

            35* Leaflets opposite.   PAP-**Millettieae**
                    e.g. *Poecilanthe* (AM), *Tephrosia*

28* Serial buds (F12 a). (Rhythmic growth noticeable due to the presence of short internodes). See36a, 36b, and 36c:
    36a Vertical serial buds (F12 a). (Rachis grooved, young internodes angular.   CSL-**Sclerolobieae**
                    e.g. *Campsiandra, Vouacapoua* (AM)

    36b Vertical serial buds (F12 a). Rachis not grooved.
        (Leaflets opposite).   PAP-**Sophoreae**
e.g. *Clathrotropis, Diplotropis,* (AM), *Ormosia* (AM,AS,AU)

    36c Buds collaterals (F12 c). L. paripinnate.
                        PAP-**Robinieae**
                e.g. *Sesbania* (AM)

25* Leaves compound. Rachis not ending in a terminal leaflet (ending in a mucro), (F14 g,j).
    37 Leaflets opposite (F14 g).
        38 Plant twining. (Small liana).   PAP-**Abreae**
                              Abrus

- **38*** Plant not twining.
  - **39** Serial buds (F12 a,c,d).
    - **40** Spines (short twigs): *Gleditsia*; subwhorled leaves disposed on very short twigs, venation secant (F17 n): *Haematoxylum*. **CSL-Caesalpinieae**
      Haematoxylum (AM), *Gleditsia* (AM,AS)
    - **40*** Plant different. (AM, vertical serial buds, F12 a; Rhythmic growth well marked by the occurrence of shorter internodes).
      - **41** Young stems grooved. Pinnatifide stipules. (Some myrmecophilous spp. with hollow petioles: *Tachigali*). (SCARRONE). **CSL-Sclerolobieae**
        e.g. *Sclerolobium, Tachigali* (AM)
      - **41*** Stems not grooved. **CSL-Dimorphandreae**
        e.g. *Mora* (AM)
  - **39*** All axillary buds solitary.
    - **42** Young internodes angular or with longitudinal ridges (F11 h). (TROLL or CHAMPAGNAT). (Cupuliform, F19 d, or nipple-like glands, F19 c; underside of leaflets glaucous). **CSL-Cassieae**
      e.g. *Cassia, Senna*
    - **42*** Stems different, erect? Rachis not glandular. **PAP-Aeschynomeneae**
      e.g. *Amicia* (AM)
- **37*** Leaflets alternate (F14 j) or subopposite. AM. (Rachis winged; spirodistichous phyllotaxy, F9 f1: *Coumarouna*, or embossed venation, F18 s: *Dipteryx*). **PAP-Dipterygeae**
  e.g. *Coumarouna* (~*Dipteryx*), (AM)

- **24*** Leaves bipinnate. (Phyllodes for some Australian *Acacia*, F10 b).
  - **43** Existence of buds in lateral or in vertical zigzagging series (F12 c,d), (possibly, only two buds located more or less vertically).
    - **44** Climbing or weakly prostrate (F5 e) plant.
      - **45** Branches with UE in monopodial series. (Spiny plant). If spiny: stem 5-sided with spines on the five sides. Leaves not glandular and leaflets thigmonastic (see F5): *Mimosa*; L. glandular: *Piptadenia, Mimosa myriadenia*. Prostrate, not spiny plant growing along steams: *Entada*. **MIM-Mimoseae**
        e.g. *Piptadenia* (AM), *Mimosa, Entada*
      - **45*** Branches markedly sympodial. Leaves glandular. Leaflets not thigmonastic. Spiny plant. Stems angular or groove, 10-sided (with spines) on only fives sided. **MIM-Acacieae**
        Acacia
    - **44*** Tree or self-supporting shrub.

46 Branches plagiotropic (TROLL).
    47 Existence of UE in monopodial series (field observation!).
        48 Leaves glandular. (Young internodes angular, no glands).   MIM-**Ingeae**
        e.g. *Samanea* (AM)

        48* Leaves not glandular.   MIM-**Mimoseae**
e.g. *Piptadenia* (AM), *Pseudoprosopis, Piptadeniastrum* (AF)

    47* Branches distinctly sympodial. See 49a, b, c, and d:
        49a Rachis grooved. Twigs angular with two decurrent ribs from petiole. Leaflets very numerous. (Spiny plant).   MIM-**Acacieae**
        Acacia

        49b Rachis grooved. Twigs not angular. Leaflets very numerous. Rachs 3-furrowed with a basal gland.
        MIM-**Mimoseae**
        Anadenathera

        49c Rachis convex in section. (Plant not spiny; leaflets not very numerous).   MIM-**Ingeae**
        e.g. *Enterolobium* (AM)

        49d Rachis convex in section. Well devloped axillary buds. (Short shoots modified into spines; leaflets not very numerous).   MIM-**Ingeae**
        e.g. *Chloroleucon* (AM)

46* Branches orthotropic.
    50 Leaflets thigmonastic (See text of F5). (Spiny plant of open places; indument of thick hairs f. s. spp.).
    MIM-**Mimoseae**
    Mimosa

    50* Leaflets not thigmonastic.
        51 Existence of UE in monopodial series.
            52a Leaves not glandular. Leaflets alternate, and buds forming a weak zizaging series: *Adenanthera*.   MIM-**Mimoseae**
            Tetrapleura, Adenanthera (AS,AU,OC)

            52b Convex glands situated somewhat below the insertion of the pinnae, leaflets quite large and almost symmetrical.
            MIM-**Ingeae**
            Cedrelinga, Pseudosamanea (AM)

            52c Cupular glands, leaflets asymmetrical.
            MIM-**Ingeae**
            Piptadenia, Stryphnodendron (AM)

        51* Branches sympodial. (L. glandular).

53 Leaflets >5mm broad. (Short shoots f. s.spp. of *Pithecellobium*). **MIM-Ingeae**
e.g. *Pithecellobium, Zygia* (AM)

53* Leaflets <5mm broad. **MIM-Mimoseae**
e.g. *Pseudopiptadenia* (AM)

43* Axillary buds solitary, or in distinct vertical series (F12 a).
  54 Buds always solitary.
    55 Rachis glandular (F19 e) or well marked rhythmic growth indicated by short internodes. (Branches erect). See 56a, 56b, and 56c:
      56a Leaflets small, opposite, many (>100). **MIM-Parkieae**
Pentaclethra (AF), *Parkia*

      56b Leaflets large, opposite, few (<30). Existence of UE in monopodial series. **MIM-Mimoseae**
Xylia (AF,MA,AS)

      56c Leaflets alternate. Internodes short. **MIM-Ingeae**
Serianthes (AS,OC)

    55* Rachis not glandular. Growth different. (Branches erect or decumbent). (Short shoots; rachis modified into a spine: *Parkinsonia*). **CSL-Caesalpinieae**
e.g. *Delonix* (AF,MA,AS, ornam.)
Parkinsonia, Schizolobium (AM, ornam.), *Caesalpinia*

  54* Buds in vertical series (F12 a).
    57 Leaflets very numerous, midrib dividing the leaflet in two very unequal parts (F14 f). **MIM-Parkieae**
Pentaclethra (AM)

    57* Leaflets can be fairly numerous. Leaflet divided by its midrib in two subequal parts.
      58 Growth weakly rhythmic. (Pinnae opposite; spiny weakly prostrate: *Caesalpinia* (in AM), *Mezoneuron*. **CSL-Caesalpinieae**
e.g. *Bussea* (AF,MA), *Gymnocladus* (AS,North AM)
Mezoneuron (Paleotrop.), *Caesalpinia, Peltophorum* (ornam.)

      58* Rhythmic growth distinctly marked by the occurrence of short internodes. (Pinnae or leaflets alternate f. s. spp.). **CSL-Dimorphandreae**
e.g. *Dimorphandra* (AM), *Pachyelasma* (AF)

Glossary, notes, and illustrations

1. Geographical distribution

Endemic families with woody species in tropical rain forests; numbers indicate genera / number of species, generally according to Mabberley [38]*. Non identified family indicated by an asterisk.

America: *Asteranthaceae, 1/1; Bixaceae 1/1 (if Cochlospermaceae are kept apart); Brunelliaceae, 1/52; Cactaceae (but *Rhipsalis* extending into Africa and Sri Lanka); Caryocaraceae, 2/25; Cobaeaceae, 1/10 (if *Cobaea* is separated from Polemoniaceae); Cyclanthaceae, 12/200; Cyrillaceae, 3/14; Dialypetalanthaceae, 1/1; *Duckeodendraceae, 1/1; Eremolepidaceae, 3/11; Goupiaceae, 1/3 (here separated from Celastraceae); Marcgraviaceae, 5/108; Pellicieraceae, 1/1; Peraceae, 1/40 (here included in Euphorbiaceae); Peridiscaceae, 2/2; Picramniaceae, 2/50 (if separated from Simaroubaceae); Picrodendraceae, 1/1 (here separated from Euphorbiaceae); Quiinaceae, 4/45; Rhabdodendraceae, 1/3; *Stylocerataceae, 1/5 (if separated from Buxaceae); Theophrastaceae, 4/90; Tovariaceae, 1/2.

Africa (including Madagascar, Mascarenes and Seychelles): Aphloiaceae, 1/1 ; Balanitaceae, 1/25 (here not placed in Zygophyllaceae); Brexiaceae, 3/11; Dioncophyllaceae, 3/3; *Hoplestigmataceae, 1/2; Huaceae, 2/3; *Lepidobotryaceae, 2/2 (if not in Oxalidaceae); *Medusandraceae, 2/9; Melianthaceae, 2/8; Monotaceae, 1/26 (here in Dipterocarpaceae); Montiniaceae, 2/4 (here not placed in Grossulariaceae); Octoknemaceae, 1/5 (here in Olacaceae); Oliniaceae, 1/8; *Pentadiplandraceae, 1/1 (if separated from Capparidaceae); *Ptaeroxylaceae, 3/10; Scytopetalaceae, 5/20, if kept apart from Lecythidaceae; Uapacaceae, 1/61 (here in Euphorbiaceae).

Madagascar and **Mascarenes**: *Asteropeiaceae, 1/5; *Didymelaceae, 1/2; *Diegodendraceae, 2/18; Foetidiaceae, 1/6 (here in Lecythidaceae); *Kaliphoraceae, 1/1; Melanophyllaceae, 1/6; Physenaceae, 1/2; *Psiloxylaceae, 1/1; Sarcolaenaceae (syn.: Chlaenaceae), 9/51.

Asia and/or **Australia, New Guinea** and **Pacific Islands**: *Amborellaceae, 1/1; *Austrobaileyaceae, 1/1; *Blepharocaryaceae, 1/2; Bischofiaceae, 1/2 (here separated from Euphorbiaceae); Casuarinaceae, 4/95; *Davidsoniaceae, 1/2; *Degeneriaceae, 1/2; Eupteleaceae 1/2; *Himantandraceae, 1/2-3; Lophopyxidaceae, 1/1; *Medusagynaceae, 1/1; *Oncothecaceae, 1/2; *Paracryphiaceae, 1/1; *Pentaphylacaceae, 1/1; *Phellinaceae, 1/12 (if separated from Aquifoliaceae); Sabiaceae, 1/19 (if separated from Meliosmaceae); Siphonodontaceae, 1/7 (here in Celastraceae); *Strasburgeriaceae, 1/1 (if separated from Ochnaceae); *Trimeniaceae, 1/5.

Asia and/or **Malay Archipelago**: Aucubaceae, 1/3 ; Bischofiaceae, 1/2; Crypteroniaceae *sensu stricto*, 3/10; Daphniphyllaceae, 1/10; Dipterocarpaceae (AS, PF, except *Monotes* and *Pakaraimea*), 14/653; Erythropalaceae, 1/1 (if separated from Olacaceae); Rhodoleiaceae, 1/(1-7) (if separated from Hamamelidaceae); Sabiaceae, 1/19 (if separated from Meliosmaceae); *Sarcospermataceae, 1/8 (if separated from Sapotaceae); *Scyphostegiaceae, 1/1; Tetramelaceae, 2/2 (here separated from Datiscaceae); Xanthophyllaceae, 1/93 (sometimes in Polygalaceae).

Australia (Queensland or Northern Territories) and/or **Pacific Islands**: *Austrobaileyaceae, 1/1; *Balanopaceae, 1/9; *Blepharocaryaceae, 1/2 (if separated from Anacardiaceae); Casuarinaceae, 4/95 (a few species in the the Malay Archipelago); Corynocarpaceae, 1/4; *Davidsoniaceae, 1/2; *Degeneriaceae, 1/2; *Eupomatiaceae, 1/2; *Lactoridaceae, 1/1 ; *Trimeniaceae, 1/5; *Xanthorrhoeaceae, 10/90.

* Numbers in brackets refer to bibliographic references.

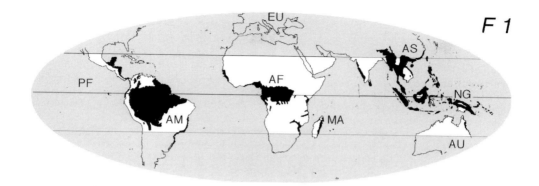

Potential distribution of the tropical rain forests

AM: tropical North, Central and South America; AF: tropical Africa; AS: tropical Asia; AU: tropical Australia; EU: Europe; MA: Madagascar, Comoro Islands, Reunion and Mauritius; NG: New Guinea; OC: Oceania.

Families with woody tropical species on two, but not three, different land masses (Dicotyledons). The specified genera have a discontinuous distribution, when the latter not mentioned.

America / Africa or **Madagascar**: Canellaceae, 5/13; Caricaceae 4/33; Humiriaceae 8/50 (only one African species); Mendonciaceae *s. str.* (*Mendoncia*), 2/60, (here in Acanthaceae); Velloziaceae, 4/250; Vochysiaceae 8/210 (only one African species).

America / Asia or **Australia**: Actinidiaceae (*Saurauia*) 3/340; Araucariaceae, 2/29; Bonnetiaceae, 3/34 (here separated from Theaceae); Clethraceae, 1/64; Chloranthaceae (*Hedyosmum*), 4/75; Elaeagnaceae (*Elaeagnus*), 3/45; Epacridaceae, 31/375; Fagaceae (*Trigonobalanus*), 8/700 (not in tropical AF); Illiciaceae, 1/42; Iteaceae (*Itea*), 1/15 (here separated from Grossulariaceae); Juglandaceae, 8/59; Magnoliaceae, 7/165; Meliosmaceae *s. str.* 2/62 (if separated from Sabiaceae); Pinaceae, 12/220; woody Santalaceae (*Jodina* in AM, *Santalum* in AS, *Eucarya* in AU); Schisandraceae 2/47; Styracaceae (*Styrax*), 11/160; Tetrameristaceae, 2/4; Theaceae (*Ternstroemia* especially in AM and AS), 21/605; Trigoniaceae, 3/26 (with *Humbertiodendron* in MA); Winteraceae (*Drimys*), 4/60. The Aquifoliaceae, Araliaceae, Cunoniaceae, Lauraceae, Monimiaceae, Santalaceae and Symplocaceae are, in addition, poorly represented in Africa (without Madagascar).

Africa or **Madagascar / Asia** or **West Pacific** area: Ancistrocladaceae, 1/12; Ctenolophonaceae, 1/3 (if separated from Linaceae); Hymenocardiaceae, 1/6-7 (if separated from Euphorbiaceae); Irvingiaceae, 3/8 (*Irvingia*), (here separated from Ixonanthaceae and Simaroubaceae); Leeaceae, 1/34; Moringaceae, 1/12; Pandaceae, 4/18; Pandanaceae, 3/875; Pittosporaceae, 9/200 (*Pittosporum*); Sonneratiaceae, 2/8 (*Sonneratia*), if excluded from Lythraceae; Thunbergiaceae, 4/105 (*Thunbergia*), (if excluded from Acanthaceae).

2. Outer bark and lenticels

For definitions of morphological terms used in the following glossary, see also [3].
For description and terminology of bark and rhytidome, see [4], [5], [6], [7], [8], [9], [10], [11], [12].
Outer bark (or rhytidome): layer of dead sheet of periderm external to the living bark [5], [5b].
Periderm: The secondarily developed protective bark tissue replacing the epidermis, or built up during rhytidome formation; consists of phellem, phellogen and phelloderm [8]. The periderm can peel off periodically, it is rarely permanent and single. The phelloderm forms a thin, but obvious layer, and its colour is in some cases very different to that of the remaining bark [6], [7].
Lenticels: Restricted areas of relatively loosely arranged cells, suberized or not, in the periderm [8]. The lenticels result from the transformation of stomata; they are much developed in the areas of tangential expansion of the bark, at the periphery of funnel-shaped rays [7].
Branch collar (eye mark, [9]): A peculiar conformation of the rhytidome appearing on trunks (and old erect branches) at the level of insertion of lateral branches. These structures result from secondary growth and from a peculiar mode of rupture of the rhytidome.

Sloughing off and rupture of the rhytidome
The following characters have to be examined especially in young branches.
- **a** Suberization forming well-anchored scales (Buxaceae: *Buxus*; Ixonanthaceae: *Ixonanthes*; Sapotaceae: *Sideroxylon*; Rubiaceae: *Ixora*), (Branch, x 2).
- **b** Peeling where the dead layers slough off in the form of straw-like scales (Myrtaceae; Melastomataceae; Onagraceae).
- **c** Peeling in sheets (Combretaceae: *Combretum*; Myrtaceae: *Eucalyptus*; Sapindaceae: *Pometia*). (Trunk, x 1/8).
- **d** Peeling producing longitudinally-torn strips (Caprifoliaceae: *Lonicera*; Hydrangeaceae: *Hydrangea*; Rubiaceae: *Aiouea*; Lythraceae; Melastomataceae).
- **e** Rupture of the periderm or the rhytidome forming thin longitudinal slits (x 1), the upper internodes not yet suberized. This characteristic of cracking eventually disappears as it becomes hidden by other phenomena (*Conceveiba* and other Euphorbiaceae; Elaeocarpaceae: *Sloanea*; Apocynaceae: *Cerbera*, *Himatanthus*).
- **f** Well-marked longitudinal slits (Euphorbiaceae: *Baccaurea*, *Hura*, *Phyllanthus*), (x 3); Phytolaccaceae: *Phytolacca*.

Branch collar
- **g** Branch collar associated with bases of rising branches (Fagaceae; Moraceae: *Ficus*; Sapindaceae: *Melicoccus*); (x 1/10).
- **h** Branch collar relatively flat (Fagaceae); (x 1/10).

Lenticels
- **j** Lenticels retaining a circular form during the thickening of the rhytidome (Apocynaceae; Burseraceae; Connaraceae; Dichapetalaceae; Humiriaceae; Malpighiaceae; Meliaceae; Sapindaceae).
- **k** Lenticels elongating transversely during the thickening of the rhytidome (Chrysobalanaceae; Leguminosae; Moraceae; Urticaceae).
- **m** Lenticels elongating longitudinally during the thickening of the rhytidome, possibly with a longitudinal furrow (Flacourtiaceae; Passifloraceae; some Euphorbiaceae).
- **n** Protruding lenticels, partially covered by the old periderm (Bignoniaceae; Oleaceae).

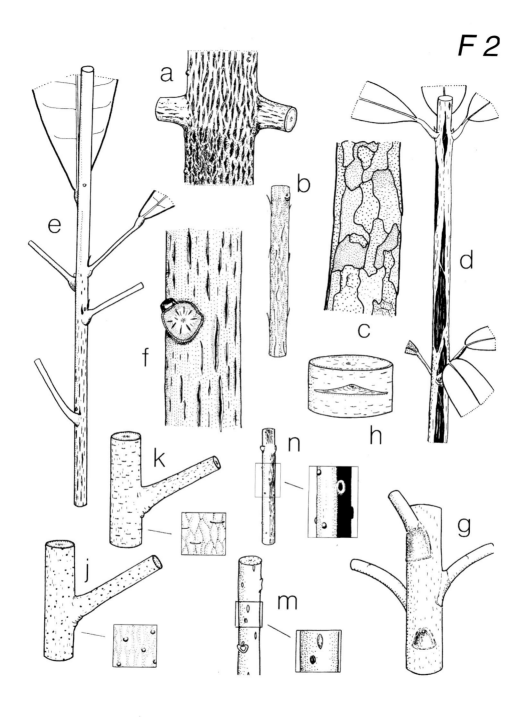

3. Macroanatomy I : inner bark, rays and exudates

Inner bark (living bark or external phloem): Layer formed by the phloem and its rays (living layers) and including in many cases fibres [5], [6]; the inner bark spreads to the periphery of the cambium [6], [7]. Lianas may have several vascular cambia (see *F 4*) bordered by internal phloems.
Fibres: Networks of fibres pertain to the bark and are often observable with the naked eye. The "arches" contain fibres (with series of sclerified cells). The fibres grow round the prolongation of the medullary rays. The occurrence of a network is an important taxonomic character [9], [10], [11].
Rays: Series of parenchymatous cells spreading radially in the wood and the bark. Multiseriate rays ending in some cases in furrows grooved in the periphery of the wood.
Phloem fibres: When they occur in the secondary phloem, they are conspicuous elements because of their thick walls and characteristic distribution patterns [9]. Presence of cortical fibres does not always produce a network observable with the naked eye or a hand lens (Fagaceae and various other families).

Exudates

Latex consist of polyisomers of isoprene cis, [13], (Euphorbiaceae; Moraceae; Apocynaceae). Latex is a colloidal suspension or emulsion of water-insoluble substances; it is typically white (milky), but may be yellow, or colourless [9]. **Gutta** are polyisomers of isoprene trans (Sapotaceae), they always contain resin, [13]. **Gums** are derived from polymers of pentoses or hexoses, they are not crystallizable but soluble in water (*Prunus, Sterculia*); gums are viscid secretions, generally colourless and tasteless; they become hard, clear, glassy masses on dessication [9]. **Resins** are gums associated with variable proportions of essential oils. When essential oils exist in limited quantity, hard resin or **copals** are produced (Caesalpinioideae). When the fraction of essential oils is high, soft resins, **damars** (*Agathis*, Dipterocarpaceae) or **balsams** (Burseraceae, Styracaceae) are produced [13].

Networks of fibres

a Network with relatively short meshes; detail of the net (x 8), (*Theobroma* and other Malvales: Bombacaceae, Malvaceae, Sterculiaceae, Tiliaceae; Annonaceae; Bixaceae; Caricaceae; Flacourtiaceae: *Pangium*; Lecythidaceae: *Barringtonia*; Monimiaceae; Thymelaeaceae). Cortical fibres forming networks with very small meshes occur in the Lecythidaceae (*Eschweilera*).
b Deep purplish or black network contrasting with the cortical parenchyma (Annonaceae: *Guatteria*).
c Network of rather loose meshes, distinctly longer than broad (Boraginaceae: some species of *Cordia*; Malvales: *Bombax*; *Dendrocnide* and other Urticaceae).
d Network with broad strips of fibrous layers (several mm to a few cm), (Bombacaceae; Sterculiaceae: *Pterospermum*; a few Leguminosae: e.g. *Inga* pp.).

Furrowed wood (at its periphery)

e Thin and elongated furrows (Rosaceae: *Prunus arborea*); (x 1). Some woody Asteraceae do possess long parallel furrows: *Mikania*).
f Wood (w) markedly furrowed. Large rays (r) protruding out of the wood; the bark has been peeled off (an unidentified Fagaceae of Sumatra); (x 3).
g Furrows shallow (Casuarinaceae; Fagaceae; Nyctaginaceae: *Pisonia*; Platanaceae); (x 2).
h Furrows (f) marked in such a way that the external wood appears as a anastomosed structure (Proteaceae; Rhizophoraceae: *Sterigmapetalum*); (x 2).
j External wood furrowed. Wood traversed by numerous rays (r) penetrating the bark where they are surrounded by the phloem. The "arches" of the network are readily visible and prolong the poles of phloem to the rhytidome (Proteaceae and Platanaceae); (x 3).

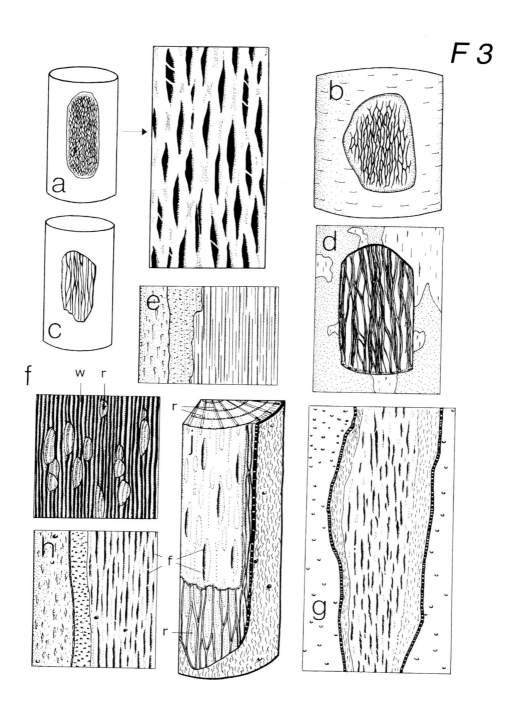

F 3

4. Macroanatomy II : internal phloem, wood and pith

This section applies especially to lianas: phloem is figured in black, rhytidome and bark are dotted. The families or genera of woody lianas are often identifiable by mere observation of their trunk macroanatomy [15].
Internal phloem: Phloem appearing surrounded by wood, hence separated from the bark (in a transversal cut of the stem). This situation occurs frequently with woody lianas.
Wood (or xylem): The whole set of vascular bundles destined, at least initially, for the transport of water taken up by roots. In old stems of arborescent Dicotyledons and Gymnosperms, almost all the xylem consists of dead cells.
Patterns of xylem disposition: The xylem (or wood) may appear:
- entire (simple xylem, more or less circular in section);
- not divided and not entire (xylem more or less lobate or with rays or strands of phloem);
- divided, i.e. arranged in isolated strands separated by layers of phloem and parenchyma (xylem in rings or in blocks).
Pith (or medulla): Parenchymatous central cylindrical part of many stems of Dicotyledons, Ferns, and Gymnosperms. The medulla does not contribute to water transport.

Xylem and phloem

a Typical structure of a self-supporting Dicotyledon trunk where the phloem is produced between the wood and the rhytidome. This structure occurs also in the twining lianas (Apocynaceae; Connaraceae: *Jaundea*; Sterculiaceae: *Byttneria*) and the non-twining lianas (Ancistrocladaceae; Annonaceae; Leguminosae: *Machaerium*).
b The primary wood has a quadrangular section (Rubiaceae: *Uncaria*).
c Xylem in concentric rings (Connaraceae; Leguminosae-Papilionoideae: *Machaerium*, *Millettia*).
d Xylem bilobate with two poles of phloem (Polygonaceae: *Coccoloba*).
e Xylem pentalobate (Leguminosae-Mimosoideae: *Acacia*; Caesalpinioideae: *Senna*).
f Xylem quadrilobate (Leguminosae-Mimosoideae: *Pithecellobium* sp.).
g Two rows of xylem in blocks (Icacinaceae: *Iodes*).
h Xylem in excentric rings (Connaraceae).
k Xylem with rays of phloem (Menispermaceae; Dilleniaceae; Vitaceae) or in rings (Menispermaceae).
m Xylem with scattered strands of phloem (Combretaceae: *Combretum*; Loganiaceae: *Strychnos*).
n Xylem lobate with much protruding phloem (Bignoniaceae; Convolvulaceae: *Neuropeltis*; Celastraceae: *Salacia*).
p Xylem in blocks (Celastraceae-Hippocrateoideae).
q Xylem in blocks or rings (Celastraceae-Hippocrateoideae).

Pith

r Sympodial character of a stem noticeable by its discontinuous pith.
s Rhythmic growth of a stem noticeable by the variable width of its pith.

Bark

t Sclerenchymatous, often orange or yellow-coloured, rod-like inclusions in bark (Flacourtiaceae: *Laetia*; Icacinaceae: *Iodes*, *Gonocaryum*; Lauraceae: *Aniba*; Meliaceae: *Dysoxylum*).

Latex

u Laticiferous threads in leaves (or even in bark), (Celastraceae, in part; Cornaceae: *Cornus*; Gnetaceae; Ixonanthaceae: *Ochthocosmus*; Monimiaceae: *Tambourissa*).

F 4

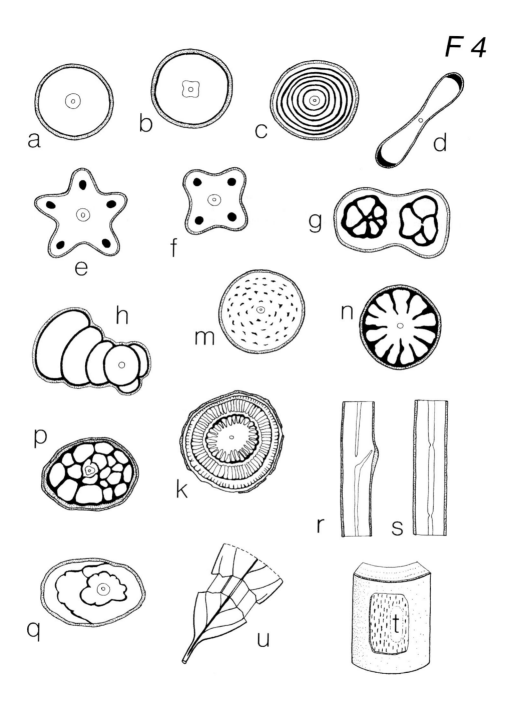

5. Climbing systems

For a synopsis of this subject, see [14]. The duration of the self-supporting stage of a liana is highly dependent upon the species or the situation (open land or forest, for example), [15].
Twining: Winding of an axis around a potential support (volubile or twining axis). This phenomenon can affect a stem, a leaf, a peduncle or a root.
Thigmonastism: Motion of an organ triggered by contact with an object. The direction of motion is independent of the contact point (i.e. closing of the leaflets in Mimosa), [16].
Thigmotropism: Motion of an organ triggered by contact with an object or even by the close proximity of an object [17]. The motion acts towards the object [16]; (i.e. tendrils move and/or grow towards a potential support).
Weakly prostrate plant: A plant which climbs by leaning on its stems, often helped by spines or hairs or non-prehensile hooks.
Hook: Distal part of a thigmonastic stem becoming woody and persistent or short shoot allowing the plant to climb.
Tendril: Twining, short lived and non-thickening thigmonastic(-tropic) axis.

- **a** A long portion of the stem twines, growth monopodial (Apocynaceae; Asteraceae: *Mikania*; Celastraceae: *Celastrus*; Convolvulaceae; Icacinaceae: *Phytocrene*; Lardizabalaceae; Leguminosae: Phaseoleae; Malpighiaceae).
- **b** Stem twining only at the extremity, growth sympodial (Connaraceae).
- **c** Twigs thigmonastic and leaves opposite (Celastraceae: *Salacia* ; Apocynaceae-Willughbeeae: *Landolphia, Pacouria, Willughbea*).
- **d** Twigs spiny and thigmonastic (Leguminosae: *Machaerium*).
- **e** Weakly prostrate shrub with decumbent branches (Apocynaceae: *Allamanda*; Oleaceae: *Jasminum*; Verbenaceae: *Petrea*). This habit can be also acquired by means of plagiotropic branches (Thymelaeaceae: *Dicranolepis*; Icacinaceae: *Rhaphiostylis*).
- **f** Extremity of stem modified into a hook (Rhamnaceae: *Gouania, Ventilago*; Leguminosae: *Machaerium*).
- **g** Short twig modified into a hook (Linaceae: *Hugonia, Indorouchera*; Loganiaceae: *Strychnos*; Rubiaceae: *Uncaria*). Illustration: *Strychnos*.
- **h** Extremity of a stem module transformed into a hook (Annonaceae: *Artabotrys*).
- **j** Thigmonastic twig, more or less twining (Annonaceae: *Uvaria*; Polygalaceae: *Bredemeyera*).
- **k** Persistent petiolar bases, turning woody and offering a passive means of climbing (Verbenaceae: *Clerodendron*).
- **m** Module of the main stem ending in a tendril (Icacinaceae: *Iodes*).
- **n** Tendril tightly coiled (or circinnate) of *Paullinia* (Sapindaceae), or not tightly coiled (Passifloraceae: *Passiflora*), subtended by a mature leaf.
- **p** Twig-tendril subtended by a scale-leaf (Lophopyxidaceae: *Lophopyxis*).
- **q** Tendril oppositifoliate (Vitaceae).
- **r** Rachis ending in one or more tendrils (Bignoniaceae; Leguminosae: *Entada*; Polemoniaceae: *Cobaea*). Illustration: leaf of a Bignoniaceae.
- **s** Twining petiole (Ranunculaceae: *Clematis*; Hernandiaceae: *Illigera*).
- **t** Thigmonastic petiole (Icacinaceae: *Desmostachys*; Solanaceae: *Solanum*), modified into a hook.
- **u** Twining stipular tendrils (Smilacaceae: *Smilax*).
- **v** Midrib ending in a bifid hook (Dioncophyllaceae).
- **w** Root modified into a tendril (Araceae).

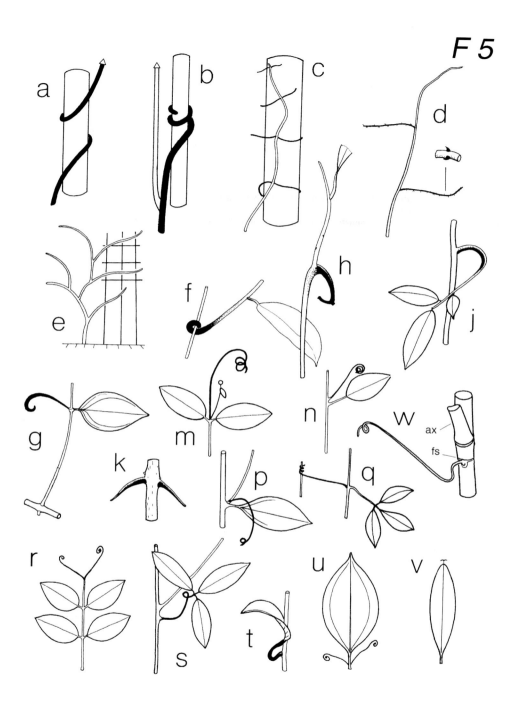

6. Unit of extension, monopodium and sympodium

Meristem: Permanent or temporary zone of actively dividing cells from which mature tissues differentiate.

Unit of extension: Portion of stem which has elongated in an uninterrupted fashion [18]. In practice, the limits of the unit of extension (UE) are observable by the presence of shorter internodes, relatively small leaves, a narrower width of the pith (**4 s**) or a variation in the bark texture. (Limits of UE are figured by double-bars in the illustrations).

Monopodium: Shoot unit developed by the vegetative extension of one apical meristem.

Sympodium: Linear series of shoots, each new distal shoot unit developing from an axillary bud sited on the previous shoot unit. The sympodial character of a stem can be easily recognised by its interrupted pith (**4 r**). A sympodium is possibly formed by two or more UE in a « monopodial » series.

Ramification: Process leading to the formation of a 'lateral' stem inserted on a 'main' stem.

Modules (cf. modular architecture): Simple and non-varying morphogenetic units produced in a linear sequence by a sympodial mechanism (cf. modular architecture), [19], [20]. In a much less restricted sense: a shoot unit developed from one apical meristem.

Plagiotropic growth: Growth tending away from the vertical. Plagiotropic growth is generally associated with distichous phyllotaxy, twisting of internodes (when leaves are opposite), or reorientation of the leaves to the horizontal plane (see F 9). Plagiotropy may also occur in shoots with spiral phyllotaxy.

Apposition growth: The displacement of a terminal meristem by an axillary meristem promoting the further extension growth of the branch complex; the evicted terminal bud continues its vegetative growth, usually as a short shoot [20].

Substitution growth: The replacement of a terminal meristem by a lateral meristem, after the former has either aborted or has become differentiated as a terminal flower or an inflorescence [20].

UE and ramification (examplified by monopodial growth)

- **a** Ramification of the UE does not occur at the time of extension, i.e. ramification is **delayed** (Aquifoliaceae; Euphorbiaceae: *Hevea*; Sapindales: Anacardiaceae, Burseraceae, Meliaceae, Sapindaceae, Simaroubaceae, Rutaceae; *Quercus* and other Fagaceae). c = scale-leaf.
- **b** The unit of extension produces one or several lateral axes during its development, i.e. ramification is **immediate** (Ebenaceae: *Diospyros*; Myristicaceae; Lauraceae: *Ocotea*; etc.).
- **c** Hypersyllepsy. If Δ equals the difference between internodes number of the main axis and internodes number of a lateral axis and if Δ is taking negatives values for a certain time, the development of a lateral axis is stronger than that of the main axis (Cupressaceae: *Chamaecyparis*, *Thuja*, etc., but not *Cupressus*, or possibly in taxa conforming to ROUX-ATTIM's models.
- **d** **A**. sectorial anisoclady: the dominant branches develop preferentially in one of the four sectors of the cylinder centred on the trunk (frequent for the Oleaceae, Acanthaceae, and Bignoniaceae: *Parmentiera*), (top-view). **B**. helicoidal anisoclady, the dominant twigs lie in all the sectors around the trunk (frequent for the Rubiaceae: *Lasianthus*; Loganiaceae: *Antonia*), (top-view).

Sympodial growth (death of apical meristems indicated by "x")

- **e** Branch plagiotropic by apposition growth (c = scale-leaf). Each UE is at first plagiotropic but soon acquires orthotropic growth (Bonnetiaceae: *Ploiarium*; Combretaceae: *Terminalia*; Loganiaceae: *Fagraea*; Rhizophoraceae: *Bruguiera*; Sapotaceae: *Manilkara*).
- **f** Branch plagiotropic by substitution growth (Euphorbiaceae: *Excoecaria*; Icacinaceae: *Lasianthera*; Chrysobalanaceae; Violaceae: American *Rinorea*).
- **g** Branch with sympodial units consisting of two (or more) units of extension (UE1, UE2, ... delimited by double bars). (Lecythidaceae: *Barringtonia*, Leguminosae: *Piptadenia*).

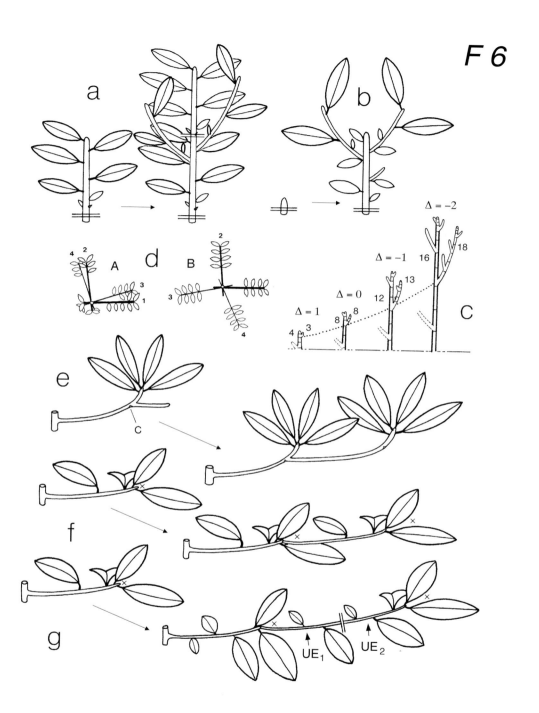

F 6

7. Architectural models (trunk monopodial)

Node: Insertion zone on the stem at which one or more leaves are inserted.
Internode: The portion of stem between two consecutive nodes.
Continuous ramification: Ramification leading to the formation of lateral shoots at each node of an axis [20]. This mode of ramification is associated with an absent or weak delimitation of the UE.
Rhythmic ramification: Ramification leading to the formation of lateral shoots regularly separated by several non-ramified nodes. Rhythmic ramification is frequently correlated with the rhythmic production of UE's, in such a way that the lateral shoots form tiers [20]. (See **6 a, 6 b**).
Diffuse ramification: Ramification not continuous such that the internodes maintain a relatively constant length and the lateral shoots do not form tiers.
Anisoclady: Specific mode of ramification, associated with opposite or whorled phyllotaxy, leading to an unequal development of the lateral shoots at the same node. Anisoclady is maximal when only one stem of several develops at the node.
N.B. Architectural models were named after patronyms of various botanists [20], [21]; to make text shorter, Roux's model would simply be designated by 'ROUX', Rauh's model by 'RAUH', etc.).

Patterns of trunk ramification

a Continuous ramification, branches monopodial. **A**. branches plagiotropic, 'ROUX' (*Casearia, Coffea, Dipterocarpus, Durio, Garcinia, Rhizophora*, etc.); **B**. branches orthotropic, 'ATTIMS' (*Avicennia, Eucalyptus, Laguncularia*, etc.).
b A pattern very similar to 'ROUX', where lower branches are shedding, 'COOK' (Euphorbiaceae: *Phyllanthus mimosoides*; Monimiaceae: *Glossocalyx longicuspis*; Moraceae: *Castilla elastica*).
c Continuous ramification, branches sympodial. **A**. branches plagiotropic, 'PETIT' (*Gossypium*). **B**. branches orthotropic, 'STONE' (*Pandanus*)
d Rhythmic and immediate ramification, branches plagiotropic, 'MASSART' (*Alangium, Aniba, Ceiba, Diospyros, Duabanga, Napoleonaea*, etc.).
e As in d, but trunk sympodial, 'NOZERAN', this pattern is included in this plate because a taxon conforming to 'NOZERAN' (e.g. *Theobroma*) is usually close to some taxa with monopodial trunks (many other Sterculiaceae).
f Rhythmic, immediate ramification, branches orthotropic, 'RAUH' (*Ficus* pp. *Ocotea* pp.).
g Rhythmic and delayed ramification, branches orthotropic, 'RAUH' (*Ficus* pp., *Artocarpus, Entandophragma, Hevea, Quercus*, etc.).
h As in g, but stems at first orthotropic, secondarily bending (probably by gravity) 'CHAMPAGNAT' (*Caesalpinia, Crescentia, Jasminum*).
j Immediate ramification leading to formation of basal plagiotropic branches (still not observed!).
k Rythmic ramification, branches sympodial. **A**. branches orthotropic, 'SCARRONE' (*Anacardium, Mangifera, Simarouba*); **B**. branches sympodial by apposition, apical meristems lost, 'FAGERLIND' (*Fagraea, Genipa, Magnolia*).
m Rythmic ramification, branches monopodial. **A** branches orthotropic, 'RAUH' (*Ficus, Ocotea, Sterculia*, etc.); **B**. branches sympodial by apposition, apical meristems maintained, 'AUBREVILLE' (*Baccaurea, Elaeocarpus, Manilkara, Ocotea, Persea, Sterculia, Terminalia*, etc.).

Location of the ramification on the unit of extension

n Basal, immediate ramification (*Cinnamomum, Endlicheria, Ficus, Eugenia malaccensis*).
p Intermediate, immediate ramification (Clusiaceae: *Symphonia*; some Lauraceae and Podocarpaceae).
q Distal, immediate ramification (*Diospyros*; Myristicaceae). See also 'PREVOST' and 'NOZERAN'.

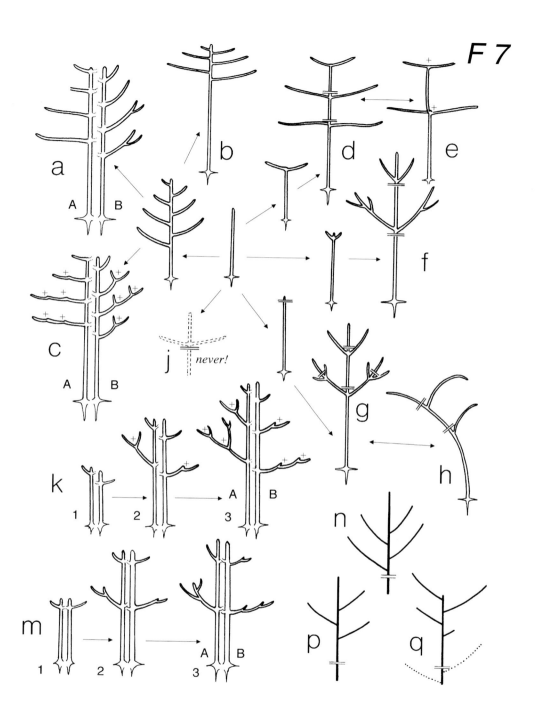

8. Architectural models (trunk sympodial)

Not ramified trunk

a   Initial stages of development. **a1**, apex resuming growth, i.e. ramification will be delayed; **a2**, immediate ramification.
b   Not branched trunk, with terminal inflorescence, 'HOLTTUM', (Agavaceae: *Agave*; Arecaceae: various species of *Caryota*, *Corypha*, *Metroxylon*; Rutaceae: *Spathelia*).

Ramification delayed in trunk

c   Monocaulous and sympodial trunk consisting of the stacking modules, always with orthotropic growth, 'CHAMBERLAIN' (Leeaceae: *Leea*; Quiinaceae: *Quiina*; Sapindaceae: *Talisia*).
d   Branches modular, orthotropic, one leader becoming distinctly erect (Leguminosae-Papilionoideae: *Bowdichia*).
e   A special case linked with opposite phyllotaxy. One of the new modules becoming erect, the other plagiotropic (Myrtaceae: *Eugenia*).

Ramification immediate in trunk

f   Branches modular, orthotropic, modules all equal, 'LEEUWENBERG' (*Alstonia*, *Anthocleista*, *Schefflera*).
g   Branches modular, orthotropic, modules initially equal, all apparently branches, but later unequal, one becoming a trunk, 'KORIBA' (Nyctaginaceae: *Pisonia*; Pittosporaceae: *Pittosporum*). Ramification is generally immediate.
h   Branches sympodial, modular, plagiotropic, 'PREVOST' (Apocynaceae: *Geissospermum*; Boraginaceae: *Cordia*; Euphorbiaceae: *Dichostemma*).

Plant bearing mixed axes

j   Growth direction of trunk modules changing by primary growth, at first (proximally) orthotropic, later (distally) plagiotropic, 'MANGENOT' (Apocynaceae: *Funtumia* ; Icacinaceae: *Rhaphiostylis*; Loganiaceae: *Strychnos*; Melastomataceae: *Mouriri*; Monimiaceae: *Matthaea*; Thymelaeaceae: *Dicranolepis*).
k   Sympodial trunk consisting of the stacking of modules. The growth of a module is initially plagiotropic but its basal part reorientates orthotropically before onset of secondary thickening, most often after leaf-fall, 'TROLL' (Annonaceae: *Annona*, *Monodora*; Chrysobalanaceae; Leguminosae: *Cynometra*, *Inga*, *Swartzia*).
m   Monopodial version of TROLL's model, trunk apex remaining plagiotropic, ramification is monopodial. This pattern, however, is not favourised in large, emergent trees (Annonaceae: *Anaxagorea*).

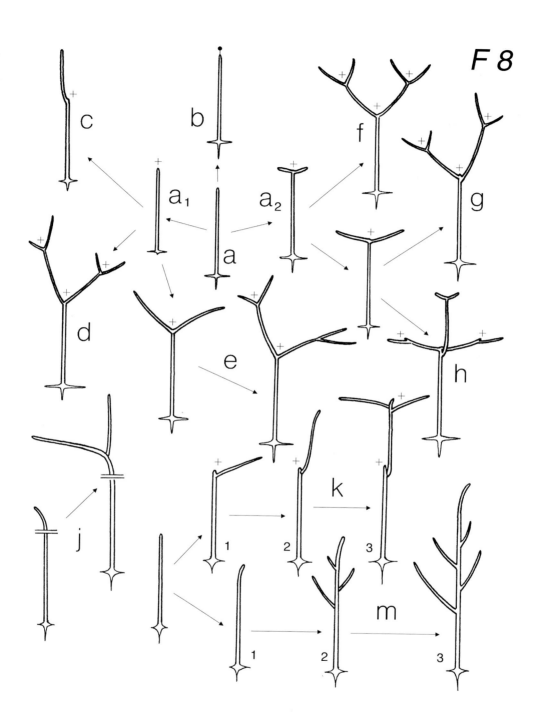

F 8

9. Phyllotaxy and torsion

Phyllotaxy: Sequence of origin of leaves on a stem [24]. When there are only one leaf per node, phyllotaxy is generally distichous or spiral; when there are two (or more) leaves per node, phyllotaxy is opposite (or whorled), [24].
N.B. (concerning the woody Dicotyledons with distichous or spiral phyllotaxy): an individual can produce some shoots with distichous phyllotaxy or produces only shoots with non-distichous (spiral) phyllotaxy. Here are two mutually exclusive taxonomic characters, very stable at the genus or even the tribal level [22].

Phyllotaxy
a Distichous phyllotaxy: branches of the Annonaceae; Chrysobalanaceae; Dichapetalaceae; Leguminosae: *Swartzia*; Myristicaceae; Sterculiaceae: *Theobroma*; Ulmaceae; etc.
b Spiral phyllotaxy with leaves somewhat grouped at the end of the UE: (Elaeocarpaceae: *Sloanea*, Fagaceae: *Quercus*, Pittosporaceae: *Pittosporum*, etc.).
c Spiral phyllotaxy becoming pseudowhorled at the end of the module (Illiciaceae: *Illicium*; Pittosporaceae: *Pittosporum*; Rutaceae: *Pilocarpus*; Theophrastaceae: *Jacquinia*).
d Opposite-decussate phyllotaxy, successive pairs of petiolar bases forming angles of approximately 90° (Chloranthaceae; Clusiaceae; Monimiaceae; Verbenaceae; etc.). For some Acanthaceae (*Ruellia*), the successive pairs of leaves (dotted in the illustration) are more or less oriented in the same half-cylinder (sectorial anisophylly).
e Bijugate phyllotaxy; successive pairs of petiolar bases forming angles distinctly different from 90° (*Cassipourea* and other Rhizophoraceae; some Rubiaceae).
f **f1**: Spirodistichous phyllotaxy, successive petiolar bases forming angles slightly less than 180° (Anacardiaceae: *Campnosperma*). **f2**: Spirotristichous phyllotaxy, leaves in three rows with 120° between twisted rows (Pandanaceae: *Freycinetia*, *Pandanus*).

Modes of disposition of leaves in the same plane
This item concerns plagiotropic shoots or shoots growing in a more or less horizontal plane (decumbent orthotropic shoots).
g Spiral phyllotaxy, the petioles twist and bend moving the laminas into the same plane; plagiotropic branches of many Lauraceae; Boraginaceae: *Cordia*; Icacinaceae: *Rhyticaryum*.
h Distichous phyllotaxy, the petioles neither twist nor bend; plagiotropic branches of the Alangiaceae; Ulmaceae; Lauraceae: *Cryptocarya* pp.; Ebenaceae: *Diospyros*; Lecythidaceae: *Eschweilera*; Icacinaceae: *Discophora*, *Gomphandra*; Thymelaeaceae: *Gonystylus*; Olacaceae: *Heisteria*; Chrysobalanaceae; Leguminosae: *Swartzia* and many other genera; Rhamnaceae: *Ziziphus*.
j Opposite phyllotaxy (1st case): the petioles twist and bend moving the laminas into the same plane. (This phenomenon occurs in horizontal segments of decumbent branches in *Combretum* (Combretaceae), Malpighiaceae; Verbenaceae; for plagiotropic branches in *Duabanga* (Lythraceae) and *Gnetum* (Gnetaceae).
k Opposite phyllotaxy (2nd case): the internodes twist alternatively clockwise and counterclockwise moving the laminas into the same plane (Celastraceae: *Cheiloclinium*; Clusiaceae: *Vismia*; Gentianaceae: *Tachia*; Melastomataceae: *Mouriri*; Monimiaceae: *Siparuna*; Myrtaceae: *Eugenia*, *Psidium*; Rubiaceae: *Coffea*; Vochysiaceae: *Vochysia*). Bending of petioles can also occur.

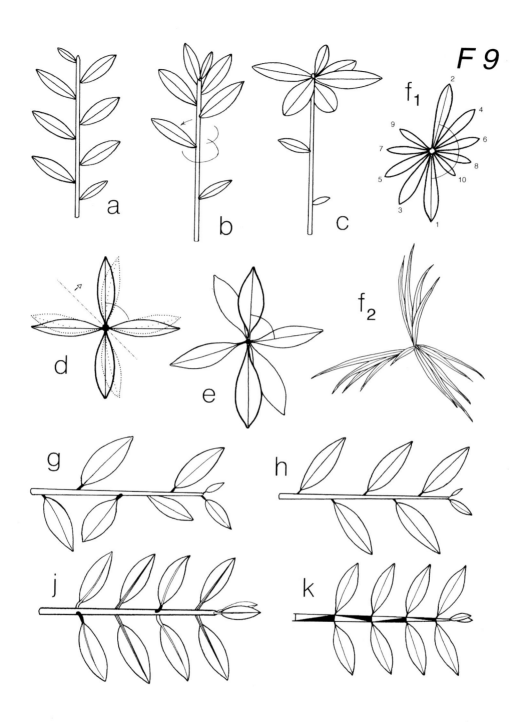

10. Heterophylly

Heterophylly: Important variation of the shape and/or size of leaves during the development of a plant. Different kinds of heterophylly can be recorded:
1) **Foliar dimorphism**: Leaves of the young stage differ from those of the adult stage [23, 24].
2) **Heterophylly between different kind of shoots**: Two or more leaf types differing in shape or size depending on whether they are inserted on trunk, branches or short shoots.
3) **Heteroblasty**: A developmental sequence of differing leaves along a shoot, typically occurring along the same unit of extension or the same module.
4) **Anisophylly**: Two leaf types, typically differing in size, occurring on the same shoot; this variation being correlated with the dorsiventral character of the shoot [25]. Anisophylly can be associated with opposite, whorled or spiral phyllotaxy.

Foliar dimorphism (juvenile/adult leaf = leaf of the juvenile/adult individual)
- **a** **Juvenile** leaf of simpler form than the **adult** leaf :
  For many Bignoniaceae, the developmental sequence starts with simple leaves, then bifoliate leaves and ends with tendril bearing adult leaves. Generally, in Connaraceae, Meliaceae and Sapindaceae, the juvenile leaves are simple and the adult leaves are compound; adult leaves of *Cecropia* (Moraceae) are more deeply lobate than sapling leaves.
- **b** **Adult** leaf of simpler form than the **juvenile** leaf :
  Phyllode-bearing *Acacia* (Leguminosae), the juvenile leaf is compound whilst the adult leaf is reduced to its petiole and rachis (phyllode). For some Proteaceae (*Dilobeia*, *Roupala*), the juvenile leaf is compound, the tree later bearing simple leaves.
- **c** For some Quiinaceae, the shape of the juvenile deeply lobed leaf foreshadows the shape of the simple adult leaf.

Heterophylly between different shoots
- **d** The trunk bears small leaves whilst the branches bear "normal" leaves (Ebenaceae: *Diospyros*; Lythraceae: *Duabanga*). This kind of heterophylly is quite common for lianas. The figure shows the heteroblastic trunk of *Diospyros* for which each unit of extension bears scale-leaves (or cataphylls = c) and a terminal rosette of small leaves.

Heteroblasty (UE delimited by "//")
- **e** Monopodial trunk with spiral phyllotaxy; the shape and size of the leaves varies progressively according to their location in the unit of extension (*Sterculia*). Examples with pronounced variation: *Altingia* (Hamamelidaceae), *Erythroxylum* (Erythroxylaceae), *Litsea* (Lauraceae), *Ouratea* (Ochnaceae).
- **f** Lateral twig bearing scale-leaves (c) at its base; abrupt transition from scale-leaves to photosynthetic leaves (Anacardiaceae: *Bouea*; Fagaceae: *Quercus*; Vochysiaceae: *Callisthene*; and for many other species or genera for which lateral axes undergo delayed ramification).

Anisophylly
- **g** Typical anisophylly: each node bears a large and a small leaf (Melastomataceae : *Sonerila*; Monimiaceae: *Glossocalyx*; Urticaceae: *Pilea*).
- **h** Anisophylly and pseudodistichous phyllotaxy on a plagiotropic twig: the small leaves are inserted above the twig and the large ones are inserted below (Anisophylleaceae: *Anisophyllea*).
- **j** Twig with opposite-decussate phyllotaxy, the petioles inserted in a vertical plane are those of the small leaves, e.g. *Faramea* (Rubiaceae). Hence, at first sight, all leaves appear located in the same plane.

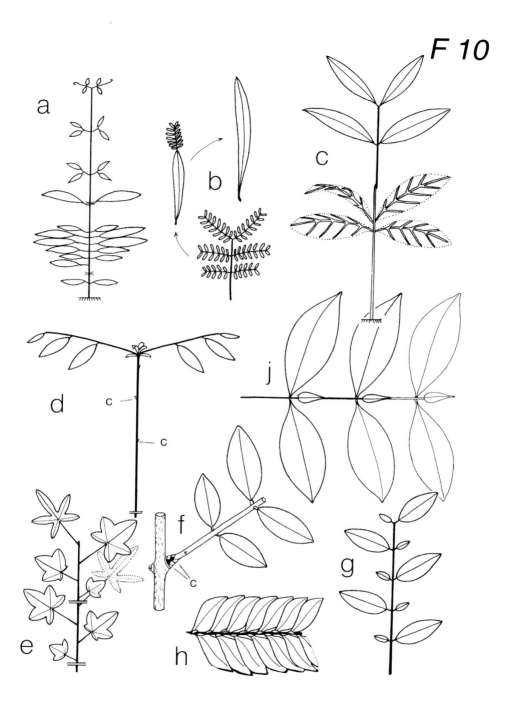

11. Shape of stems

Angular or grooved stem: The development of leaves induces modifications of the typically cylindrical form of the stem. The stem is often slightly thickened below the node. This thickening decreases progressively while forming a rib. The disposition of these structures is dependent upon phyllotaxy.

Short shoot: A short shoot can be a spine (Balanitaceae; Flacourtiaceae: *Flacourtia*, *Scolopia*; Nyctaginaceae: *Bougainvillea*) or an axis with short internodes (Bombacaceae: *Ceiba*; Lauraceae: *Aniba*; Lecythidaceae: *Couratari*; Rosaceae: numerous species in temperate zones). The distal parts of the modules of branches that are plagiotropic by apposition could be interpreted as terminal short shoots (***F6* e**). Well-developed axillary buds of many temperate species are usually not considered as short shoots, even when these bear several scale-leaves (Aceraceae, Betulaceae, Fagaceae, etc.).

Nodes and internodes
a Extremity of an angular twig (frequent in the Sapindaceae).
b Extremity of a grooved twig (Lauraceae; Sapindaceae; Leguminosae: *Sclerolobium*).
c Internodes longitudinally striate (Aristolochiaceae; Hernandiaceae: *Hernandia*).
d Stem with a more or less quadrangular section (Verbenaceae), winged for some species (Vitaceae: *Cissus quadrangularis*; Lythraceae: *Lagerstroemia*; Myrtaceae: *Eugenia* pp.).
e Stem grooved on two opposite sides (Bignoniaceae).
f Flattened twig or trunk (Leguminosae: *Bauhinia*).
g Base of the internode swollen; this phenomenon is more obvious on old stems (Acanthaceae; Gnetaceae; Nyctaginaceae: *Mirabilis*).
h Enlarged nodes (n) of a twig (Aristolochiaceae; Piperaceae). Illustration: *Piper* spp.
i Small cushion below the point of insertion of leaf (Convolvulaceae; Solanaceae: *Cestrum*).
j Young internodes in which the periderm exhibits thin longitudinal ribs. Each petiolar insertion is the origin of two ribs, these attenuating progressively towards the proximal part of the stem. Alternate phyllotaxy (Passifloraceae: *Barteria*; Dichapetalaceae: *Dichapetalum*; Flacourtiaceae: *Laetia*; Olacaceae). Opposite phyllotaxy (Buxaceae: *Buxus*; Oleaceae: *Olea*; Oliniaceae: *Olinia*).

Petiolar bases
k Petiolar base prolonging itself into a wing of the periderm (Tetramelaceae: *Octomeles*).
m Petiole (p) "fused" to the stem (base decurrent), without abscission joint at its base (Aristolochiaceae; Sapotaceae; usually in Sapindales: Anacardiaceae, Burseraceae, Meliaceae, Rutaceae, Sapindaceae, Simaroubaceae).
n Petiole with an abscission joint marked by a small annular furrow (Ericaceae; Leguminosae; Menispermaceae).

Short shoots
p Short shoot modified into a spine (sp); (Apocynaceae: *Carissa* (but leaves and spines opposite); Euphorbiaceae: *Adelia*; Flacourtiaceae: *Flacourtia*, *Xylosma*; Rubiaceae: *Randia*; Rutaceae: *Citrus*; Sapotaceae: *Sideroxylon*.
q Short shoot bearing leaves (s = scar of the axillary leaf); Upper node: lateral axis developed, Erythroxylaceae : *Erythroxylum*; Lecythidaceae: *Couratari*; Leguminosae: *Calliandra*; Meliaceae: *Turraea*; Rosaceae: *Prunus*. Lower node (inf): lateral axis very short, Bignoniaceae: *Crescentia*.

Scars
r Oval or almost circular scars. These scars appear in the rhytidome after the shedding of the twigs (Myrsinaceae: *Ardisia*, *Oncostemum*); (fs = foliar scar).

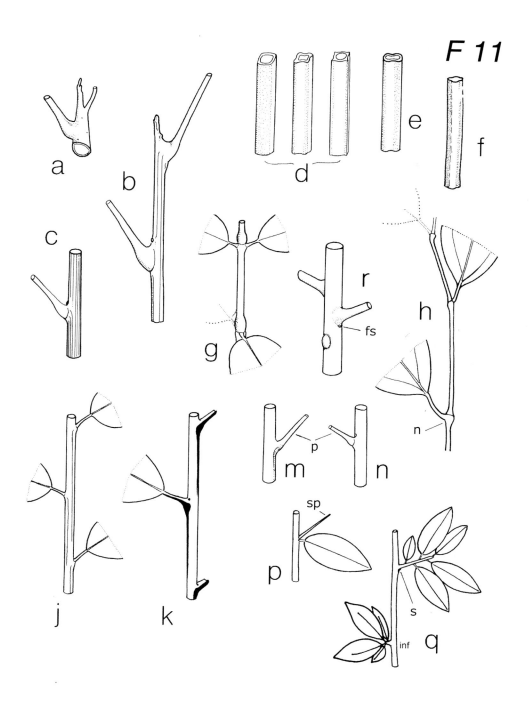

F 11

12. Axillary buds and prophylls

(Vegetative) bud: Group of weakly differentiated cells, formed in the axil of a leaf, which may develop into a twig. In some cases tropical plants possess vegetative buds covered with several conspicuous scale-leaves (Fagaceae: *Lithocarpus*, *Quercus*; Leguminosae: *Cynometra*, *Inga*, *Lonchocarpus*; etc.).

Serial buds: More than one bud or meristem formed separately in the axil of a single leaf (Annonaceae, Araceae, Aristolochiaceae, Caesalpinieae, *Piper*). An axillary bud can develop into a much condensed branch system (Bamboos, Mimosoideae), [21], but the buds of a vertical serie develop independently [26].

Prophylls: The first two leaves formed by an axillary bud or by a lateral twig of a Dicotyledon. The prophylls are usually designated by the symbols a and b. Lateral twigs or buds of a Monocotyledon bear only one prophyll, the latter usually appressed to the stem.

Scale-leaf (cataphyll): Reduced leaf usually existing at the base of a lateral twig or of a unit of extension (UE). Prophylls are very often scale-leaves.

Serial buds

a Buds disposed longitudinally (in one row), the most developed bud lies here in the distal position (Bignoniaceae: *Tabebuia*; Caesalpinieae: *Caesalpinia*, *Gymnocladus*, *Peltophorum*; Mimoseae: *Pentaclethra*; Sophoreae: *Myroxylon*, *Ormosia*; Connaraceae: *Connarus*; Malpighia-ceae: *Banisteriopsis*, *Malpighia*; Verbenaceae).

b Buds disposed longitudinally (in one row), the most developed bud lies here in the proximal position (Caprifoliaceae: *Lonicera*).

c Buds, in a zigzagging series (or in two rows), forming a typical condensed branch system (Mimosoideae: *Inga*, *Acacia*, *Pithecellobium*).

d Buds in a lateral or zigzaging series (or in two rows), the largest lying in a distal position (Papilionoideae-Swartzieae: *Swartzia*).

e Buds in a vertical series, the distal one modified into a spine (Nyctaginaceae: *Bougainvillea*).

Prophylls

f The two prophylls α/β of a Dicotyledon, their petioles inserted perpendicularly to the plane formed by the main and the lateral stems. The prophylls here are developed leaves. This occurs frequently when ramification is not delayed (numerous Ebenaceae; Lauraceae; Myristicaceae; Sapotaceae). For some cases of immediate ramification, α is scale-leaf but β is a developed leaf (Hamamelidaceae: *Exbucklandia*), for others the reverse holds (Piperaceae: *Piper*).

g The two prophylls α/β of a Dicotyledon inserted perpendicularly to the plane formed by the main and the lateral stems. The prophylls here are scale-leaves. This occurs frequently when ramification is delayed (most of the Rosaceae; Leguminosae; Celastraceae; Sapindales: Meliaceae, etc.).

h Prophyll appressed to the stem or slightly shifted (Araceae; Annonaceae: *Xylopia*; Aristolochiaceae; Schisandraceae: *Schisandra*).

j Prophyll appressed and annular as for some Araceae (*Heteropsis*).

k Two equal, leaf-like prophylls (Bignoniaceae: *Anemopaegma*).

m Two sessile, unequal, leaf-like prophylls (*Solanum* and other Solanaceae, possibly β after a long internode in *Cestrum*).

n The two prophylls modified into a spine (Phytolaccaceae: *Seguieria*).

p Concrescence of organs: for some species of the Solanaceae the prophylls are fused to the petiole; another petiole (drawn in black) is adnate to the stem; flowering is terminal (*Atropa*, *Cyphomandra*, *Datura*).

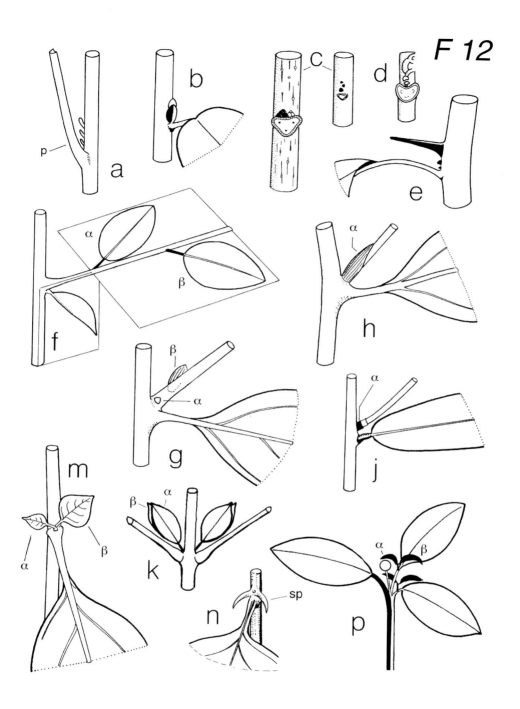

F 12

13. Stipules and interpetiolar ridge

Stipule (in Dicots): Pair of outgrowths associated with the base of a leaf developing from part of the leaf primordium in the early stages [24].
N.B. Some stipuliform expansions may not be stipules (petiole of the Dilleniaceae [27]; basal leaflets of a few Sapindales [28]).
Interpetiolar ridge: Ridge or crest observable on the nodes of some plants with opposite or whorled phyllotaxy. This trait usually disappears with the thickening of the rhytidome. When a taxon does possess interpetiolar ridges (e.g. Gentianaceae, Loganiaceae), the probability is high that a taxon of the same order is stipulate (e.g. Rubiaceae). When leaves are subopposite, stipules are oftenly missing.

Stipule and stipuliform expansion
- **a** Stipules lateral. **A.** Stipules inserted on the stem; distichous or spiral phyllotaxy (Dipterocarpaceae; Euphorbiaceae; Flacourtiaceae; Moraceae; Sterculiaceae; etc.); opposite phyllotaxy (Celastraceae: *Salacia*, *Lophopetalum*; Malpighiaceae; Quiinaceae; Rhamnaceae: *Maesopsis*; Trigoniaceae: *Trigonia*). **B.** Stipules inserted on the petiole (Chrysobalanaceae: *Licania*, Malpighiaceae: *Hiraea*).
- **b** Stipules intrapetiolar; alternate phyllotaxy (Erythroxylaceae; Leguminosae: *Saraca*; Urticaceae: *Urera*); opposite phyllotaxy (Malpighiaceae: *Byrsonima*).
- **c** Interpetiolar stipules fused into a ring (Loganiaceae: *Logania*).
- **d** Interpetiolar stipules free from the petiole (*Coffea*, *Ixora* and other Rubiaceae; Rhizophoraceae; Dialypetalanthaceae).
- **e** Interpetiolar stipules sheathing the apical bud (*Weinmannia* and other Cunoniaceae).
- **f** Sheathing petiolar bases giving the appearance of a swollen node; stipules emerge from the margin of the vaginate sheath on either side of the petiole (Chloranthaceae).
- **g** Stipules free from the stem and adnate to the petiole (Loganiaceae: *Neuburgia*).
- **h** Stipules fused into a flattened hood adnate to petiole (Leeaceae: *Leea*; Hamamelidaceae: *Exbucklandia*).
- **j** Stipuliform expansions adnate to the petiole (Dilleniaceae: *Dillenia philippensis*).
- **m** Stipules hood-like and conical-shaped, leaving an annular scar (Moraceae: Cecropioideae; Dipterocarpaceae: *Dipterocarpus*; Magnoliaceae: *Magnolia*; Sterculiaceae: *Triplochiton*). L = leaf.
- **n** Stipules hood-like, narrow and elongated (Irvingiaceae; Sarcolaenaceae).
- **p** Petiolar base swollen and protecting the stem apex (Apocynaceae: *Couma*; Clusiaceae: *Garcinia*; Loganiaceae: *Fagraea*; Rutaceae: *Metrodorea*).
- **q** Petiolar base winged (Dilleniaceae: *Curatella*, *Dillenia*).
- **r** Stipules (stipuliform petiolar expansions) adnate to a grooved petiole (numerous Araliaceae).
- **s** Stipules fused into a cylinder, leaving an annular scar (ochrea of the Polygonaceae).
- **t** Ochrea reduced to an annular ridge (Polygonaceae: *Afrobrunnichia*).
- **u** Ochrea trumpet-like (Platanaceae: *Platanus*). L = leaf.
- **v** Petiolar base swollen and associated with an interpetiolar ridge (Apocynaceae: *Odontadenia*).
- **w** Stipules modified into spines (Capparidaceae: *Capparis*; Rhamnaceae: *Ziziphus*; also for some Leguminosae, but leaves compound: *Acacia*, *Robinia*). Illustration: leaves opposite as in *Ziziphus*.

Interpetiolar ridge
- **y** Interpetiolar ridge (Caprifoliaceae: *Sambucus*; Clusiaceae; Gentianaceae; Loganiaceae: *Fagraea*, *Potalia*; Melastomataceae; Avicenniaceae: *Avicennia*).

Cactaceae
- **z other** Areole, i.e. group of hairs inserted on a swelling in the axil of a shed leaf (*Epiphyllum* and other Cactaceae).

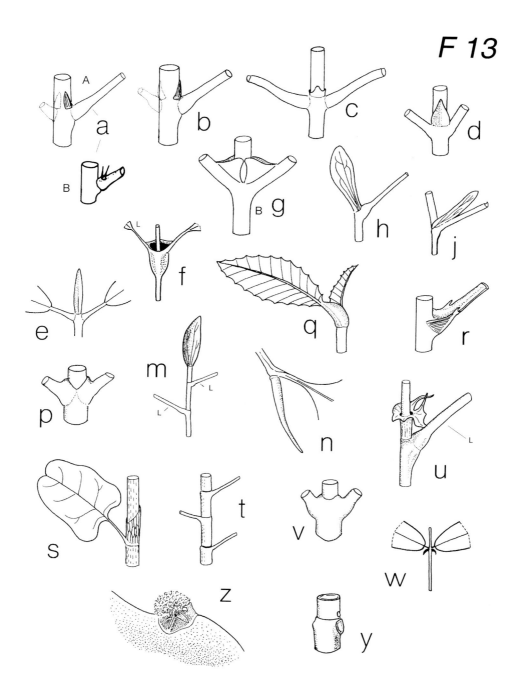

14. Leaf : blade and rachis

Blade (lamina): The flat expanded portion of a leaf.
Rachis: Segment of leaf axis bearing the pinnae or leaflets.
Pinna: A primary subdivision of a compound leaf. Pinnae may be further divided into pinnules.
Rachilla: Segment of axis bearing higher order pinnae.
Stipel: A stipule-like appendage at the base of a petiolule of a leaflet.
N.B. Pinnate leaves belonging to closely-related species, or even to the same individual, can be very variable in number and location of leaflets along the rachis. The presence of a rachis ending in a mucro, or of a rachis ending in a leaflet, is nevertheless an important taxonomic character that is quite stable at the genus or even the tribal level. Pinnately lobate leaves are much less common than palmately lobate leaves in tropical woody plants (e.g. Aralidiaceae: *Aralidium*; Moraceae: *Artocarpus incisa*; leaves of juvenile Sapindaceae).

Blade shape

a Leaf cordate (black), palmately lobate (stippled) or palmate; these three kinds of leaves may exist in the same family (Sterculiaceae: *Dombeya, Triplochiton, Cola*; Bombacaceae: *Quararibea, Ochroma, Bombax*; Euphorbiaceae: *Croton, Manihot, Hevea*).
Families in which simple and palmate leaves, but not palmately lobate leaves occur, are in fact mostly pinnate-leafed (Rutaceae: simple leaves for *Citrus*, palmate leaves for *Cusparia, Vepris* pp., and pinnate leaves for most of other species).

b Leaf auriculate (Bonnetiaceae: *Ploiarium*).

c Leaf notched (or retuse), mucronate (Capparidaceae); m = mucro.

d Leaf or leaflet with base of lamina inserted slightly above the petiole or the petiolule (Connaraceae).

e Leaf peltate (Hernandiaceae: *Hernandia*; Euphorbiaceae: *Macaranga*; some Menispermaceae).

f Leaflet asymmetrical, divided into two unequal parts by its midrib (Leguminosae-Mimosoideae: *Pithecellobium*).

Pinnate leaves

g Paripinnate leaf: leaflets opposite, rachis ending in a mucro (m), (Caesalpinioideae: *Afzelia*, Cassia, *Cynometra, Saraca*; Meliaceae: *Carapa*; Sapindaceae: *Nephelium*; Simaroubaceae: *Simaba*). m = mucro.

h Imparipinnate leaf: leaflets opposite (or alternate) and rachis ending in a terminal leaflet (Papilionoideae: *Dalbergia, Hymenolobium, Sophora, Swartzia*; Rutaceae: *Fagara*; Anacardiaceae; a few Sapindaceae and Meliaceae). Stipels (= s) are for Papilionoideae.

j Leaflets more or less alternate and rachis ending in a mucro (m), (Caesalpinioideae: *Crudia, Tessmannia*; Meliaceae: *Carapa*; numerous Sapindaceae).

k Rachis ending in a bud, this bud developing into additional rachis and leaflets; the leaf twig-like (Meliaceae: *Guarea, Chisocheton*).

Rachis

m Rachis swollen at point of leaflet insertion (Bignoniaceae).

n Rachis not swollen at point of leaflet insertion (generally for the Sapindales), the leaf seems stipulate due to the presence of a basal pair of leaflets (Burseraceae: *Canarium* pp., *Pachylobus edulis*; Meliaceae: *Trichilia schomburgkiana*; Sapindaceae: *Pometia, Lepisanthes, Toulicia*; Simaroubaceae: *Picrasma javanica*).

P Rachis winged. Leaflets opposite (Bignoniaceae: *Crescentia*; Cunoniaceae: *Weinmannia*; Leguminosae: *Swartzia*; Verbenaceae: *Peronema*); leaflets alternate (Sapindaceae: *Filicium*).

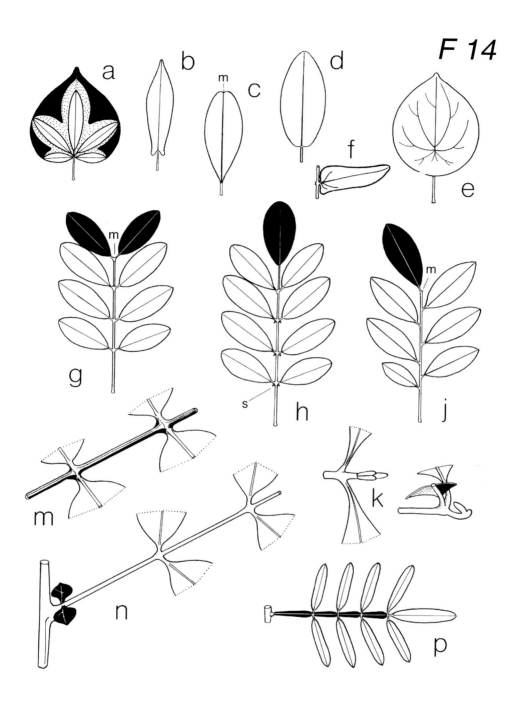

15. Leaf : petiole and petiolule

Petiole/petiolule: The stalk attaching a leaf/leaflet blade to a stem/rachis.
Pulvinus: The enlarged base or apex of a petiole or a petiolule, which functions in the movement of a leaf or a leaflet.
Abscission joint: Joint or zone of articulation where a leaf or a leaflet will break off. Such joints are frequently swollen and usually also bear an annular constriction groove marking the location of future breakage [24].
N.B. The leaves (excluding scale-leaves) of the tropical forest Dicotyledons are almost always stalked ; sessile leaves are quite uncommon (Bonnetiaceae, Loranthaceae, Viscaceae).

Petiolar bases

a **a1**: Petiolar base sheathing the stem (Araliaceae; Piperaceae ; Monocotyledons: Araceae, Cyclanthaceae). **a2**: Non-petiolate leaf sheathing the stem (numerous Monocotyledons: Agavaceae, Pandanaceae, etc.). **a3**: Petiolar base decurrent to the stem (Araucariaceae; Podocarpaceae, and other Coniferae); leaves opposite for *Agathis* (Araucariaceae).
b Petiolar base canaliculate or grooved (Annonaceae; Burseraceae; Dilleniaceae).
c Petiolar base biconvex in section (Meliaceae; Sapindaceae).
d Petiole winged and articulated with the blade (Rutaceae: *Citrus*).

Petiolar scars

e Scar broad and circular (Euphorbiaceae: *Uapaca*; Hernandiaceae).
f Scar shield-like (Anacardiaceae; Burseraceae; Meliaceae; Sapindaceae).
g Phyllotaxy opposite-decussate and scars almost encircling the stem (Cornaceae: *Cornus*; Loganiaceae: *Fagraea*).

Pulvini, swellings, and abscission joints

h Petiolule separated from the rachis by an abscission joint (a); the leaflet is generally able to move autonomously when its petiolule is differentiated into a pulvinus, as illustarted here: (Connaraceae; Leguminosae; Oxalidaceae; Zygophyllaceae); abscission joint uncommon for the Sapindales (e.g. *Fagara*); pulvini for *Picramnia* (Simaroubaceae-Picramniaceae).
j Petiolule united to the rachis without abscission joint: Balanitaceae, Cunoniaceae, Leeaceae, Melianthaceae, Rosaceae, Sapindales: Anacardiaceae, Burseraceae, Hippocastanaceae, Meliaceae, Sapindaceae, Simaroubaceae.
k Petiolules bearing a distal pulvinus (p):
Pinnate leaves: numerous Burseraceae, uncommon for the Meliaceae (*Walsura*).
Palmate leaves: Bignoniaceae (*Tabebuia*).
l Petiolar base pulvinate (Connaraceae; Leguminosae) allowing the leaf to move autonomously. s = stipule.
m Petiole weakly enlarged distally or inconspicuously pulvinate (Olacaceae: *Coula*; a few Dipterocarpaceae; some climbing Icacinaceae).
n Petiole short and enlarged at its base (Humiriaceae; Ixonanthaceae; Linaceae).
p Petiole distally pulvinate: Palmate venation (Euphorbiaceae: most of the Crotonoideae-Acalyphoideae; Menispermaceae; Malvales: Bombacaceae, Tiliaceae, but petioles of Malvaceae and some Sterculiaceae are not pulvinate). Pinnate venation (Capparidaceae: *Stixis*; Elaeocarpaceae: *Sloanea*; Flacourtiaceae: *Ryparosa*; Rutaceae: *Erythrochiton*, *Tetractomia*, unifoliolate *Evodia*; Verbenaceae: *Teijsmanniodendron*, unifoliolate *Vitex*).
q Petiole with an abscission joint (a), (unifoliolate Bignoniaceae: *Tabebuia* spp ; unifoliolate Connaraceae: *Ellipanthus*; unifoliolate Leguminosae : *Baphia, Bauhinia, Bocoa*).

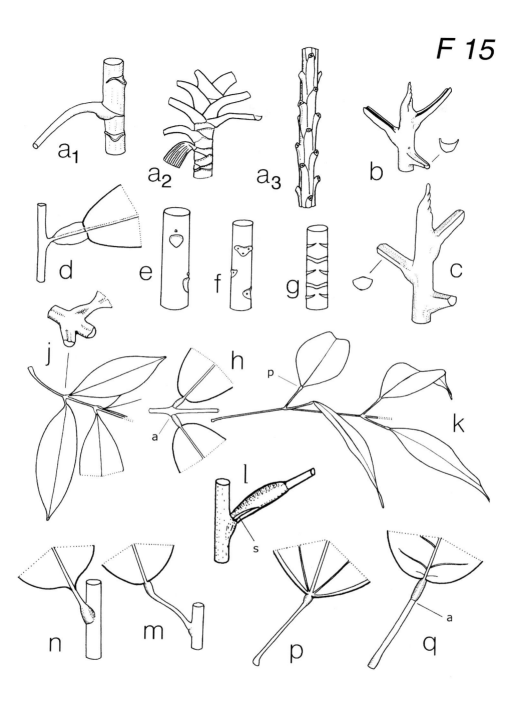

16. Leaf folding and aestivation

Leaf folding: In the strictest sense, the mode of folding of the young leaf or the young leaflet. The term "ptyxis" is sometimes employed [24], [29]. Thus leaf folding defines a transitory stage of the shape of the leaf (leaflet); this term should not be used to describe the variations of packing of different leaves (leaflets) in bud development.
Aestivation: Relative modes of packing of the young leaves (leaflets) in the bud.
N.B. Observations of leaf folding and related characters can be impossible if all the apices are in a resting stage and/or the leaves are too small. Hence these characters are not always permanent, even if the whole individual is examined.

Modes of leaf folding
a Plane or in a flattened "V" (Apocynaceae; Aristolochiaceae; Capparidaceae; Clusiaceae; Convolvulaceae; Euphorbiaceae; Myrtaceae: *Eucalyptus*).
b Conduplicate: **b1** (Annonaceae; Chrysobalanaceae; Dilleniaceae; Leguminosae; *Omphalea* and other Euphorbiaceae; Polygalaceae: *Bredemeyera*; Rosaceae: *Prunus*; Sapotaceae; Vochysiaceae: *Qualea*, *Vochysia*).
Plicate: **b2** (Rhamnaceae: *Ziziphus*); **b3** (Arecaceae, Cyclanthaceae).
c Convolute (Theales: Bonnetiaceae, Marcgraviaceae, Pellicieraceae); convolute-involute (Theaceae: *Camellia*).
d Involute, i.e. with margins rolled adaxially (Caprifoliaceae: *Viburnum*, leaflets of *Sambucus*; Celastraceae; Elaeocarpaceae; Euphorbiaceae (uncommon): *Mabea*; Staphyleaceae: *Turpinia*; Linales: Humiriaceae, Irvingiaceae, Ixonanthaceae, Linaceae; Lecythidaceae; Rosaceae: *Spiraea*; Ochnaceae, Theaceae).
e Revolute, i.e. with margins rolled abaxially, (Ericaceae: *Bejaria*; Goodeniaceae: *Scaevola*; Magnoliaceae: *Magnolia*; Platanaceae; Polygonaceae).
f Two false longitudinal veins appearing as a consequence of involute folding (Erythroxylaceae: *Erythroxylum*; Sarcolaenaceae).

Modes of aestivation
g Young (conduplicate) leaves overlapping in bud (Annonaceae: *Duguetia*, *Xylopia*; Polygalaceae: *Bredemeyera*).
h Young leaves appressed one to another for a stage of their growth (Clusiaceae: *Clusia*, *Harungana*, *Vismia*, etc.).
i Young leaves appressed one to another for a stage of their growth, but one involute, the other revolute (Buxaceae: e.g. *Buxus citrifolius*).
j Opposite leaves in a figure with two planes of symmetry (Apocynaceae; Clusiaceae; Myrtaceae; Rubiaceae).
k Opposite leaves in a skewed figure (Acanthaceae: *Aphelandra*; Hydrangeaceae: *Hydrangea*; Myrtaceae: *Eugenia*; Oleaceae: *Ligustrum*).
m Opposite leaves in a figure with only one plane of symmetry (anisophylly), (frequent for the Melastomataceae).

F 16

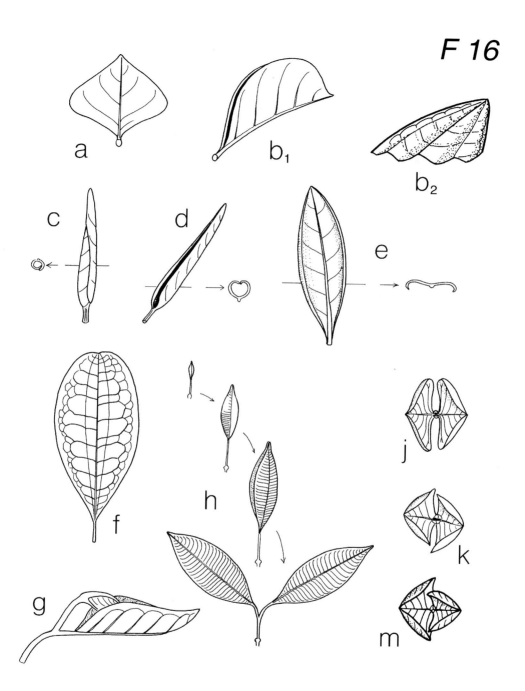

17. Venation I

References: For a typology of venation, see [30] (the source of the terms "brochidodromous", "camptodromous", "campylodromous" and "parallelodromous").
N.B. The two main types of venation, mutually exclusive, are pinnate and palmate. The palmate type is understood here in its widest sense, including trinerved blades (**18 d**). The supratrinerved blades, (**18 e**) are interpreted here as of pinnate type. See also *F18* for distinction between pinnate and palmate types of venation.

Pinnate venation
- **a** Brochidodromous type: **a1**, forming arches regularly curved (Apocynaceae; Clusiaceae; Dipterocarpaceae: *Dipterocarpus*); **a2**, forming arches in broken-lines (Linales: Humiriaceae, Linaceae; Laurales: Monimiaceae; Magnoliales: Annonaceae, Magnoliaceae, Myristicaceae).
- **b** Brochidodromous-camptodromous type (many Lauraceae).
- **c** Camptodromous type (Alangiaceae: *Alangium*; Chrysobalanaceae: *Couepia*; Combretaceae: *Combretum*; Cornaceae: *Cornus*; Rhamnaceae: *Ventilago*; Rubiaceae: frequent).
- **d** Camptodromous type with secondary veins ending almost on the margin (Dipterocarpaceae: *Dipterocarpus, Isoptera, Parashorea, Shorea*; Fagaceae: *Fagus, Lithocarpus*; Moraceae: *Cecropia*; Sapotaceae: *Chrysophyllum* pp., *Ecclinusa*).
- **e** Campylodromous type (Dioscoreaceae; Piperaceae: *Pothomorphe*).
- **f** Parallelodromous type (Araucariaceae; Podocarpaceae; Bambusaceae).
- **g** Numerous parallel secondary veins (Clusiaceae: *Clusia, Calophyllum*; Sapotaceae: *Micropholis*; Vochysiaceae: *Qualea*).

Pseudopalmate venation (venation fundamentally pinnate)
- **h** Left half of a trinerved leaf with one lateral vein ascending to the apex. Venation not, or weakly, scalariform (Rhamnaceae: *Gouania, Ziziphus*; Ulmaceae).
- **j** Right half of a leaf with two lateral veins ascending uto the apex. Venation distinctly scalariform (Melastomataceae: Melastomatoideae; Menispermaceae).

Pinnate venation
- **k** Venation scalariform: the tertiary veins are roughly parallel (Dipterocarpaceae: *Cotylelobium, Dryobalanops*; Leeaceae; Melastomataceae; Sapotaceae: *Palaquium*).
- **m** Leaf trinerved. Venation brochidodromous and not scalariform. Two ascending lateral veins (specimen of an unidentified family from French Guiana).

Relation between veins and margin, false veins (resiniferous ducts)
- **n** Entire leaf or leaflet with secondary veins ending at the margin (secondary veins secant), (Anacardiaceae: *Ozoroa, Astronium*; Brexiaceae: *Brexia*).
- **p** Toothed leaf with secondary or tertiary veins ending in the teeth (Actinidiaceae: *Saurauia*; Brunelliaceae; Dilleniaceae: *Curatella, Dillenia*; Myricaceae: *Myrica*).
- **q** Fimbrial vein, of a leaf or a leaflet, formed by the ending of the secondary veins along the leaf edge (Araliaceae: American *Schefflera* (or *Didymopanax*); Leguminosae-Papilionoideae: *Dussia, Machaerium, Piscidia*; Thymelaeaceae: *Aquilaria*).
- **r** Leaf with resiniferous ducts (Clusiaceae: *Garcinia*; Myrsinaceae: *Maesa*).

F 17

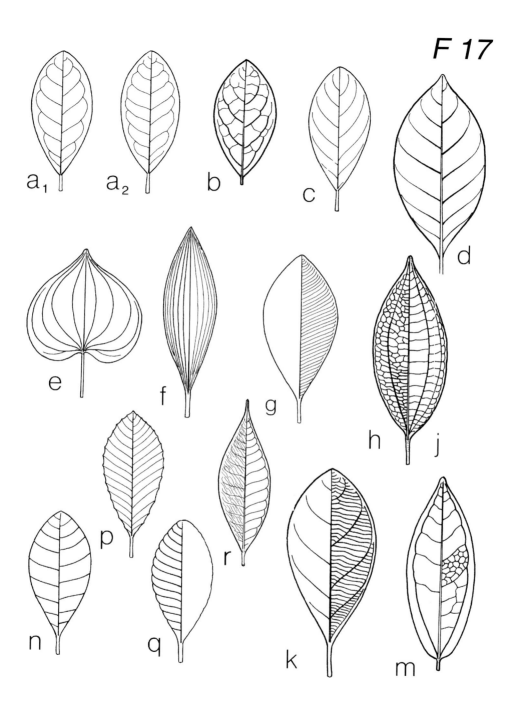

18. Venation II

Order of ramification: Order 1 is attributed to the midrib, the main lateral veins are of order 2, etc. Figure a shows a leaf with venation ramified up to the 6th order.

N.B. When considering vegetative characters, there is a marked convergence between *Gnetum* (Gymnosperm) and the Dicotyledons. However Gnetum does not possess blind veinlets, but exhibits a closed network of veinlets. All Dicotyledons and very few Monocotyledons (e.g. *Smilax*) possess at least some blind-veinlets [31].

a Venation with numerous blind-veinlets (Anacardiaceae: *Anacardium*). a1/a2 densely / not densely reticulate venation.

b Venation without blind-veinlets (Gnetaceae: *Gnetum*).

c Typical palmate venation (Aristolochiaceae: *Aristolochia*; Bixaceae: *Bixa*; Bombacaceae: *Ochroma*; Euphorbiaceae: *Croton, Macaranga, Mallotus*; Flacourtiaceae: *Caloncoba, Pangium*; Hernandiaceae: *Hernandia*; Malvaceae: *Hibiscus*; Sterculiaceae: *Triplochiton*; Tiliaceae: *Berrya*; etc.).

d Venation palmate: leaf trinerved (Alangiaceae: *Alangium*; Celastraceae: *Bhesa*; Euphorbiaceae: *Aporusa, Croton, Ptychopyxis*; Flacourtiaceae: *Hasseltia, Scolopia*; Hernandiaceae: *Sparattanthelium*; Moraceae: *Ficus*; Sterculiaceae: *Sterculia*; Tiliaceae: *Apeiba, Grewia*; Ulmaceae: *Celtis*; Urticaceae: *Boehmeria*; etc.).

e Venation not palmate (the two main lateral veins are inserted above the very base of the main vein) thus leaf is supratrinerved (Anisophylleaceae: *Anisophyllea*; Lauraceae: *Cinnamomum*; Loganiaceae: *Strychnos*; Melastomataceae: *Medinilla, Miconia*).

f,g Typical pinnate venation, the two basal, main secondary veins, not inserted at the same level on the main vein (Annonaceae; Apocynaceae; Celastraceae: *Salacia*; Chrysobalanaceae; Combretaceae; Erythroxylaceae; Euphorbiaceae: *Phyllanthus*; Flacourtiaceae: *Hydnocarpus, Laetia*; Humiriaceae; Magnoliaceae; Malpighiaceae; Monimiaceae; Myristicaceae; Ochnaceae; Rubiaceae; Symplocaceae; Theaceae; Violaceae; etc.).

h Venation pinnate with basal secondary veins grouped (Convolvulaceae: *Dicranostyles*).

j Leaf or leaflet with two ascending basal secondary veins (frequent for the Connaraceae; Oxalidaceae: *Sarcotheca*).

k Typical brochidodromous venation with secondary veins regularly spaced and ending in an intramarginal vein (Annonaceae; Clusiaceae: *Harungana, Tovomita, Vismia*; Dichapetalaceae; Loganiaceae: *Fagraea*; Myristicaceae; Myrtaceae: *Eugenia*; Sapotaceae: *Payena, Pouteria*; Lythraceae: *Duabanga*; Thymelaeaceae: *Gonystylus*).

m Brochidodromous venation forming an intramarginal vein very close to the margin (Anacardiaceae: leaflets of *Spondias*; Apocynaceae: *Himatanthus, Macoubea, Neocouma*; Clusiaceae: *Garcinia*; Myrtaceae: *Calyptranthes, Eugenia*).

n Brochidodromous venation forming two intramarginal veins (Myrtaceae: *Eugenia*).

p Secondary veins abruptly curved close to the margin (Anacardiaceae). Small secondary veins oriented towards the base of the lamina (Anacardiaceae: *Sorindeia*).

q Midrib ramified at its end (Bignoniaceae: *Deplanchea*).

r Raised venation (protruding slightly on the upper side of the lamina), here with midrib (V I) and secondary veins (V II) striate (Ochnaceae: *Ochna, Ouratea*; Flacourtiaceae; Violaceae).

s Venation embossed on the upper side of blade (Connaraceae, Myrtaceae, Prunaceae, Theaceae); ad: adaxial side (i.e. upper side) of blade, ab: abaxial side (i.e. underside) of blade.

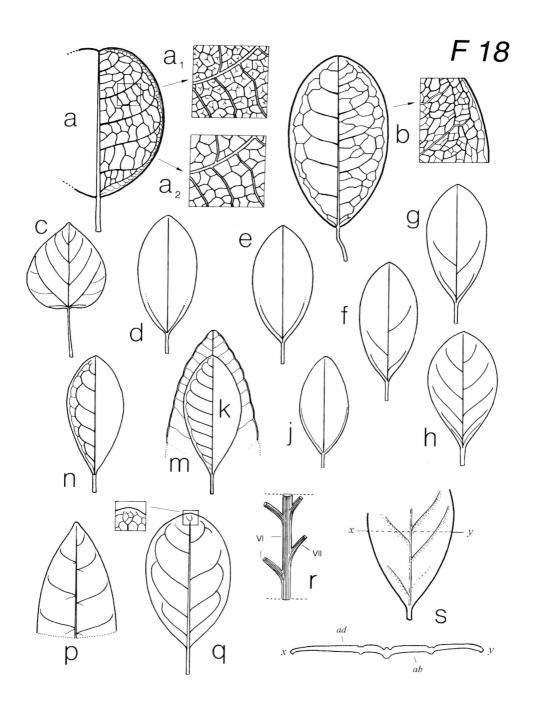

F 18

19. Glands and translucent dots

Glands: These are here confined to extrafloral nectaries. Glands, often situated on the leaves (foliar nectaries), are secretory organs involved in the feeding of insects, principally ants [32]. See also [33].
Translucent dots: These are also glands, and contain essential oils. These dots are usually scattered over the whole lamina (Canellaceae: *Cinnamodendron*, Flacourtiaceae, Leguminosae: some Detarieae and Sophoreae; Myrtaceae, Rutaceae).

Caulinary glands

a Glands (gl) located close to the petiolar insertions of the leaves (L); distichous or spiral phyllotaxy: *Bredemeyera*, *Securidaca* (Polygalaceae); opposite phyllotaxy: *Sonneratia* (Lythraceae). *Barteria* (Flacourtiaceae) has stipules modified into glands.
b Glands accompanied by stipules (st): *Qualea* (Vochysiaceae).

Rachis glandular

c Nipple-like glands (Leg-Caesalpinioideae: *Senna*, *Erythrophleum*; Moringaceae: *Moringa*).
d Cupuliform glands (Leg-Mimosoideae: *Inga*, *Pithecellobium*; Caesalpinioideae: *Chamaecrista*).
e Convex-shaped glands (Leguminosae-Mimosoideae: *Parkia* spp.).

Petiolar glands

f Lateral glands (Passifloraceae: *Dilkea*, *Passiflora*).
g Adaxialglands (Leguminosae: *Vouacapoua*; Chrysobalanaceae: a few spp. of *Parinari*).
h Distal glands, venation pinnate (Chrysobalanaceae: *Parinari*; Euphorbiaceae: *Sapium*; Passifloraceae: *Passiflora*).
j Distal glands, venation palmate (Euphorbiaceae: *Aleurites*, *Croton*, possibly one solitary gland in *Jatropha*, *Macaranga*, *Omalanthus*; Passifloraceae: *Adenia*, *Passiflora*).
k Glands inserted at the junction of the petiole with the lamina (Euphorbiaceae: *Omphalea*, *Macaranga*; *Pausandra*; Malpighiaceae: *Stigmaphyllon*).

Laminar glands

m Basal glands, (Bignoniaceae: *Newbouldia*; Chrysobalanaceae: *Chrysobalanus*; Dipterocarpaceae: *Pakaraimea*; Lecythidaceae: *Napoleonaea*; Malpighiaceae: *Acridocarpus*; Rosaceae: *Prunus*).
n Glands between the veins, at base and underside of lamina (Euphorbiac': *Alchornea*, *Ptychopyxis*).
p As above, but located on the upper side of the lamina (Euphorbiaceae: *Alchorneopsis*, *Micrandra*; Flacourtiaceae: *Hasseltia*).
q Glands located on the margin of the lamina, in some cases at its very base (Chrysobalanaceae: *Licania*; Combretaceae; Flacourtiaceae: *Banara*; Rhamnaceae: *Colubrina*; Rosaceae: *Prunus*; Verbenaceae: *Citharexylum*). Glands under the lamina, along the margin (not illustrated), (Euphorbiaceae: *Mabea*).
r Glands located on the margin of the lamina, at the extremity of a veinlet (Caesalpinioideae: *Gilbertiodendron*).
s Glands located on the veins at the base and the lamina (on the upper side, Euphorbiaceae: *Glycydendron*, *Nealchornea*; on the underside, Sterculiaceae: *Byttneria*).
t A solitary gland on the midrib, on the upper side of the lamina (Dipterocarpaceae: *Monotes*; Euphorbiaceae: *Elaeophora*).
u Waxy spots on basal veins, under the lamina (Moraceae: many Asiatic species of *Ficus*).
v Glands scattered on the underside of the lamina (Xanthophyllaceae: *Xanthophyllum*).
w Glands on the underside of the lamina, scattered primarily near the midrib (Dichapetalaceae: *Dichapetalum*; Euphorbiaceae: *Pera*); or evenly scattered (Chrysobalanaceae: *Hirtella*).

F 19

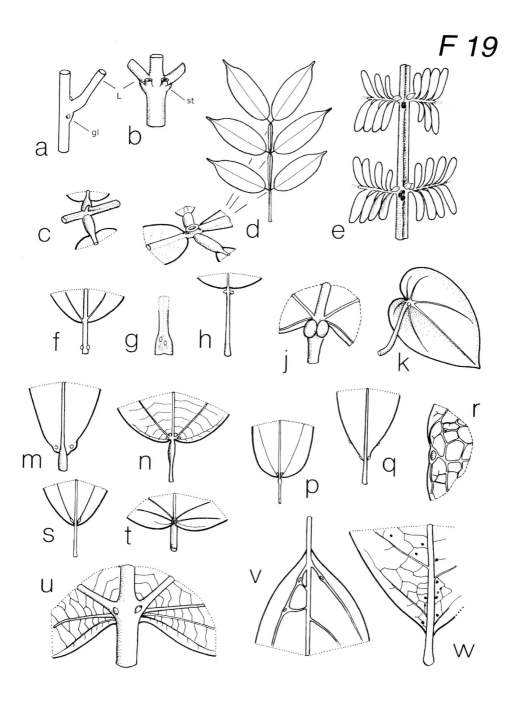

20. Epidermal structures of leaves

References: See [34] for a typological classification of foliar teeth.

Teeth
a Simple tooth, not enlarged at its apex (Brunelliaceae; Sterculiaceae: *Dombeya*).
b Glandular tooth (Euphorbiaceae: *Acalypha, Epiprinus*; Flacourtiaceae: *Abatia, Hemiscolopia, Hasseltia, Scottelia*; Passifloraceae: *Smeathmannia*).
c Glandular tooth innervated by a main veinlet and two lateral veinlets (rosoid tooth), (Rhamnaceae; Vitaceae).
d Large gland, of rosoid type, located on a small lobe of the lamina (Simaroubaceae: *Ailanthus*).

Indumentum
e Apex or bud with appressed hairs (Annonaceae; Dichapetalaceae; Ebenaceae; Lacistemataceae; Malpighiaceae; Theaceae; Thymelaeaceae).
f Apex or bud with erect or oblique hairs (Actinidiaceae; Asteraceae; Boraginaceae; Goupiaceae: *Goupia*; Lauraceae: *Eusideroxylon*; Rhamnaceae: *Lasiodiscus*; Solana-ceae; Trigoniaceae).
g Hairs simple, erect, thin and long (Brunelliaceae; Euphorbiaceae: *Omphalea*; Icacinaceae: *Pyrenacantha*; Lauraceae: *Eusideroxylon*; Rhamnaceae: *Lasiodiscus*).
h Hairs fasciculate, or a single stellate hair without pedicel (Annonaceae: *Cyathocalyx*). Fasciculate hairs can be inserted on small epidermic protuberances (not illustrated), (Actinidiaceae: *Saurauia*; Dipterocarpaceae: *Anisoptera, Shorea*).
k Stellate hair, with arms free up at its base and inserted on a very short pedicel (Huaceae: *Afrostyrax*; Sterculiaceae: *Sterculia*; Tiliaceae: *Tilia, Triumfetta*).
m Stellate hair, with arms inserted on a small disc (or scale), (Annonaceae: *Duguetia*; Styracaceae: *Styrax*).
n Stellate hairs with numerous arms inserted on a disc (Euphorbiaceae: *Croton tenuissimus, C. tessmannii*).
p Scaly, peltate hair (Annonaceae: *Duguetia*; Bignoniaceae: *Pyrostegia, Tabebuia*; Bixaceae: *Bixa*; Bombacaceae: *Durio*; Elaeagnaceae; Euphorbiaceae: *Hymenocardia*; Sterculiaceae: *Heritiera*; Tetramelaceae: *Octomeles*).
q Peltate hair, cupule-shaped (Styracaceae: *Styrax agrestis*).
r Scaly, triangular hairs (Actinidiaceae: *Saurauia*).
s Simple hairs, swollen at the base, inserted on epidermic protuberances (Boraginaceae: *Cordia*; Verbenaceae: *Nyctanthes*).
t Simple hairs, grouped into a domatium in the axil of a secondary vein (Alangiaceae: *Alangium*; Apocynaceae: *Forsteronia*; Icacinaceae: *Citronella*; Meliaceae: *Turraea*; Rubiaceae: *Chomelia*; Sapindaceae: *Arytera, Mischocarpus, Toechima*; Tiliaceae: *Apeiba, Tilia*).

Cavities, black dots
u Small cavities on underside of blade (*Osmanthus* and some other Oleaceae).
v Black dots on underside of blade (Apocynaceae: *Couma* pp; Clusiaceae: *Vismia* pp.).

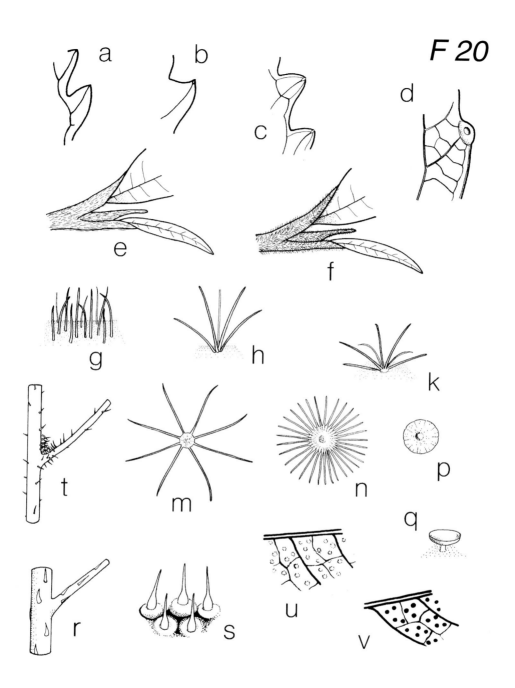
F 20

21. Key to architectural models

For relations between architecture and systematics, see [20], [21], [35], [36], [37].

Plant monocaulous (lateral flowering). .... CORNER
Plant polycaulous (several stems or a single sympodial stem), see **A**, **B** and **C**:
   A Vegetative shoots all orthotropic.
        One branch per module only. .... CHAMBERLAIN
        At least two branches per module. LEEUWENBERG
   B Vegetative axes differentiated into orthotropic and plagiotropic shoots.
      1 Trunk and branches with sympodial ramification.
          Modules initially equal, all apparently branches, but later unequal,
          one becoming a trunk. .... KORIBA
          Modules unequal from the start, trunk module appearing later than
          branch modules. .... PREVOST
      1* Trunk or branches with monopodial ramification.
        2 Trunk sympodial, bearing monopodial plagiotropic branches. .... NOZERAN
        2* Trunk monopodial.
          3 Ramification continuous.
            4 Branches monopodial.
              Branches orthotropic. .... ATTIMS
              Branches plagiotropic, (if phyllomorphic: COOK). .... ROUX
            4* Branches sympodial.
              5 Branches orthotropic. .... STONE
              5* Branches plagiotropic.
                Branches not modular. .... ROUX
                Branches modular. .... PETIT
          3* Ramification rhythmic.
            6 Branches sympodial.
              7 Branches orthotropic. .... SCARRONE
              7* Branches plagiotropic.
                8 Branches without apposition growth. .... MASSART
                8* Branches plagiotropic by apposition growth
                (F6 e).
                  Apical meristems lost. .... FAGERLIND
                  Apical meristems maintained. .... AUBREVILLE
            6* Branches monopodial.
              Branches orthotropic. .... RAUH
              Branches plagiotropic. .... MASSART
   C Growth direction of vegetative stems clearly changing during development.
      Direction changing by primary growth, at first (proximally) orthotropic,
      later (distally) plagiotropic. .... MANGENOT

      Stems at first orthotropic, secondarily bending (probably by gravity). ... CHAMPAGNAT

      Stems all plagiotropic, secondarily becoming erect, most often after leaf-fall. .... TROLL

F 21

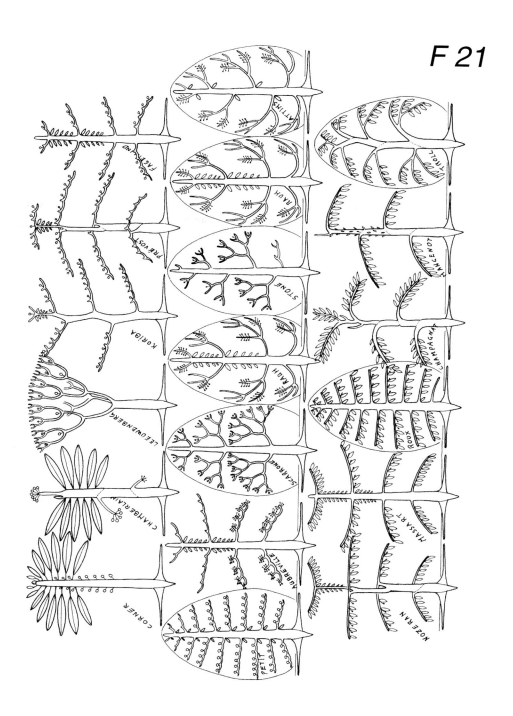

PART II

The principal families of tropical woody Dicotyledons
illustrated by means of their vegetative characters

A classification of the families identified by the key

Nomenclature follows Mabberley [38], ordering of families in reference to recent studies in cladistic and molecular systematics (i.e. [39], [40]), and opinion of the author. Figures in brackets refer to p.171. The letter-number combinations refer to the plate where the family is illustrated.

Gymnosperms

Araucariaceae
Pinaceae
Podocarpaceae
Gnetaceae

Angiosperms (Monocots)

Agavaceae / Liliaceae
Arecaceae P1 (1)
Asparagaceae
Bambusaceae
Cyclanthaceae
Dioscoreaceae
Pandanaceae
Smilacaceae

Angiosperms (Dicots)

Chloranthaceae group
  Chloranthaceae P4B

Canellaceae group
  Canellaceae P2A
  Illiciaceae P2A
  Winteraceae

Magnoliaceae group
  Annonaceae P3B
  Magnoliaceae P4A
  Schisandraceae
  Myristicaceae P4A

Lauraceae group
  Hernandiaceae P2A
  Lauraceae P2B
  Monimiaceae P3A

Piperaceae group
  Aristolochiaceae P5A
  Piperaceae P4B

Buxaceae group
  Buxaceae (2)

Sabiaceae group
  Meliosmaceae P5B (3)
  Sabiaceae

Proteaceae group
  Platanaceae
  Proteaceae P6A (4)

Menispermaceae group
  Berberidaceae
  Lardizabalaceae P5B
  Menispermaceae P5A
  Ranunculaceae P5B

Dilleniaceae group
  Dilleniaceae P6B (5)

Nyctaginaceae group
  Ancistrocladaceae
  Cactaceae
  Dioncophyllaceae
  Nyctaginaceae P7A
  Phytolaccaceae P7A
  Polygonaceae P6B

Santalaceae group
  Erythropalaceae
  Olacaceae P7B
  Opiliaceae P7B
  Loranthaceae P7B
  Santalaceae

Hamamelidaceae group
  Altingiaceae (6)
  Aphloiaceae (7)
  Daphniphyllaceae
  Grossulariaceae
  Hamamelidaceae P8A
  Leeaceae P19B
  Picramniaceae (8)

Vitaceae P19B (4)
Zygophyllaceae P8B (9)

Fagaceae group
  Casuarinaceae
  Fagaceae P8A
  Juglandaceae
  Myricaceae

Myrtaceae group
  Combretaceae P13B
  Crypteroniaceae
  Lythraceae P14A (10)
  Melastomataceae P14B
  Myrtaceae P13B
  Oliniaceae
  Onagraceae
  Vochysiaceae P14A (11)

Celastraceae group
  Brexiaceae (12)
  Celastraceae (13)

Anisophylleaceae group
  Anisophylleaceae P27B (14)
  Corynocarpaceae
  Tetramelaceae P17B (15)

Leguminosae group P20
  - Caesalpinioideae (16)
  - Mimosoideae
  - Papilionoideae
  Connaraceae P21B (17)
  Oxalidaceae P21B
  Polygalaceae P26A
  Surianaceae
  Xanthophyllaceae P26A

Clusiaceae group
  Caryocaraceae
  Clusiaceae P15A (18)
  Ochnaceae P15B
  Quiinaceae P15B

Malpighiaceae group
  Dichapetalaceae P25B
  Malpighiaceae P25B
  Trigoniaceae

Linaceae group
  Erythroxylaceae P22B
  Hugoniaceae P22B
  Humiriaceae P22B
  Irvingiaceae P22A
  Ixonanthaceae P22B
  Linaceae P22B

Cunoniaceae group
  Brunelliaceae P26B
  Cunoniaceae P26B
  Rhizophoraceae P26B

Rosaceae group
  Chrysobalanaceae P21A (19)
  Elaaeagnaceae
  Greyiaceae P8B
  Melianthaceae P8B
  Moraceae P9B (20)
  Rhamnaceae P19A
  Rosaceae P8B, P21A
  Staphyleaceae P8B
  Ulmaceae P9A
  Urticaceae P9A

Capparidaceae group
  Capparidaceae P16B
  CaricaceaeP17B
  Moringaceae
  Tovariaceae

Malvaceae group
  Bixaceae P17B
  Bombacaceae P18B
  Dipterocarpaceae P18A
  Malvaceae P18B (21)
  Sarcolaenaceae
  Sterculiaceae P18B
  Thymeleaceae P25A (22)
  Tiliaceae P148B

Flacourtiaceae group
  Flacourtiaceae P16A (23)
  Goupiaceae
  Huaceae
  Passifloraceae P16B
  Peridiscaceae
  Violaceae P17A

Euphorbiaceae group (24)
  Bischofiaceae
  Euphorbiaceae P19
  Lophopyxidaceae (25)
  Pandaceae
  Picrodendraceae
  Putranjivaceae(26)

Sapindaceae group
  Anacardiaceae P24A
  Balanitaceae
  Burseraceae P24A
  Meliaceae P23B
  Rhabdodendraceae (27)
  Rutaceae P24B
  Sapindaceae P23A
  Simaroubaceae P23B

Cornaceae group
  Cornaceae P27B
  Hydrangeacaeae

Ericaceae group
  Actinidiaceae 11A
  Bonnetiaceae 11A (28)
  Clethraceae
  Cyrillaceae P12A
  Ebenaceae P12B
  Ericaceae P12A
  Lecythidaceae P13A (29)
  Marcgraviaceae P11A
  Myrsinaceae P11B
  Pellicieraceae P11A
  Theaceae P11A (30)
  Theophrastaceae P11B
  Sapotaceae P12B
  Styracaceae
  Symplocaceae

Gentianaceae group
  Apocynaceae P31B
  Asclepiadaceae (31)
  Caprifoliaceae
  Gentianaceae P31A
  Loganiaceae P31A (32)
  Rubiaceae P32 (33)

Bignoniaceae group
  Acanthaceae P30B (34)
  Avicenniaceae
  Bignoniaceae P30A
  Boraginaceae P29A
  Buddlejaceae (35)
  Cobaeaceae
  Convolvulaceae P29A
  Hydrophyllaceae
  Lamiaceae
  Montiniaceae
  Myoporaceae
  Oleaceae P30A
  Polemoniaceae
  Solanaceae P29B
  Verbenaceae P30A (36)

Icacinaceae group
  Aquifoliaceae P27A
  Icacinaceae P28A (37)

Araliaceae group
  Alangiaceae P27B
  Aralidiaceae
  Aucubaceae
  Araliaceae P27A
  Kaliphoraceae
  Melanophyllaceae
  Pittosporaceae P27A

Asteraceae group
  Asteraceae P28B
  Goodeniaceae P28B

Remarks

1. General systematic treatment of Palms [41].
2. When considering vegetative features (i.e. a peculiar leaf estivation), Buxaceae are also apart.
3. The genus *Meliosma* is placed here in the separate family Meliosmaceae, *Sabia* (Sabiaceae) and *Meliosma* differing vegetatively in many respects, DNA sequencing also in favour [42].
4. Bark and external wood of Platanaceae and Proteaceae are very similar, see also [39], [40].
5. Dilleniaceae could be much closer to Vitaceae than one might think [39].
6. Altingiaceae often merged into Hamamelidaceae, e.g. [43].
7. *Aphloia* is now excluded from Flacourtiaceae, rhytidome characters are in favour of this opinion.
8. A monogeneric family segregated from Simaroubaceae, *Picramnia* species differ from other Sapindales by their pulvinate petiolules.
9. Zygophyllaceae and Geraniaceae conform to a same modular architecture.
10. *Sonneratia* being part of the expanded Lythraceae, see also [44].
11. According to molecular data, Vochysiaceae are separated from Polygalales [39], [40].
12. Brexiaceae are vegetatively very different from the ex Saxifragales, see also [39].
13. Including Hippocrateaceae, but excluding *Bhesa* and *Goupia* [45].
14. Anisophylleaceae are definitely separated from Rhizophoraceae [46].
15. Tetramelaceae is here separated from Datiscaceae, *Datisca* and *Tetrameles/Octomeles* differing vegetatively in many respects, DNA sequencing also in favour [47].
16. The order Fabales is synonymous with Leguminosae and thus receives the family rank [48], [49], hence Cronquist's Mimosaceae, Caesalpiniaceae and Fabaceae are given the subfamily rank.
17. The Connaraceae are kept near Leguminosae (climbing species very similar to Papilionoideae).
18. Maesaceae could be placed in a monogeneric family of its own [50].
19. Vegetatively, a Sumateran *Prunus* is very similar to American Chrysobalanaceae.
20. In this key, Moraceae includes the Cecropiaceae at the subfamily rank.
21. Malvaceae *s.l.* includes Bombacaceae, Malvaceae, Sterculiaceae, and Tiliaceae [51].
22. This opinion (see [39], [40]) is suported by the fibrous bark of the Thymeleaceae.
23. Including Lacistemaceae, but excluding *Aphloia* [40], see [52] for systematic treatment.
24. See systematic treatment of Webster [53] for the Euphorbiaceae-Pandaceae.
25. According to [45], *Lophopyxis* is close to the Euphorbiaceae.
26. A new family [40], enclosing the large *Drypetes* which is anomalous in Euphorbiaceae when considering fruits, lenticels, and periderm fissuring.
27. *Rhabdodendron* is vegetatively very similar to an unbranched Rutaceae (e.g *Erythrochiton*).
28. In this key, the Bonnetiaceae are treated as a separate family, linked to the Theaceae-Ternstroemiodeae.
29. Including Scytopetalaceae, if this family is merged into Lecythidaceae, the odd rarity of the latter in Africa does not exist any more. It is noteworthy that in *Key F* these two families appear close together.
30. Theaceae *sensu stricto* (without Ternstroemioideae) do not exhibit a very rhythmic growth.
31. Asclepiadaceae is included in Apocynaceae in recent studies [54].
32. Acording to molecular studies [55], *Anthocleista* should be merged into Gentianaceae, and *Mostuea* should be included in the new family Gelsemiaceae.
33. See Robbecht's systematic treatment of the Rubiaceae [56].
34. In the present work, Mendonciaceae are part of Acanthaceae.
35. Buddlejaceae (*Buddleja*, *Nuxia*) are placed near Lamiaceae because of resembling rhytidomes.
36. The monogeneric family Avicenniaceae is kept apart Acanthaceae and Verbenaceae, see also [57].
37. Some Icacinaceae (e.g. *Leptaulus*) are vegetatively very similar to *Ilex* species.

Legends to plates 1A/1B (Arecaceae)

Classification according to Uhl & Dransfield [41].
CORYPHOIDEAE: **Cor** (Corypheae), **Pho** (Phoeniceae), **Bor** (Borasseae);
CALAMOIDEAE: **Cal** (Calameae), **Lep** (Lepidocaryeae); NYPOIDEAE: *Nypa*;
CEROXYLOIDEAE: **Cer** (Ceroxyleae), **Hyo** (Hyophorbeae); ARECOIDEAE: **Car** (Caryoteae), **Iri** (Iriarteeae), **Are** (Areceae), **Coc** (Cocoeae), **Geo** (Geonomeae).

1A

a Palmate and induplicate leaf (i.e. divisions are 'V' shaped, while reduplicate divisions are 'Λ' shaped), here folds are crossing divisions and veins (e.g. **Cor**, *Rhapis*).
b Costapalmate induplicate leaf (e.g. *Sabal*, **Cor**), briefly costapalmate, but reduplicate: **Lep** (e.g. *Mauritia*).
c Pinnate and induplicate leaf (e.g. *Phoenix*), also in all Phoeniceae and Caryoteae.
d Pinnate and reduplicate leaf (e.g. *Hyophorbe*), also in NYPOIDEAE and all ARECOIDEAE (except Caryoteae), i.e. Calameae, Ceroxyleae, Hyophorbeae, Caryoteae, Iriarteeae, Areceae, Cocoeae, Geonomeae.
e Pinnate and reduplicate leaf, its leaflets half-spirally arranged (e.g. *Syagrus*, **Coc**).
f Old leaves persisting on trunk before shedding (e.g. **Cor**, *Pritchardia*).
g When leaves shed completely, the trunk is smooth (e.g. *Coccothrinax*, **Cor**).
h Climbing rattan (e.g. *Calamus*, **Cal**).
j Spiny stem and leaf rachis in *Calamus*.
k Palmately compound leaves (e.g. *Licuala*, **Cor**).
m Small Palm of the genus *Geonoma* (**Geo**), adult leaves like young leaves of large Palms.

1B

a Pinnate leaf with praemorse (i.e. jaggedly toothed, as if bitten) leaflets (e.g. *Arenga*, **Car**), also in many other Caryoteae.
b Praemorse leaflets in *Caryota* (**Car**), the leaves of this genus are twice compound.
c Fibrous basal leaf expansion (e.g. *Cocos*, **Coc**).
d Typical habit of an Areceae, with its infrafoliar inflorescence (e.g. *Areca*).
e Columnar basal sheaths (e.g. *Hyophorbe*, *Veitchia*) also in many Hyophorbeae and Areceae. Inflorescence position is infrafoliar.
f Spliting leaf base (e.g. *Hyophorbe*, **Hyo**, but unusual for this genus). Spliting old leaf bases, but thicker and woody, are well examplified in *Sabal*, **Cor**.
g Trunk with persisting and diverging basal sheaths, e.g. **Coc**-Attaleinae: *Attalea*, *Scheelea*.
h Tangential cut in bark showing wood with sclerenchymatic fibres (e.g. *Euterpe*, **Are**).
j Slender trunk of a dwarf palm (e.g. *Chamaedorea*, **Hyo**); its leaf nodes distinctly spaced.
k Distincly ribbed trunk (e.g. *Attalea*, **Coc**).
m Leaf bases persisting and short, (e.g. *Butia*, **Coc**); the pineapple pattern appearing because of an high phyllotactic index (see [22] for an explanation of Fibonacci's series).
n Spiny stem and leaves (e.g. *Bactris*, *Acrocomia*, **Coc**), also in many other Cocoeae-Bactridinae.
p Adventicious aerial roots, this feature being, however, associated with swampy ecology (e.g. *Iriartea*, *Socratea*, **Iri**).

Legends to plates 2A/2B

2A Hernandiaceae, Illiciaceae and Canellaceae

a *Hernandia*: **A**. spiral phyllotaxy with short internodes, leaves trinerved (peltate for some species); **B**. large and protruding petiolar scars, petiole decurrent to the stem.
b *Illigera*: Trifoliate leaf with prehensile petiole, petiole decurrent to the stem.
c *Gyrocarpus*: Tree with palmately lobate leaves.
d *Sparattanthelium*: Liana with nastic twigs, these oriented backwards.
e *Illicium*: Orthotropic, sympodial branches with leaves arranged in rosettes (pseudo-whorled phyllotaxy). "+" denotes the death of an apical meristem.
f *Canella*: SCARRONE's model: Monopodial trunk, rhythmic branching, branches sympodial and orthotropic.
g *Cinnamosma*: MANGENOT's model: Sympodial trunk consisting of the stacking of modules, each module firstly orthotropic, then becoming abruptly plagiotropic.
h *Cinnamodendron*: Coriaceous, pellucid-dotted, cuneate leaf with brochidodromous venation.

2B Lauraceae

a An architectural model similar to that of ROUX (branching is continuous in trunk), but UE are distinguishable (reminiscence of MASSART' model); branches plagiotropic and monopodial (*Cryptocarya, Eusideroxylon*).
b MASSART's model: Monopodial trunk, rhythmic branching, branches monopodial and plagiotropic. No short shoots (*Aniba* pp., *Ocotea*).
c An erect branch of an individual conforming to RAUH's model (e.g. *Nectandra* sp. monopodial trunk, rhythmic branching, branches orthotropic and monopodial). These characters occur in many species of various genera of the Lauraceae in part because many species do undergo an architectural metamorphosis (i.e. MASSART when young, and RAUH when getting older).
d AUBREVILLE's model: Monopodial trunk, rhythmic branching, branches plagiotropic by apposition growth (Aniba pp., *Ocotea* pp.).
e Young stem with grooved or angular internodes (typical for Lauraceae).
f *Aniba* pp.: Very rhythmic growth of an individual conforming to MASSART's model; cataphylls (c) at the base of an UE; rhytidome (rh) becomes soonly suberised (e.g. *Aniba, Licaria*), the reverse situation (i.e. a periderm remaining green or black-green for a long time) do exist for most of genera (*Aiouea, Beilschmiedia, Cinnamomum, Ocotea, Persea*, etc.).
g Apical and lateral buds covered with scale-leaves (e.g. *Cinnamomum*), hence growth is strongly rhythmic.
h Venation densely reticulate (typical for Lauraceae) and brochidodromous (e.g. *Ocotea*).
j Venation camptodromous (*Cryptocarya* pp.).
k Venation brochido-camptodromous (e.g. *Aiouea*), with domatia (dom).
m Supratrinerved leaf (*Cryptocarya* pp., *Cinnamomum*).
n Young axis distinctly lenticellate (e.g. *Ocotea* pp.), also in old trunks of several species.

N.B. All these families have in general immediate branching in trunk.

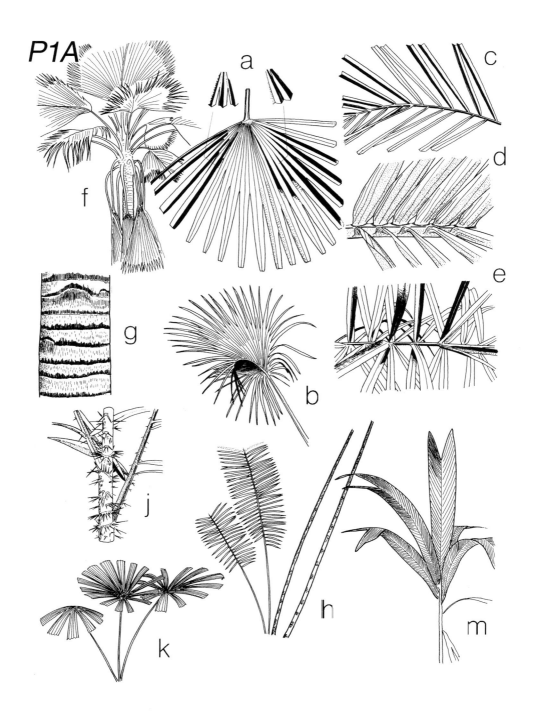

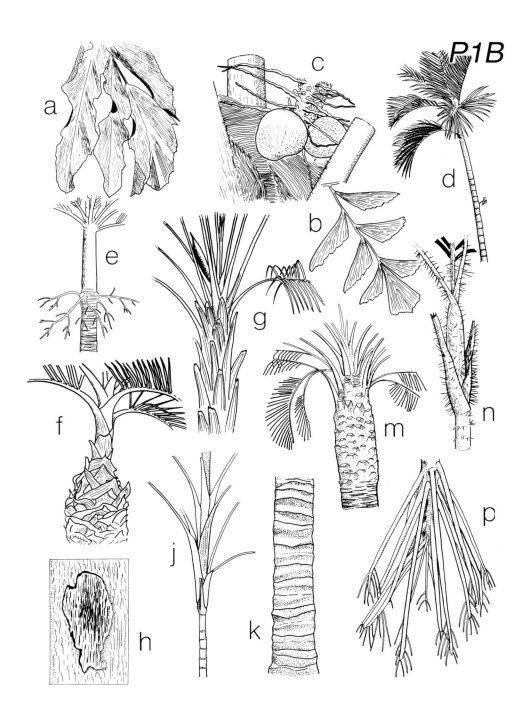

P1B

P2A

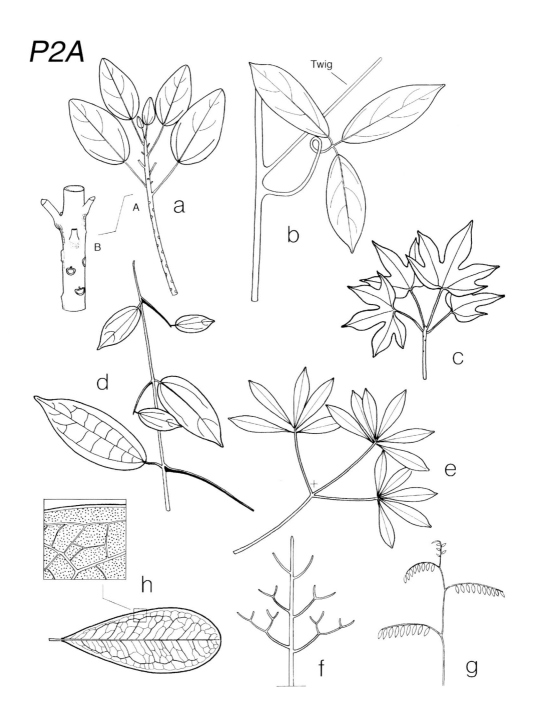

176

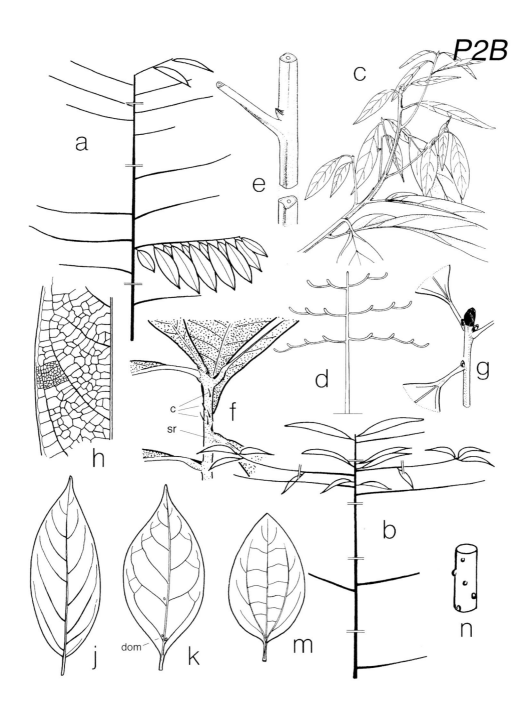

Legends to plates 3A/3B

3A Monimiaceae

a KORIBA's model: Trunk and branches modular, modules initially equal, all apparently branches, but later unequal, one becoming a trunk (*Kibara, Matthaea, Siparuna*). For these genera, the branches are often decumbent.

b MANGENOT's model: Trunk formed by the stacking of modules, each module firstly orthotropic, then becoming abruptly plagiotropic (*Mollinedia, Tambourissa*).

c MASSART's model: Monopodial trunk, rhythmic branching, branches monopodial and plagiotropic (an unknown species from Seram (Indonesia).

d ATTIMS's model: Trunk monopodial with continuous branching, branches orthotropic and monopodial (*Nemuaron*).

e *Glossocalyx*: Very pronounced anisophylly, where one leaf of each pair is much reduced (rl) and soon shed. **A**. monopodial trunk, continuous branching, branches (br) plagiotropic and monopodial (ROUX's or COOK's model), only one branch develops at each node, so anisophylly is associated with anisoclady. **B**. top view of a plagiotropic branch with distichous phyllotaxy (reduced leaves in black).

f Internodes flattened apically, opposite phyllotaxy (characters typical for the Monimiaceae).

g Typical shape of a serrulate leaf of a Monimiaceae.

h Laticiferous threads in the leaf (*Tambourissa, Xymalos*).

3B Annonaceae

a TROLL's model: Sympodial trunk consisting of the stacking of modules, each module firstly plagiotropic, then becoming orthotropic at its base (very common for *Annona, Isolona, Monodora, Rollinia*).

b ROUX's or PETIT's model: Monopodial trunk, plagiotropic, monopodial (**A**) or sympodial (**B**) branches (very common for *Duguetia, Guatteria, Polyalthia, Xylopia*). Death of apical meristems indicated by "+".

c Bark with network of fibres (a very typical character for the Annonaceae).

d Plagiotropic twig with drooping extremity (typical).

e Zigzagging extremity of a twig (typical).

f Conduplicate young leaves bearing appressed hairs (typical).

g Section of a trunk of a lianescent species, the bark is relatively very thick, its fibrous structure forms a network of phloem strands associated with sclerenchymatous cells (Artabotrys).

h *Uvaria*: Liana climbing by means of prehensile twigs.

j *Artabotrys*: Liana climbing by means of hooks.

k *Duguetia*: Stellate or peltate hairs.

N.B. Monimiaceae and Annonaceae have immediate branching in trunk..

Legends to plates 4A/4B

4A Magnoliaceae and Myristicaceae

a TROLL's model: Sympodial trunk consisting of the stacking of modules, each module firstly plagiotropic, then becoming orthotropic at its base (some species of Asiatic *Magnolia* subgenus *Talauma*).

b ROUX's model: Monopodial trunk, continuous branching, branches plagiotropic and monopodial (*Elmerrillia*).

c FAGERLIND's model: Monopodial trunk; rhythmic branching; branches plagiotropic by apposition growth; flowering is terminal (*Magnolia*).

d Young trunk showing annular stipular scars, its apex covered by the stipular hood (sh); detail of a leaf base with the scar (sc) of the caducous, adnate to petiole, stipule (*Michelia*).

e Stipule (s) adnate to the petiole, axillary bud (b) covered by the stipule (*Talauma*)

f MASSART's model: Monopodial trunk, rhythmic branching, branches monopodial and plagiotropic (for all genera of the Myristicaceae, e.g. *Horsfieldia, Iryanthera, Knema, Pycnanthus, Virola*, etc.).

g The plagiotropic branches bear distichous leaves (the leaves of pioneers species are often chewed by insects).

h Leaf entire with brochidodromous venation (typical for the Myristicaceae).

j Apex of stem covered with very short, erect hairs (typical).

4B Piperaceae (*Piper*) and Chloranthaceae

a PETIT's model: Trunk monopodial, continuous branching, branches plagiotropic and modular (*Piper*).

b A modular branch of *Piper*, each module consists of one long internode (li) bearing a developed leaf and one very short internode (si) bearing a scale-leaf (sl). Inflorescence is terminal. fs = scar left by a shed inflorescence.

c The terminal inflorescence is a long-lived spike.

d Some species of *Piper* possess stipules inserted on the petiole.

e ATTIMS's model: Trunk monopodial, continuous branching, branches orthotropic and sympodial (*Hedyosmum*).

f Very short stipular crown around a swollen node (*Sarcandra*).

g Serrulate leaves, stipules inserted on the crown formed by petiolar expansions (e.g., *Hedyosmum*).

N.B. All these families have in general immediate branching in trunk.

P3A

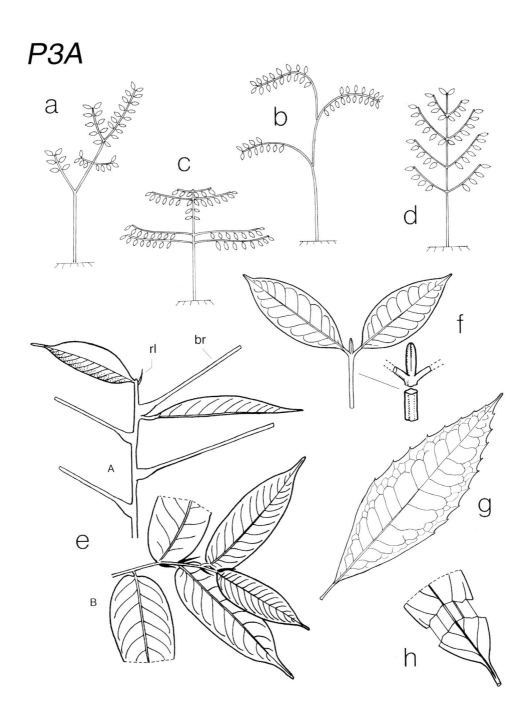

180

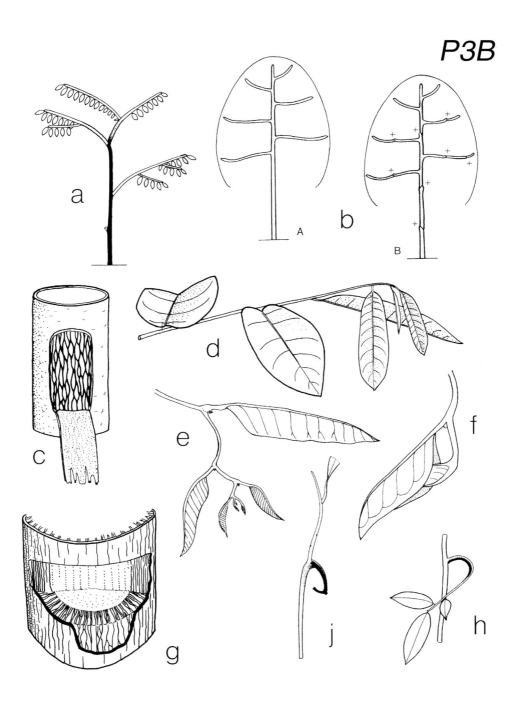

P3B

P4A

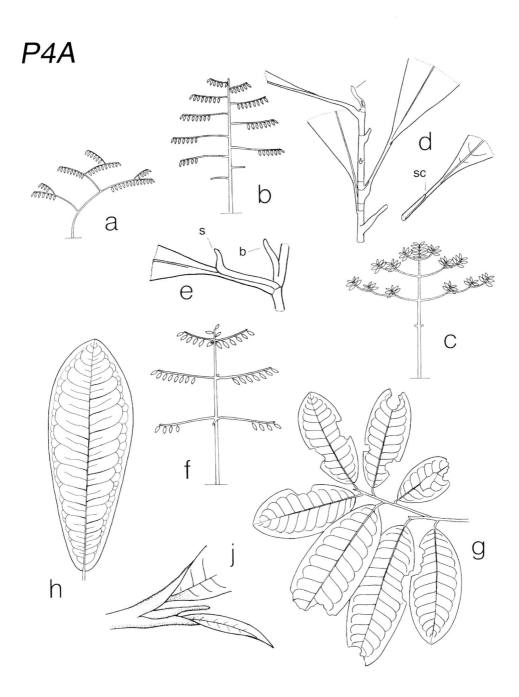

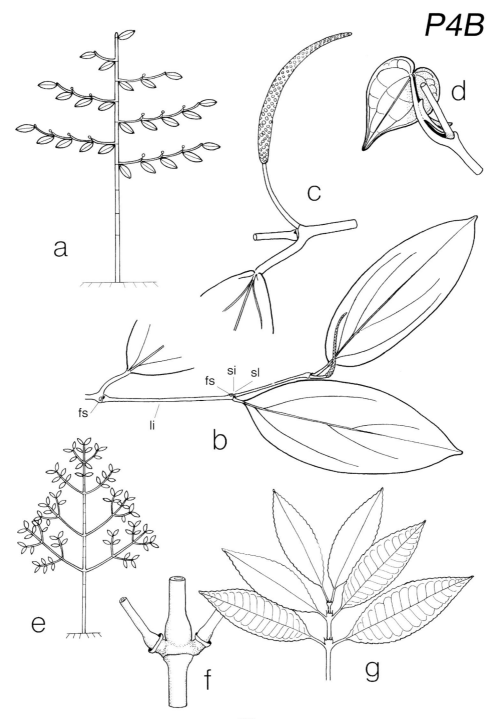

P4B

Legends to plates 5A/5B

5A Menispermaceae and Aristolochiaceae

a   Twining trunk of a lianescent individual (the commonest case for the Menispermaceae).

b   An arborescent individual conforming to LEEUWENBERG's model, i.e. stems orthotropic and sympodial, not distinctly differentiated into trunk and branches (all stems equivalent), (*Abuta* pp.).

c   An arborescent individual conforming to MANGENOT's model, i.e. sympodial trunk consisting of the stacking of modules, each module firstly orthotropic, then becoming abruptly plagiotropic (e.g. *Cocculus*).

d   Xylem in rings and with rays of phloem (almost all Menispermaceae).

e   A typical leaf: the petiole is pulvinate at both ends, blade with scalariform and palmate venation, petiolar base not decurrent to the stem.

f   Lobate leaf (rather uncommon, e.g. *Burasaia* pp., *Tinospora* pp.).

g   Peltate leaf (rather uncommon, e.g. *Cyclea*, *Stephania*).

h   Liana with twining trunk, petiolar base decurrent to the stem. Detail: radiate wood and well developed cork of a non-herbaceous *Aristolochia*.

j   Typical leaf of an Aristolochiaceae: the petiolar base is decurrent to the stem, blade is cordate with palmate venation.

5B Meliosmaceae (*Meliosma*), Lardizabalaceae and Ranunculaceae (*Clematis*)

a   ROUX's (or COOK's?) model: Monopodial trunk, continuous branching, monopodial and plagiotropic (phyllomorphic) branches. For the Asiatic *Meliosma* conforming to this model, phyllotaxy is distichous and the leaves are simple.

b   KORIBA's model: Trunk and branches modular, modules initially equal, all apparently branches, but later unequal, one becoming a trunk. For the Asiatic *Meliosma* conforming to this model, phyllotaxy is spiral, leaves are pinnate and compound.

c   Serrulate leaf or leaflet of an Asiatic *Meliosma*.

d   **A**. orthotropic branch with spiral phyllotaxy (model unknown). **B**. grooved stem, leaves simple with petioles enlarged at their bases (an American *Meliosma*).

e   *Mahonia*: sheathing petioles leaves paripinnate and serrate, spinescent.

f   *Mahonia*: fissuring periderm, petiole stipulate.

g   *Parvatia* (Lardizabalaceae): plant twining, its leaves bipinnate.

h   *Clematis* (Ranunculaceae): plant climbing with its petioles or petiolules. Leaves are oppposite.

N.B. All these families have delayed branching in trunk.

Legends to plates 6A/6B

6A Proteaceae

a   *Grevillea*: trunk with numerous transversely elongated lenticels.
b   *Helicia*: **A**. lenticellate twig; **B**. Young lateral twig with supplementary bud (sb); **C**. Young internode angular (int ang).
c   *Helicia*: Leaf serrulate, shortly petiolate.
d   Typical bark of the Proteaceae, with its external wood furrowed, lenticellate rhytidome and fibrous bark.
e   *Stenocarpus*: Rhytidome without lenticels.
f   Leaf simple, but deeply lobate (e.g. *Heliciopsis, Stenocarpus*).
g   Leaf simple and serrulate (e.g. juvenile leaves of *Roupala*).
h   Pinnate compound leaf (e.g. adult leaves of *Roupala*).
j   Phyllotaxy alternate to subopposite, a double bars is placed between the two adjacent UE (e.g. *Panopsis*), missing leaves (f).
k   Stem apex with angular internodes (int) (e.g. *Panopsis*).
m   Nearly opposite or whorled phyllotaxy, leaves entire (e.g. *Panopsis*). Detail: vertical serial buds (sb).

6B Dilleniaceae and Polygonaceae

a   Stipuliform petiolar expansions, hairs appressed, young leaf folding conduplicate (all these characters are typical for the Dilleniaceae), (e.g. *Dillenia*).
b   Twining liana (numerous Dilleniaceae). The venation of leaves here is brochidodromous, which is not typical for the family (e.g. *Doliocarpus*).
c   *Curatella*: Base of petiole surrounding the apical bud, venation secant on the margin of the blade.
d   Scabrous stem with hairs directed downwards, allowing the plant to grip its support.
e   A lianescent and twining species of *Coccoloba*. The leaves inserted at the base of suckers are reduced to petiolar sheaths.
f   *Coccoloba*: Section of a trunk showing a bipolar arrangement of the phloem (frequent for the lianescent species).
g   MASSART's / RAUH's model of an arborescent *Coccoloba*. The branches bear short twigs, phyllotaxy is spiral.
h   A well-developed ochrea (typical character of the Polygonaceae).
j   Young leaf folding revolute, i.e. margin of the blade rolled abaxially (a typical character in this family). Short shoots are typical in twining *Coccoloba*.
k   Stem ending in a tendril (*Afrobrunnichia, Antigonon*).

N.B. All these families have delayed branching in trunk.

P5A

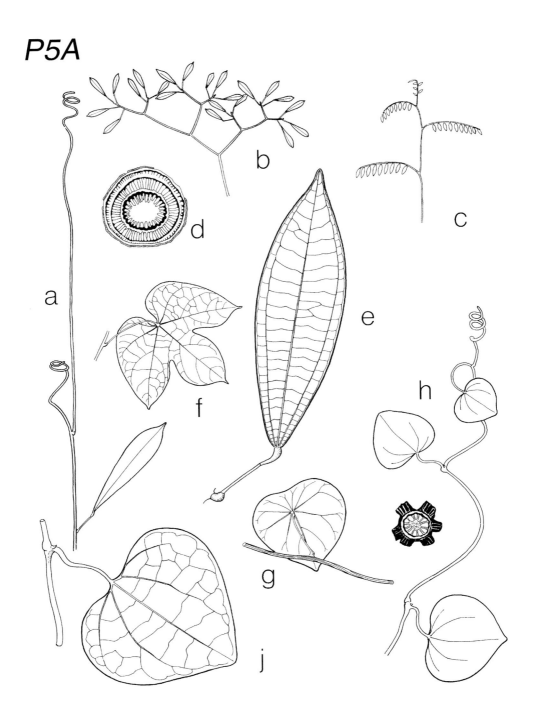

P5B

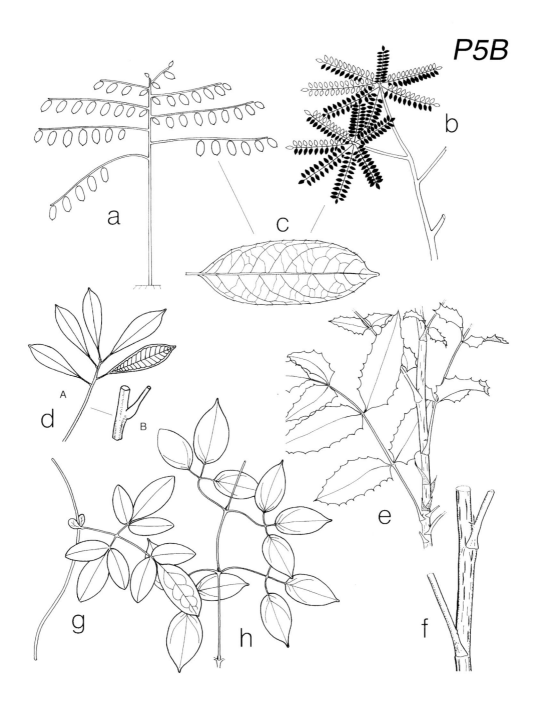

P6A

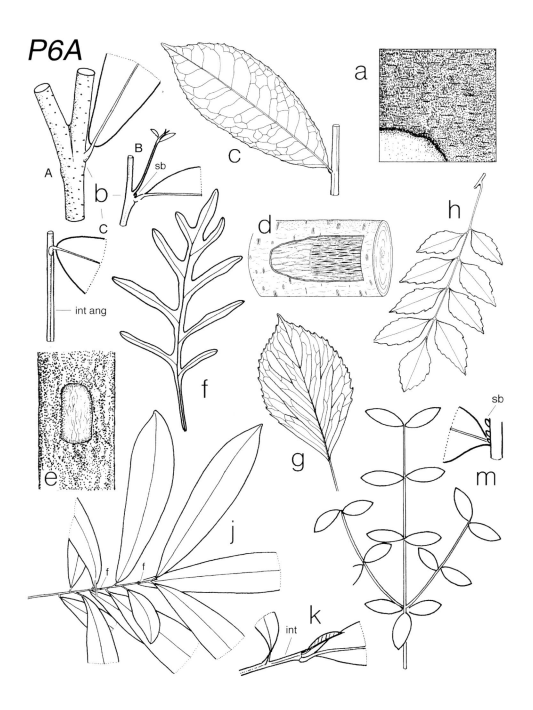

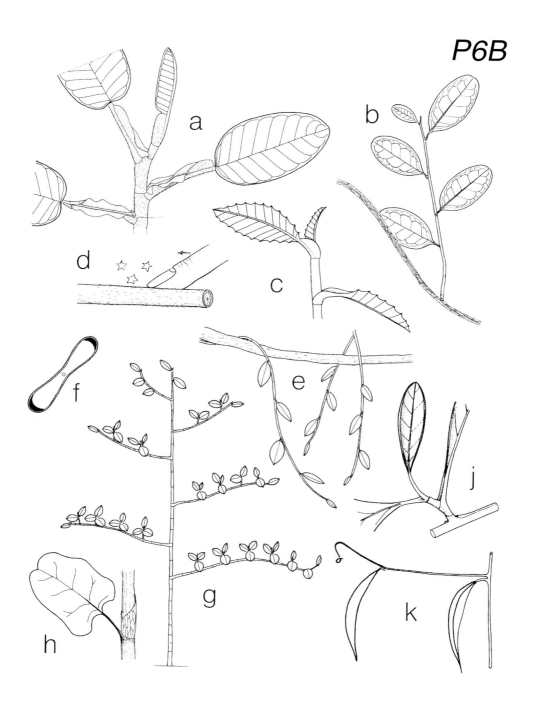

P6B

Legends to plates 7A/7B

7A Nyctaginaceae and Phytolaccaceae

a   *Bougainvillea*: shrub with decumbent branches.
b   *Bougainvillea*: short shots modified into a spine.
c   *Pisonia*: modular architecture (KORIBA's model); «+» indicates a death of an apical meristem.
d   *Neea*: external wood furrowed.
e   *Neea, Pisonia*: two different kind of sympodial branches, in both cases sympodial units consist of only 2-3 long internodes.
f   *Pisonia*: anisophylly, phyllotaxy is opposite or subopposite.
g   *Petiveria* (Phytolaccaceae): chlorophyllous periderm (p) in black, the older suberised periderm becomes fissured.
h   *Petiveria*: stipules (s) thin, but not spiny.
j   *Seguieria*: two spiny prophylls in stipular position.
k   *Phytolacca americana*: sympodial architecture with oppositifoliate inflorescence, the two prophylls (a, b) are foliaceaous.

7B Olacaceae, Opiliaceae and Loranthaceae

a   ROUX's model: Monopodial trunk, continuous branching, branches plagiotropic and monopodial (Olacaceae: *Ochanostachys, Scorodocarpus, Strombosia*).
b   MANGENOT's model: Sympodial trunk consisting of the stacking of modules, each module firstly orthotropic, then becoming abruptly plagiotropic (*Heisteria*).
c   *Minquartia guianensis*: flutted trunk.
d   Young twig angular (typical for the Olacaceae), leaves entire with camptodromous venation.
e   Petiole forming a pulvinus (pu), its distal half weakly enlarged or at least able to twist or bend (*Coula, Octoknema*).
f   *Agonandra*: Leaf venation poorly reticulate and corky rhytidome.
g   LEEUWENBERG model's and opposite phyllotaxy (e.g. *Loranthus, Viscum*).
h   Alternate phyllotaxy (e.g. *Antidaphne, Nuytsia*: an Australian parasitic tree).
j   Serial buds, «colateral» in main axis A (i.e. t1, t2, t3), and vertical in lateral axis t1 (i.e. r1, r2, r3), (e.g. *Phthirusa*); ls is the scar of the fallen leaf axillary to t1, t2, t3.
k   Two opposite leaves in *Psittacanthus*, venation poorly reticulate.
m   Leaf in *Viscum*, venation poorly reticulate.

N.B. All these families have delayed branching in trunk.

Legends to plates 8A/8B

8A Hamamelidaceae and Fagaceae

a *Altingia* (Hamamelidaceae-Altingioideae): Rhythmic and orthotropic growth of branches. Details: the apical buds possess several scale-leaves, phyllotaxy is always spiral; some species bear small stipules on their petioles.

b *Exbucklandia* (syn.: *Symingtonia*) *populnea* (Hamamelidaceae-Exbucklandioideae): A1. trunk; A2. and A3. branches. Branching is continuous. Leaves are arranged on the twig in a very peculiar fashion: they are lying in the same quadrant of a cylinder, the axis of which would be the twig itself. Stipules (s) are appressed one to another and adnate to the petiole. Phyllotaxy alternate and distichous in the branches, leaves trilobate.

c *Maingaya malayana* (Hamamelidaceae-Hamamelidoideae), hood-like stipules (s).

d *Rhodoleia championii* (Hamamelidaceae-Rhodoleioideae): Sympodial architecture, all branches with spiral phyllotaxy. Death of apical meristems indicated by "+".

e Furrowed wood of the Fagaceae (entire bark must be stripped off for this character to be observed).

f Branch collar, elongated or flat (a frequent character).

g A young *Lithocarpus*, not conforming to TROLL's model, with sympodial trunk and branches. The phyllotaxy is distichous in the plagiotropic axes. TROLL's model might exist for this genus, as in *Fagus*. Death of apical meristems indicated by "+".

h Spiral phyllotaxy of a branch of *Quercus* and *Castanopsis*. "+": as before.

j Furrowed internode of a young stem in the Fagaceae, (s = stipule).

k Camptodromous venation (typical for entire leaves in the Fagaceae).

8B Zygophyllaceae, Staphyleaceae, Greyiaceae, Melianthaceae, Rosaceae, and *Ribes*

a *Guaiacum* (Zygophyllaceae): sympodial stem becoming progressively plagiotropic, each module bears only two opposite compound leaves and, at its base, two small prophylls.

b *Guaiacum*: old tree with scaly rhytidome.

c *Turpinia*: stem distinctly lenticellate.

d *Turpinia* (Staphyleaceae): stem apex showing opposite phyllotaxy, imparipinnate leaves; leaflets serrulate, young leaflets with involute folding.

e *Greyia* (Greyiaceae): leaf serrulate, venation palmate and sheathing petiolar (sh p) base (detail).

f *Greyia*: rhytidome coming off in strips, without lenticels, very different from that of *Turpinia*.

g *Melianthus* (Melianthaceae): leaves compound, leaflets serrulate; stipules (st) are foliaceous and intrapetiolar.

h *Eryobotrya* (Rosaceae): monopodial stem with distinct rhythmic growth; base of UE bearing scale-leaves (c); stipules (st, in black) are persistent. Rhytidome of *Eryobotrya* (Maloideae) is very different from these of Rosoideae. such as *Potentilla*, *Spiraea* (N. temperate), *Polylepis* (Andean), which are fibrous or papery.

j *Ribes* (Grossulariaceae): leaf in axiliary position in respect to spines (sp).

N.B. Except *Exbucklandia*, all these taxa have delayed branching in trunk.

P7A

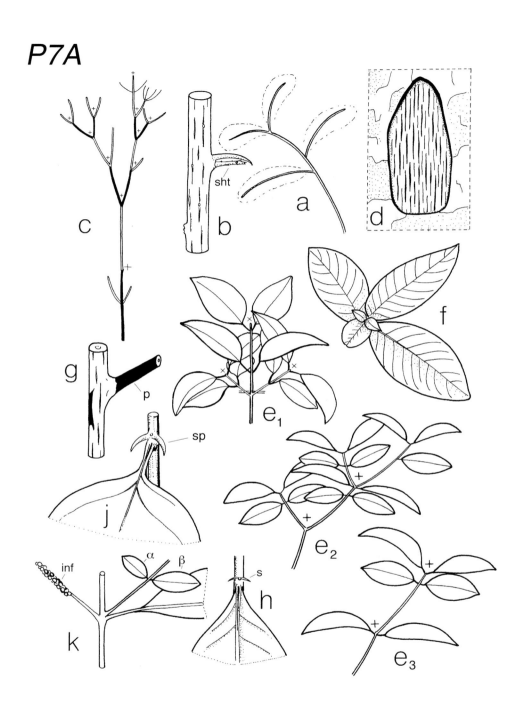

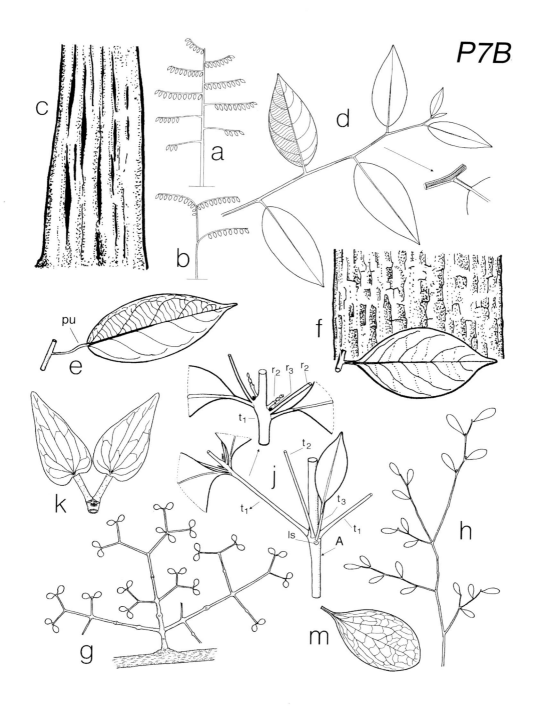

P8A

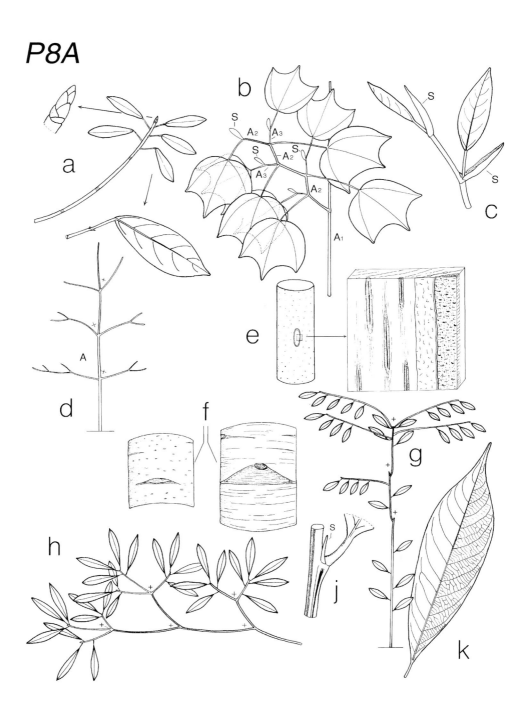

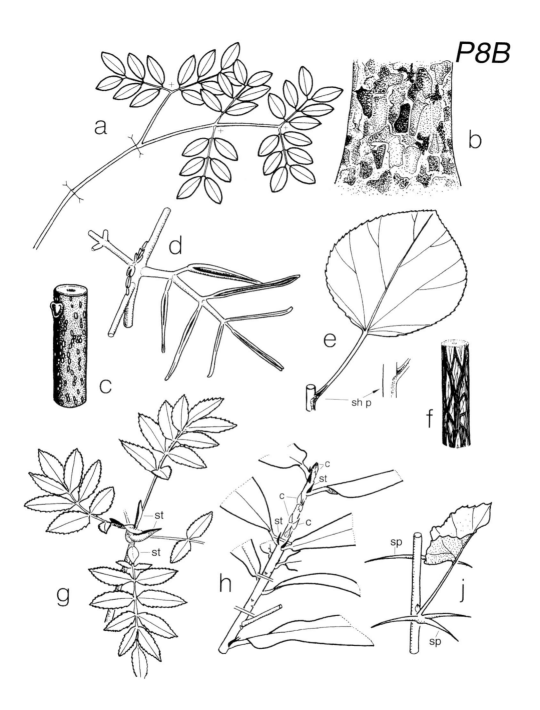

P8B

Legends to plates 9A/9B

9A Ulmaceae and Urticaceae

a ROUX's model: Monopodial trunk; continuous branching; branches plagiotropic and monopodial (*Celtis*, *Trema*).
b TROLL's model: Sympodial trunk consisting of the stacking of modules, each module firstly plagiotropic, then becoming orthotropic at its base (*Trema*).
c Plagiotropic branch with distichous phyllotaxy. Leaves are trinerved, venation is somewhat camptodromous (*Celtis*, *Trema*).
d Immediate ramification in *Celtis philippensis*, A2 and A3 the second and third branch orders. Leaves are entire and distinctly trinerved.
e Short shoots modified into spines (*Celtis*).
f Bark with network of fibres, lenticels becoming transversely elongated (e.g. *Dendrocnide*).
g Asymmetrical stipules (typical for Urticaceae with spiral phyllotaxy, e.g. *Poikilospermum*, *Urera*). s1/s2, pair of fused stipules of the leaf ls, s2 lies in an intrapetiolar position. l: another leaf.
h Leaves serrulate, trinerved with scalariform and densely reticulate venation (typical for the Urticaceae).
j Opposite phyllotaxy, stipules small, not intrapetiolar. (e.g. *Boehmeria*).
k Stem and petiole with long, rigid hairs allowing the plant to climb (e.g. *Urera*). Leaf entire (rather uncommon for the Urticaceae), with intrapetiolar stipules.
m Stinging hair (e.g. *Laportea*).

9B Moraceae

a Fluted trunk formed by anastomosing aerial roots (*Ficus*).
b TROLL's model: Sympodial trunk consisting of the stacking of modules, each module firstly plagiotropic, then becoming orthotropic at its base (e.g. *Brosimum*, *Ficus*, *Sorocea*, *Streblus*, *Trophis*).
c RAUH's model: Monopodial trunk, rhythmic branching, branches orthotropic and monopodial (e.g. *Artocarpus*, *Cecropia*, *Ficus*). Double bars indicate adjacent UE.
d COOK's model: Plagiotropic short-lived branches resembling compound leafs (e.g. *Castilla*, *Perebea*, *Pseudolmedia*).
e Stipules leaving annular scars (sc), sh = stipular hood, latex whitish (Moroideaee: *Artocarpus*, *Brosimum*, *Ficus*, etc., some Moroideae have no annular stipules, e.g. *Streblus*).
f *Musanga cecropioides*, an African Cecropiodeae with almost palmately compound leaves; sc = scar left by the stipular hood. (*Cecropia* normally bears palmately lobate leaves).
g Various leaf forms in *Ficus*: **A**. lobate; **B**. cordate with waxy spots (ws) at the axils of lateral veins; **C**. with dripped tip; **D**. cuneate.

N.B. These three families have immediate or delayed branching in trunk, depending of the species (e.g. *Ficus*) or of the genus, e.g. *Trema* is "immediate", *Cecropia* is "delayed".

Legends to plates 10A/10B

10A Rhamnaceae

a   ROUX's model: Monopodial trunk, continuous branching, branches plagiotropic and monopodial (e.g. *Alphitonia, Lasiodiscus, Maesopsis*).

b   CHAMPAGNAT's model: Trunk and branches firstly erect, than becoming decumbent, phyllotaxy not distichous (e.g. *Colubrina*).

c   TROLL's model: Sympodial trunk consisting of the stacking of modules, each module firstly plagiotropic, secondarily becoming erect, most often after leaf-fall (e.g. *Ziziphus*).

d   *Ziziphus*: zigzagging twig with distichous phyllotaxy. Leaves trinerved. Stipules are modified into spines for some species.

e   Liana climbing by means of tendrils which become woody in some species (e.g. *Gouania*).

f   Glandular trinerved leaf of *Colubrina glandulosa*. gl = gland.

g   *Maesopsis*: Plagiotropic branch with opposite-decussate phyllotaxy. Strong anisoclady: there are nodes bearing only one developed lateral twig. Foliar teeth glandular.

h   Serrulate leaf, camptodromous and scalariform venation, small stipules (typical for the Rhamnaceae).

j   *Scutia*: Scrambling liana climbing by means of short shoots modified into spines (ramification rhythmic, MASSART's model). Detail: stipules are very small, spines are subtended by a leaf.

10B Vitaceae and Leeaceae (*Leea*)

a   Section of trunk of a liana showing well developed rays, still visible in the scaly rhytidome (e.g. *Cissus*).

b   Liana of the understorey with two kinds of stems: erect and creeping. Both types of stem bear simple and compound leaves (heterophylly). Phyllotaxy remains distichous in erect stems (e.g. *Tetrastigma*). Oppositifoliate tendrils, elongating before the associated leaves reach adult size. Detail: stems swollen at nodes; a = axis, a, b = prophylls; s = stipule.

c   Margin of lamina serrulate with rosoid teeth, i.e. teeth innervated by a main veinlet and two lateral veinlets (typical for Vitaceae).

d   Stem kneed at the nodes. Leaves imparipinnate with serrulate leaflets (*Leea*).

e   Appressed stipules, leaving an annular scar.

f   Leaflet serrulate with scalariform venation.

g   Spiny trunk (in some arborescent species).

N.B. Rhamnaceae have in general immediate branching in trunk. Vitaceae and Leeaceae have probably delayed branching in trunk (except for the tendrils of the lianescent Vitaceae which develop immediately). Rhamnaceae (near Rosaceae) ought to be very remote from Vitaceae-Leeaceae, see [42].

P9A

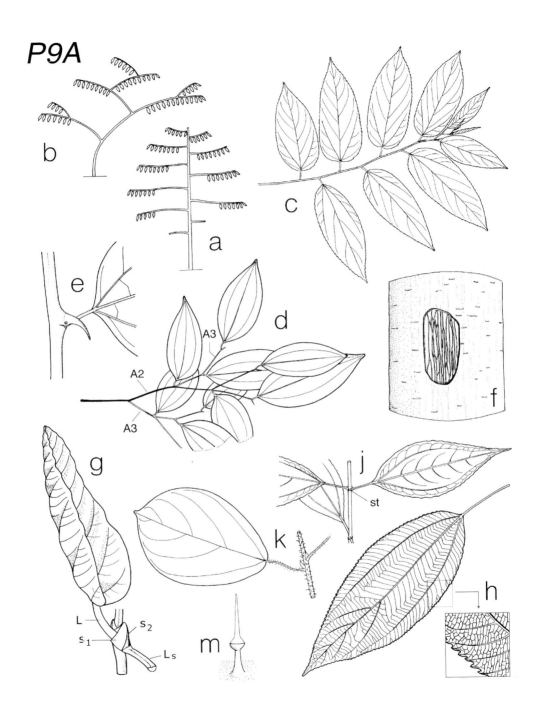

P9B

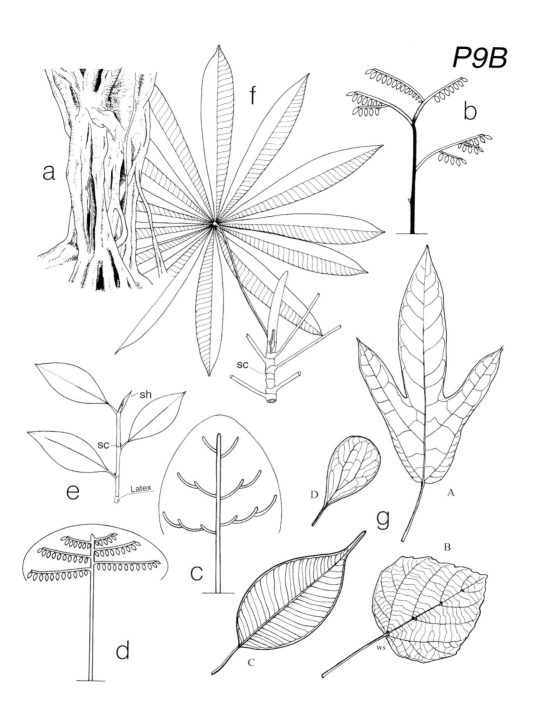

P10A

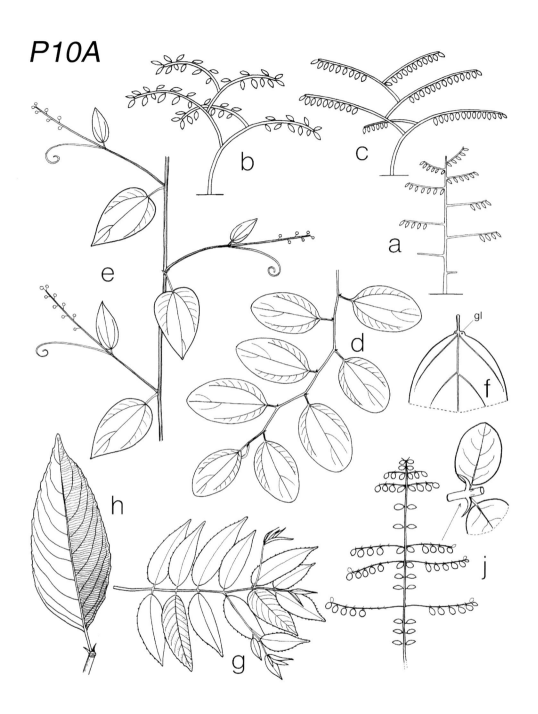

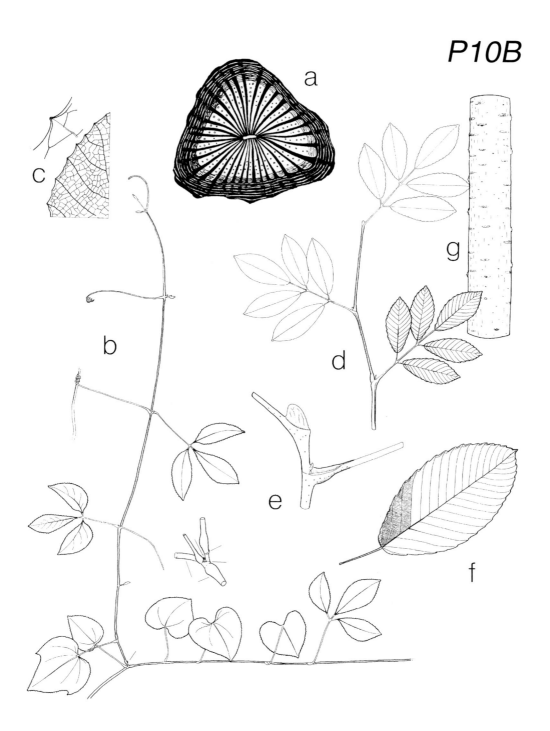

P10B

Legends to plates 11A/11B

11A Theaceae, Actinidiaceae, Marcgraviaceae, Pellicieraceae and Bonnetiaceae

a *Gordonia, Laplacea* : Branches erect and sympodial, but UE could be disposed in monopodial sequences.

b ROUX's model: Monopodial trunk, continuous branching, branches plagiotropic and monopodial (*Adinandra*).

c Serrulate leaf of a Theaceae (e.g. *Camellia*). In a tropical environment, the leaves are usually entire.

d Young leaf involute, i.e. the margins are rolled adaxially (typical for the Theaceae). Detail: petiole covered with long appressed hairs (e.g. *Freziera*).

e *Saurauia*: Apex of stem with thick, oblique or appressed hairs (typical for the Actinidiaceae). Details showing form and position of hairs.

f *Marcgravia*: Heterophylly. Liana climbing by means of clamp-roots.

g *Pelliciera*: Leaf folding convolute, its blade with marginal glands (gl).

h *Bonnetia, Ternstroemia*: Rhythmic growth and branching, branch monopodial and orthotropic (RAUH's model). Apical bud protected by an involute young leaf (ilf). Leaves coriaceous with a revolute margin (rm).

j *Ploiarium, Archytaea*: Rhythmic growth and branching. Branch plagiotropic by apposition growth (AUBREVILLE's model).

k Entire leaf, coriaceous, with well-marked secondary veins (*Ploiarium, Archytaea*).

11B Myrsinaceae and Theophrastaceae

a RAUH's or MASSART's model: Monopodial trunk, rhythmic branching, branches monopodial, orthotropic or plagiotropic; branches somewhat drooping (*Ardisia, Oncostemum*). CORNER's model in several genera (*Oncostemum, Rapanea, Tapeinosperma*, etc.)

b AUBREVILLE's model: Monopodial trunk, rhythmic branching, branches plagiotropic by apposition growth (*Ardisia*).

c Rhythmic growth of an orthotropic branch (RAUH's model), (*Cybianthus, Rapanea*). Double bars indicate adjacent UE.

d Oval scars left by the shed branches (*Ardisia, Oncostemum*). Ls = leaf scar left by a shed leaf.

e Serrulate leaf with glandular or scaly trichomes (typical in the Myrsinaceae). **A**. leaf; **B**. glandular trichomes; **C**. scaly trichomes.

f Leaf with resiniferous ducts (rd) and glandular trichomes (*Maesa*).

g Underside of the blade with glandular trichomes inserted at the bottom of small cavities.

h CORNER's model, i.e. unbranched monocaulous treelet; leaves spiny-serrulate (*Clavija, Theophrasta*).

j KORIBA's model: Trunk and branches modular, modules initially equal, all apparently branches, but later unequal, one becoming a trunk (*Jacquinia*).

N.B. Depending of the genus, Theaceae, Myrsinaceae, and Theophrastaceae have immediate or delayed branching in trunk.

Legends to plates 12A/12B

12A Ericaceae and Cyrillaceae

a Sympodial architecture, branches erect, leaves clustered at modules ends (e.g. *Rhododendron*) or not clustered (e.g. *Bejaria*). (inversed arrows are for delimitation of the modules).
b Shrub with plagiotropic branches (e.g. *Leucothoe*, Vaccinium). Detail: a toothed leaf (uncommon in tropical species). (a '+' indicates the death of an apical meristem).
c Typical sloughing of the rhytidome, small scales with transversal fissures, older rhytidome (rh1), younger rhytidome becoming (rh2) (e.g. *Arbutus* and other Ericaceae).
d Plagiotropic, sympodial branch (e.g. *Nothopora*). Entire leaves common for tropical species.
e An «ericoid» specimen (e.g. *Erica arborea*) of the family, its slendest twigs bearing very small leaves, typical in mediterraean climates.
f Desquamation pattern in anastomosing strips (e.g. *Erica*).
g Sympodial twig, with the typical abscission zone (az) at leaf insertion (e.g. *Cavendishia*).
h Monopodial twigs (with lateral inflorescence) in *Cyrilla* (double bars idicate the delimitation of the UE)
j *Cyrilla*, leaves entire with a hyaline margin (hm), periderm suberising very soonly.

12B Ebenaceae (*Diospyros*) and Sapotaceae

a *Diospyros*: **A**. Monopodial trunk, sylleptic rhythmic branching, branches sympodial (but not modular) and plagiotropic (MASSART's model); the plagiotropic branches bear developed leaves, but the trunk bears reduced leaves, even still smaller at apex of the UE (heterophylly); **B**. terminal pseudowhorl of small leaves.
b A sympodial unit of a branch. axl = axillary leaf of the last sympodial unit. Death of apical meristems indicated by "+".
c Appressed hairs and small glands (g) scattered on the underside of blade.
d A branch plagiotropic by apposition growth (AUBREVILLE's model), (*Manilkara*, *Palaquium*, old world *Sideroxylon*).
e ROUX's model: Monopodial trunk, continuous branching, branches plagiotropic and monopodial (*Chrysophyllum* pp.).
f RAUH's model: Monopodial trunk, rhythmic branching, branches orthotropic and monopodial (*Chrysophyllum* pp., *Planchonella*, *Pouteria*). Doubles bars are placed between adjacent UE.
g TROLL's model: Sympodial trunk consisting of the stacking of modules, each module plagiotropic, secondarily becoming erect, most often after leaf-fall (*Micropholis*).
h Gutta flowing out of cut bark (typical for the Sapotaceae).
j Various types of leaves: **A**. numerous parallel secondary veins (*Manilkara* pp., *Mimusops*, *Micropholis*); **B**. venation camptodromous and scalariform (*Madhuca*, *Palaquium*); **C**. brochidodromous venation (*Chrysophyllum*, *Manilkara* pp., *Pouteria*).
k Stipules protecting the apical bud (*Palaquium*), stipules very small in *Mimusops*.
m Short shoots in some American *Sideroxylon* conforming to CHAMPAGNAT's model.

N.B. Ericaceae have delayed branching in trunk (immediate in Ebenaceae and Sapotaceae).

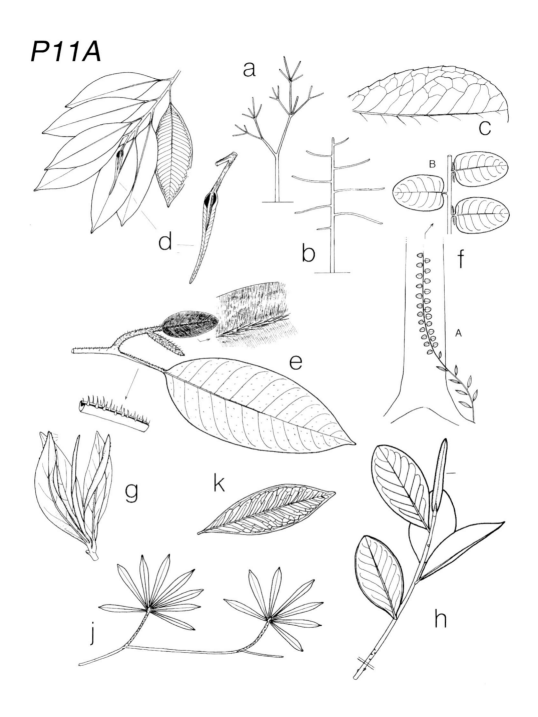

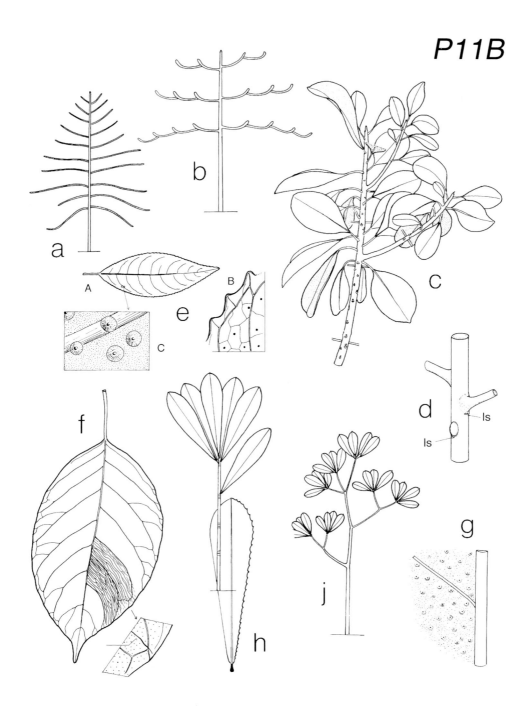

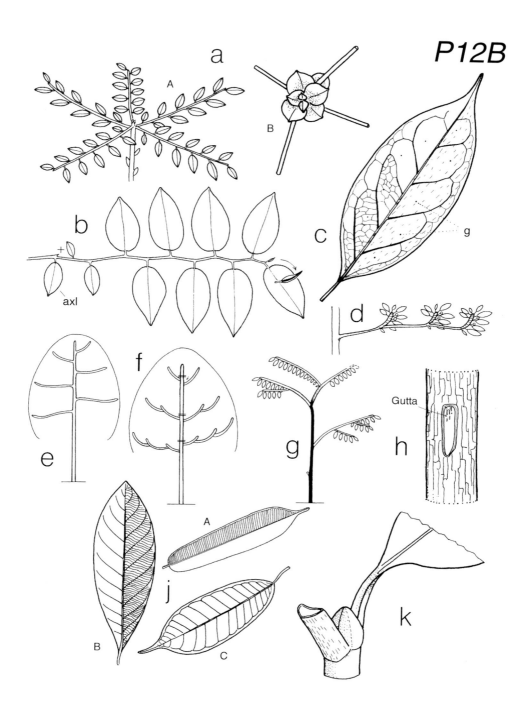

P12B

Legends to plates 13A/13B

13A Lecythidaceae

a TROLL's model: Sympodial trunk consisting of the stacking of modules, each module firstly plagiotropic, secondarily becoming erect (*Eschweilera, Lecythis*).
b RAUH's model: Monopodial trunk, rhythmic branching, branches orthotropic and monopodial (*Couroupita*). Double bars indicate adjacent UE.
c CORNER's model, i.e. unbranched monocaulous treelet (*Gustavia* pp.).
d Branch orthotropic and sympodial, but the sympodial units consist of two or more units of extension (delimited by double bars). Phyllotaxy is spiral. Branching occurs apically in the sympodial units (LEEUWENBERG, KORIBA, or SCARRONE's model), flowering is terminal, (Lecythidoideae: *Gustavia*; Planchonioideae: *Barringtonia, Planchonia*).
e Short shoots inserted distichously on a plagiotropic branch (*Couratari*); MASSART's model: monopodial trunk with rhythmic branching (*Couratari, Napoleonaea*).
f Young leaf folding involute, i.e. the margins are rolled adaxially (typical for the Lecythidaceae, but well marked in *Eschweilera, Lecythis*).
g Leaf with brochidodromous, densely reticulate venation (e.g. *Bertholletia, Eschweilera*).
h Vestigial glandular teeth (vgt), appearing as black dots on the margin of adult leaves (*Eschweilera, Lecythis*).
j Base of the lamina bearing two glands (gl), (*Napoleonaea*).

13B Combretaceae and Myrtaceae

a AUBREVILLE's model: Monopodial trunk, rhythmic branching, branches plagiotropic by apposition growth (*Buchenavia, Terminalia*).
b Terminalia-branching: development of a new module from the axil of a scale-leaf (sc). **A.** a pseudowhorl with few leaves; **B.** more leaves and a new module have developed.
c *Combretum*: Extremity of a branch, leaves opposite or subopposite. Detail: glands.
d Leaf entire with camptodromous venation (typical for the Combretaceae).
e ATTIM's model: Trunk monopodial, continuous branching, orthotropic and monopodial branches (Leptospermoideae: *Eucalyptus*). Various species of this genus show foliar dimorphism: the phyllotaxy is opposite-decussate in the young individual and becomes spiral in the adult tree.
f A typical mode of growth for *Eugenia* (Myrtoideae): each sympodial unit of the trunk bears apically two branches, one of which (b1) becomes erect to form the trunk, the other (b2) remaining plagiotropic to form a branch (a variant of a KORIBA's model?). b3: another plagiotropic branch. Branches are generally all erect in savannas or dry forests (e.g. *Ugni*: RAUH; *Myrcia, Rhodomyrtus*).
g Various types of sloughing off of the outer bark in the Myrtaceae. **A.** the dead layers slough off in the form of straw-like scales (frequently in young branches); **B.** and **C.** sloughing off in sheets (**C** is typical for old trunks of *Eucalyptus*).
h Entire leaf with translucent dots (leaf pellucid-dotted), venation is brochidodromous and forms intramarginal veins (typical for the Myrtaceae).
j Same as **h**, but venation forming two intramarginal veins.

N.B. Trunk branching delayed in these families, except Myrtaceae in part (e.g. *Eucalyptus*).

Legends to plates 14A/14B

14A Lythraceae (incl. Sonneratiaceae) and Vochysiaceae

a   *Sonneratia*: a mangrove genus with specialised aerial roots, or pneumatophores (pn).
b   *Sonneratia*: apices of internodes glandular (gl), opposite phyllotaxy.
c   *Duabanga*: strong rhythmic growth in trunk (MASSART's model), this bearing small leaves while branch leaves are large.
d   *Duabanga*: a plagiotropic branch, phyllotaxy is opposite-decussate but petiole are twisted to allow leaves to be hold in a same plane.
e   *Lagerstroemia*: for some species phyllotaxy is decussate, but not opposite, F_1-F_2, F_3-F_4, F_5-F_6 pairs of facing leaves. Camptodromous venation.
f   Scaly rhytidome in *Lagerstroemia*.
g   Leaf with intramarginal vein, secondary veins not nmerous (e.g. *Vochysia*).
h   *Vochysia*: **A**. young tree with plagiotropic axes (MASSART's model); **B**. adult tree with orthotropic branches (SCARRONE's model). Phyllotaxy is opposite or whorled.
j   Leaf with intramarginal vein, secondary veins not numerous (e.g. *Vochysia*).
k   *Qualea*: ROUX's model: Monopodial trunk, continuous branching, monopodial and plagiotropic branches.
m   Intramarginal vein, secondary veins numerous (*Qualea*).
n   Cupular glands located close to the stipules (*Qualea*). st = stipule; p = petiole.
p   Delayed ramification of a branch associated with imbricate and decussate scale-leaves at the base of the lateral twig (*Callisthene*).

14B Melastomataceae (Melastomatoideae and Memecyloideae)

a   Melastomatoideae: Sympodial orthotropic branch of an individual conforming to SCARRONE's or LEEUWENBERG's model. Detail: node with an interpetiolar ridge (ir).
b   MASSART's model: Monopodial trunk, rhythmic branching, branches monopodial and plagiotropic (Memecyloideae: *Mouriri*). (MANGENOT's model is also recorded for this American genus). l = petiole of a leaf.
c   Plagiotropic twig showing torsion of internodes allowing the opposite-decussate leaves to lie in the same plane, leaf folding is revolute (*Mouriri*, African and some Malagasian *Memecylon*).
d   KORIBA's model: some Malagasian and Asiatic *Memecylon*.
e   Anisophylly, frequent for species with trunks appressed to the support, e.g. climbing *Adelobotrys*, or to the soil, e.g. creeping *Sonerila*).
f   Anisophylly associated with myrmecophylly, the base of the bigger leaf is hollow and houses ants (Melastomatoideae: *Tococa*).
g   Leaves with their main lateral veins more or less originating from the base of the blade with scalariform venation. Melastomatoideae: **A**. e.g. *Medinilla*; **B**. e. g. *Miconia*.
h   Leaves with venation not densely reticulate, tertiary and higher order veins not visible (Memecyloideae: e.g. *Mouriri*, *Memecylon*).

N.B. Vochysiaceae have in general immediate branching in trunk; Lythraceae, except *Sonneratia*, have delayed branching, and Melastomataceae is variable in this respect..

P13A

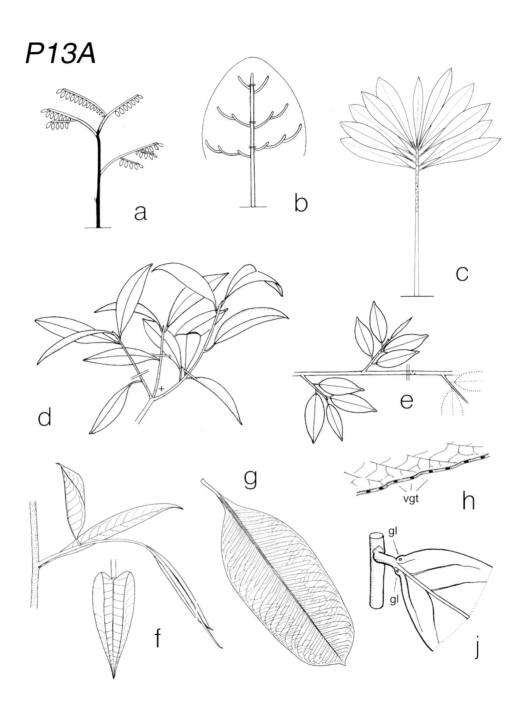

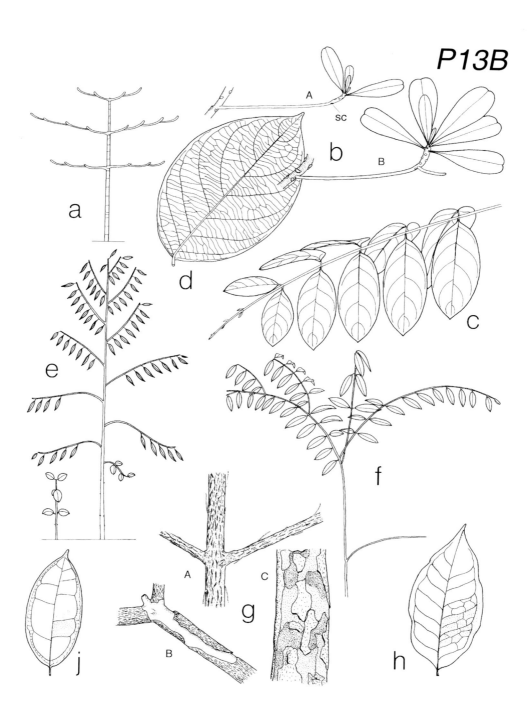

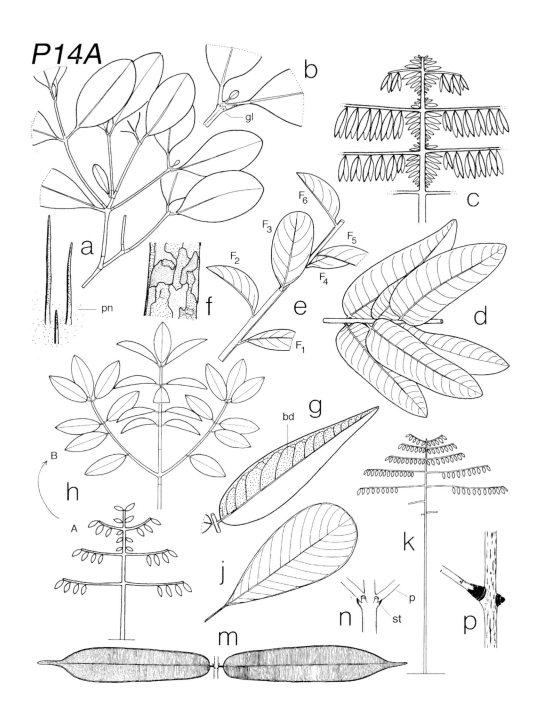

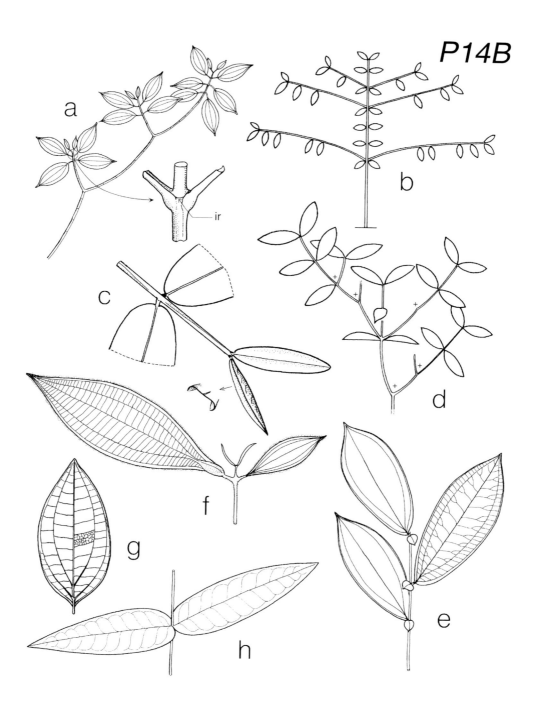

P14B

Legends to plates 15A/15B

15A Clusiaceae

a ROUX's model: Monopodial trunk, continuous branching, branches plagiotropic and monopodial (Clusioideae: *Garcinia*; Hypericoideae: *Vismia*).

b MASSART's model: Monopodial trunk, rhythmic branching, branches plagiotropic and monopodial (Clusioideae: *Platonia*, *Symphonia*; Kielmeyeroideae : *Caraipa*).

c KORIBA's model: Trunk and branches modular, modules initially equal, all apparently branches, but later unequal, one becoming a trunk (Calophylloideae: *Mesua*). Death of apical meristems indicated by "+". Branching is in general immediate in the Clusaceae.

d *Clusia*: Hemiepiphyte with its roots directed towards the soil (RAUH, also for *Chrysochlamys*, *Mammea*).

e *Moronobea*: Rhythmic growth with short shoots. Leaves with numerous parallel secondary veins. Details: rhythmic growth observable due to the presence of short internodes and scale-leaves scars (MASSART, yellow gutta is frequnt for Clusioideae).

f *Vismia*, *Harungana* (Hypericoideae): Opposite-decussate phyllotaxy, entire leaves with brochidodromous venation. Resinous, orange coloured latex in leaves.

g Clusioideae and Hypericoideae: Opposite phyllotaxy and peculiar mode of aestivation of the leaves: during the beginning of their development, the leaves are not folded and remain appressed one to another.

h Different stem apices. **A**. acute (e.g. *Chrysochlamys*). **B**, **C**. obtuse and enclosing the apical bud (e.g. *Garcinia*, *Tovomita*).

j Kielmeyeroideae (*Caraipa*, *Mahurea*): venation more or less scalariform. Young leaf flat, not conduplicate. (Resinous yellowish exudate in bark).

k Leaf with resiniferous ducts (rd) "crossing" the secondary veins (*Garcinia*, *Rheedia*).

15B Ochnaceae and Quiinaceae

a MASSART's model, with branches plagiotropic and monopodial (*Ouratea*, *Ochna*).

b KORIBA's model? A small pioneer tree bearing large leaves (*Lophira*).

c An apical bud with its scale-leaves (e.g. *Ouratea*, *Elvasia*, etc.).

d Numerous parallel secondary veins ending in a intramarginal vein (e.g. *Elvasia*).

e Large cuneate leaf with scalariform venation (*Cespedesia*).

f Rhythmic growth of a branch marked by the occurrence of short internodes. Detail: portion of axis with short internodes (sls = scale-leaves scars) and apical. A double-bar is placed between two adjacent units of extension (*Ouratea*, *Ochna*).

g *Sauvagesia*: Undershrubs with fringed stipules (st).

h Typical leaf of the Ochnaceae with serrate margin, longitudinally striate veins which protrude on the upper side of the blade and small intercostal secondary veins.

j *Quiina*, *Lacunaria*: Monocaulous plant with sympodial trunk, flowering terminal (CHAMBERLAIN's model, but *Quiina* also a branched modular treelet).

k Interpetiolar, persistent stipules (st), (typical for the Quiinaceae).

m Unique, "finger print" venation in the Quiinaceae.

N.B. Dilleniaceae, Quiinaceae and Ochnaceae have delayed branching in trunk.

Legends to plates 16A/16B

16A Flacourtiaceae

a ROUX's model: Monopodial trunk, continuous branching, branches plagiotropic, monopodial (or sympodial, but not modular), (very common for Flacourtiaceae living in the understorey of the forest, e.g. *Casearia, Hydnocarpus, Ryania, Ryparosa*).
b TROLL's model (see above for definition), (*Aphloia, Banara, Homalium, Hydnocarpus*).
c Trunk with rhythmic branching, branches orthotropic (RAUH's or SCARRONE's model, e.g. *Caloncoba, Lindackeria*).
d Branch plagiotropic with spiral phyllotaxy. Petiole distally pulvinate, e.g. *Ryparosa*.
e Leaves glandular (e.g. *Banara*) and stipulate. Detail: slightly involute leaf folding is frequent for Flacourtiaceae, gl = gland.
f *Homalium* (sympodial plagiotropic branche)
g *Ludia* (coriaceous leaves, with 'spcisl venation')
h *Neosprucea* (an uncommon type of venation, trinerved and brochidodromous).

16B Capparidaceae and Passifloraceae

a TROLL's model: Sympodial trunk consisting of the stacking of modules, each module firstly plagiotropic, secondarily becoming erect, most often after leaf-fall (e.g. *Capparis*).
b Plagiotropic branch with distichous phyllotaxy. Leaves are notched (or retuse) and possess a well-marked midrib ending in a mucro for some species (typical for *Capparis*). Detail: stipules modified into spines (*Capparis* spp.).
c Spiral phyllotaxy in an individual not conforming to TROLL's model (*Capparis* pp., *Morisonia, Steriphoma, Stixis*). Petiole distally pulvinate. Capparidaceae are not always stipulate.
d Trifoliolate leaves (*Crateva, Euadenia*).
e Different kinds of hairs in the Capparidaceae: **A**. erect and simple; **B**. stellate; C. peltate.
f Liana with circinate tendrils, glandular petiole, serial buds, trilobate leaves (*Passiflora*). gl = gland.
g Various shape of leaves, with different position of the glands in *Passiflora* and *Adenia*. **A**., **B**. trilobate; **C**. bilobate; **D**. compound; **E**. entire leaf. gl = gland, s = stipule.
h *Barteria*: **A**. COOK's model (similar to ROUX's model, but with phyllomorphic branches); **B**. the branches are inhabited by ants (see the small apertures).

N.B. Capparidaceae and Passifloraceae (except *Barteria*) have delayed branching in trunk. Flacourtiaceae is variable in this respect.

P15A

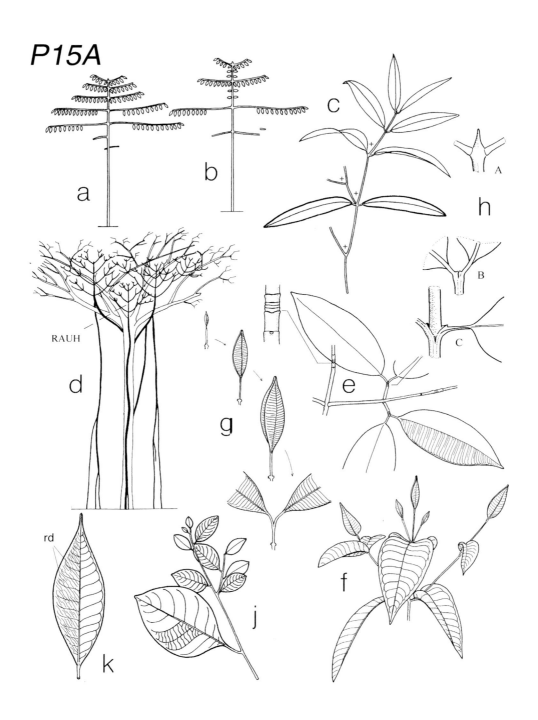

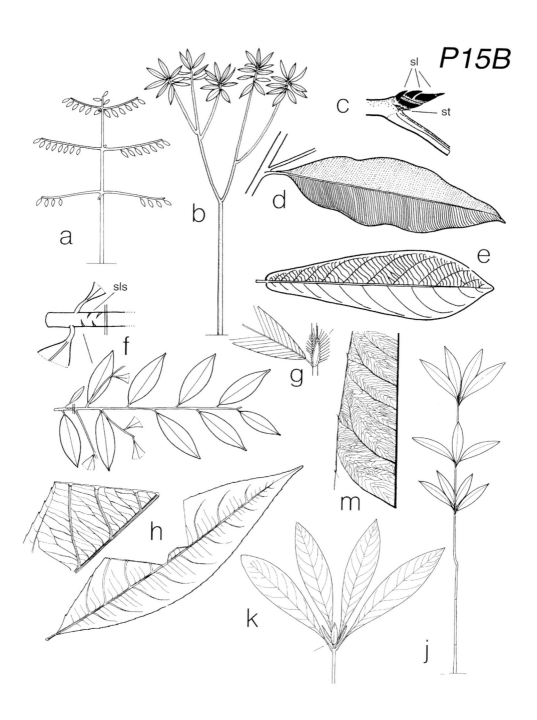

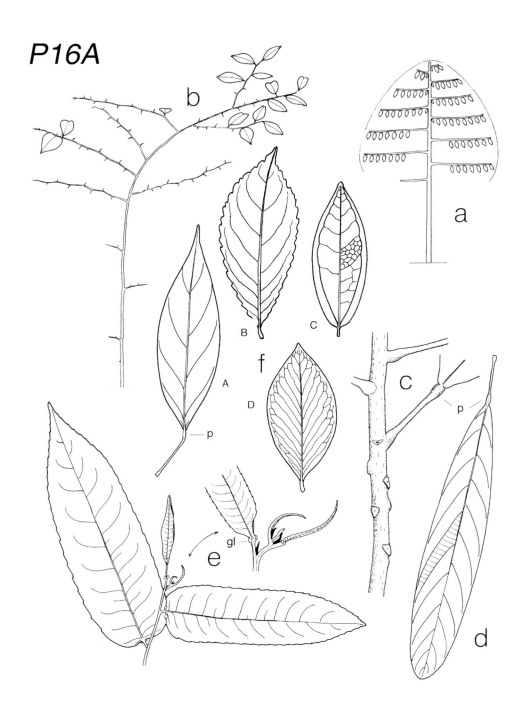

P16B

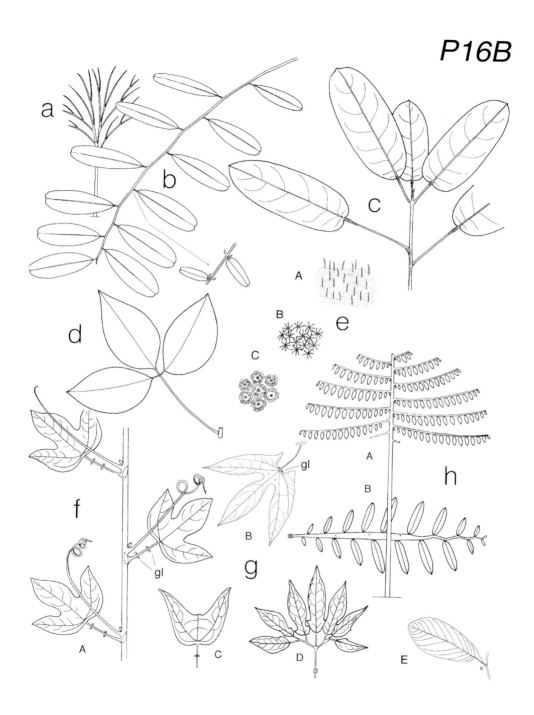

Legends to plates 17A/17B

17A Violaceae

a   A treelet with sympodial trunk and branches (immediate branching), the latter plagiotropic (frequent for the American *Rinorea*) or orthotropic (for some American *Rinorea* living in relatively sunny places. (Delimitation of sympodial units indicated by double brackets).

b   Sympodial branch with distichous phyllotaxy (e.g. *Rinorea* pp.).

c   Sympodial branch with opposite phyllotaxy (numerous American *Rinorea*).

d   Detail of twig apex of for figure c): b = axillary bud, c = cataphyll (scale-leaf), L = leaf, s = stipule; + = death of a meristem.

e   Different kind of venation, from brochidodromous (**A**. e.g. *Leonia cymosa*), to camptodromous (**B**. *Paypayrola*; **C**. *Leonia glycycarpa*).

f   Grooved petiole (gp), (e.g. *Amphirrhox, Leonia*).

g   *Leonia cymosa*: branches erect with sympodial growth and spiral phyllotaxy.

h   *Leonia glycycarpa*: branches more or less plagiotropic with sympodial growth and distichous phyllotaxy.

j   *Paypayrola*: growth monopodial and strongly rhythmic (RAUH's model). Leaves elongate and shortly petiolate.

k   TROLL's model: sympodial trunk consisting of the stacking of modules, each module firstly plagiotropic, secondarily becoming erect, most often after leaf-fall (e.g. Asiatic *Rinorea*).

m   Rhythmic growth of a branch of an Asiatic *Rinorea*, phyllotaxy is distichous and cataphylls (c) are inserted at the base of the units of extension. Branches are here monopodial. s = stipule; the double bar is placed between two adjacent UE.

17B Flacourtiaceae (*Pangium*), Bixaceae, Tetramelaceae and Caricaceae

a   *Pangium*: AUBREVILLE's model, small trees bearing cordate leaves.

b   *Bixa*: orthotropic, sympodial branches (flowering is terminal),

c   *Bixa*: long petiolate, palminerved leaves, orange coloured scales on the underside of the blade.

d   *Cochlospermum*: young lenticellate trunk, living bark fibrous with peripherical chlorophyllous layer (in black).

e   *Octomeles*: young twig with angular internodes. Leaves large and cordate. Detail: both sides of the blade bear peltate scales.

f   *Tetrameles*: indument of thick, hyaline hairs.

g   *Carica*: Unbranched treelet (CORNER's model). Leaves large, palmatilobate. Detail: bark fibrous. The African genus *Cylicomorpha* conforms to RAUH's model.

N.B. Except *Pangium* and several American *Rinorea*, all these families have delayed branching in trunk.

Legends to plates 18A/18B

18A Dipterocarpaceae and Elaeocarpaceae

a   ROUX's model: Monopodial trunk, continuous branching, branches plagiotropic and monopodial (typical for the Dipterocarpaceae, but MASSART's model for some species).
b   Stipule (s), protecting the apical bud and leaving an annular scar (*Shorea*). Other dipterocarpacean genera bear smaller, non-hood-like stipules.
c   Camptodromous and scalariform venation, petiole faintly pulvinate, able to twist or bend (*Shorea, Marquesia*). p = weakly marked pulvinus.
d   Brochidodromous venation (*Dipterocarpus*).
e   Leaf with numerous parallel secondary veins ending in an intramarginal vein (*Cotylelobium, Dryobalanops*). p = weakly marked pulvinus.
f   Base of lamina with a solitary gland (gl), (*Monotes, Marquesia*).
g   Groups of fasciculate hairs resembling stellate hairs (*Anisoptera, Shorea*).
h   RAUH's model with buttresses much developed for some species of the Elaeocarpaceae (*Elaeocarpus, Sloanea*).
j   AUBREVILLE's model: Monopodial trunk, rhythmic branching, branches plagiotropic by apposition growth (*Elaeocarpus*).
k   Leaf with pulvinate petiole and pinnate venation (*Sloanea*). s = stipule.
m   Leaf shortly petiolate, its petiole without a pulvinus (*Elaeocarpus*). **A**. developed leaf; **B**. involute young leaf.

18B Tiliaceae, Sterculiaceae, Bombacaceae and Malvaceae or Malvaceae *s.l.*, see [46].

A diagram view of the relations between architectural models and leaf-shape. The dotted lines represent the commonest associations of characters. Large dots represent terminal flowering. All woody species of this order possess a fibrous bark (fibres observable as a network).

Different shapes of leaves, clockwise, beginning with palmately compound:

a   Palmately compound (Bombacaceae: *Adansonia, Bombax, Ceiba*; Sterculiaceae: *Heritiera, Cola*, a few species of *Sterculia*).
b   Cordate, venation secant on the margin (Sterculiaceae: *Dombeya*; Tiliaceae: *Christiana, Diplodiscus*), but a non-secant venation is commoner for the cordate leaves of the Malvales (e.g. *Hibiscus, Neesia, Ochroma, Pentace, Pterygota, Pterocymbium, Scaphium*).
c   Lobate, venation secant (Sterculiaceae: *Dombeya, Firmiana, Triplochiton*; Tiliaceae: *Sparmannia*).
d   Asymmetric (Sterculiaceae: *Guazuma*; Tiliaceae: *Colona*).
e   Entire, oval or lanceolate, with scaly indumentum (Bombacaceae: *Coelostegia, Durio*).
f   Toothed, venation secant (Sterculiaceae: *Dombeya, Helicteres*).
g   Glands at the base of the lamina and between the main veins (Tiliaceae: *Apeiba* spp.).
h   Entire, more or less lanceolate (Bombacaceae: *Quararibea*; Sterculiaceae: *Cola, Heritiera, Sterculia*; Tiliaceae: *Grewia, Schoutenia*).

Abbreviations of the names of architectural models, clockwise:

A=Aubréville; **K**=Koriba; **Rh/Sc**=Rauh or Scarrone; **Ts**=Troll, with sympodial branches; **R**=Roux; **Tm**=Troll, with monopodial branches; **P**=Petit; **F**=Fagerlind; **M**=Massart.

P17A

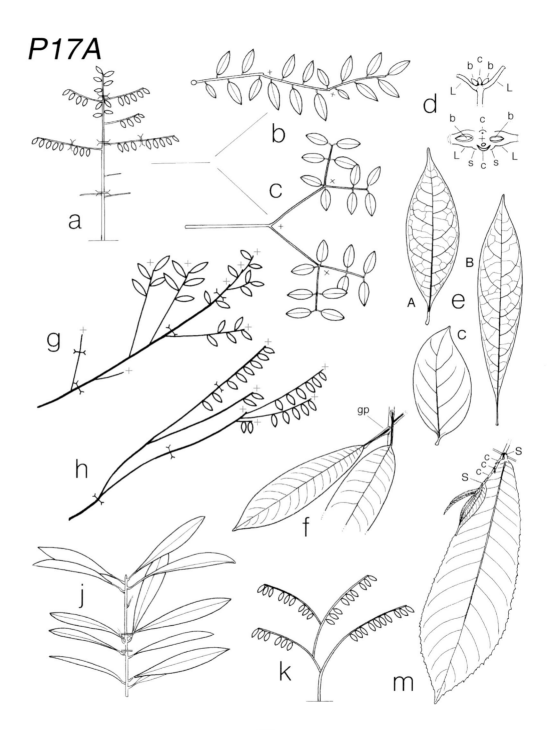

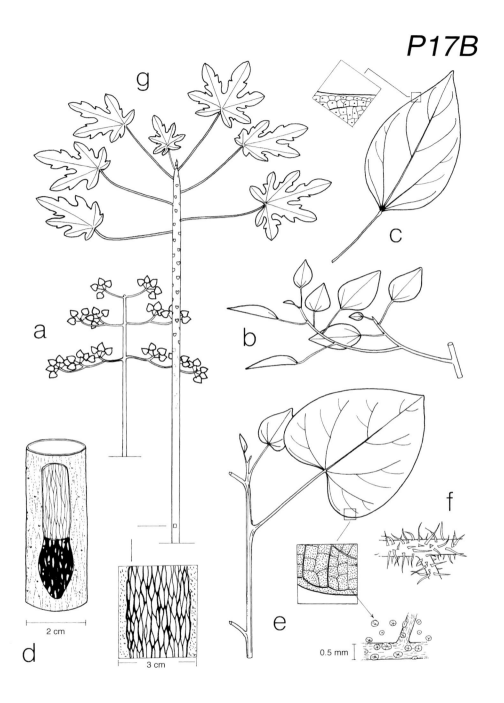

P17B

P18A

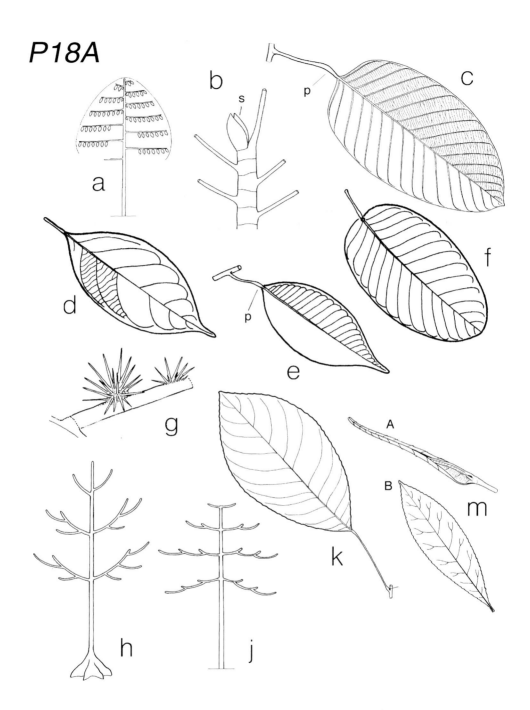

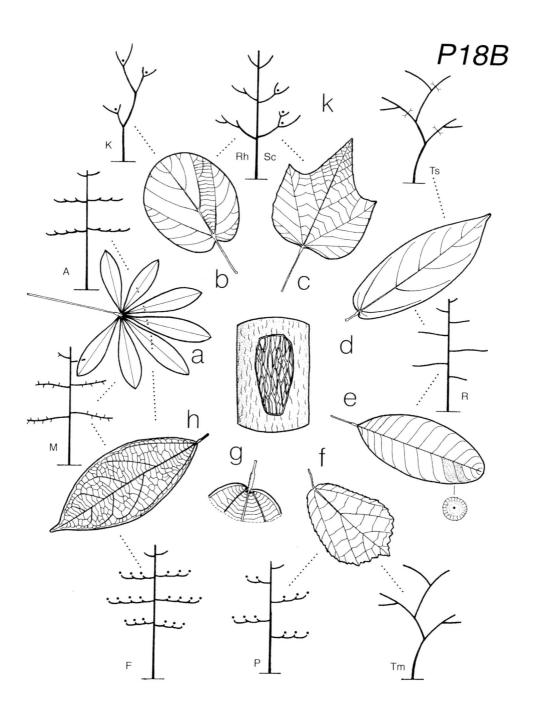

Legends to plates 19A/19B (Euphorbiaceae and Putranjivaceae)

19A mainly Phyllanthoideae

a Outer bark or periderm fissuring into thin longitudinal slits (a very typical character of the Euphorbiaceae, except in *Drypetes*, which should be placed in a family of its own: Putranjivaceae), a leaf-scar indicates approximately the size of these slits.

b COOK's or ROUX's model: Plagiotropic branches resembling compound leaves, trunk with continuous branching (e.g. *Phyllanthus*).

c Rhythmic growth for *Phyllanthus* sp. (delimitation of UE indicated by "//"). The phyllomorphic branches, bearing distichous leaves, are inserted spirally on an erect branch or trunk.

d *Uapaca*: Tree with stilt-roots.

e A twining liana bearing trilobate leaves, petiole not pulvinate (Acalyphoideae: *Dalechampia*).

f Tier of sylleptic plagiotropic branches, trunk sympodial (PREVOST's model), (Euphorbioideae: *Dichostemma*).

g Branch plagiotropic by apposition growth (*Baccaurea*, *Richeria*), bearing fruits (cauliflory).

h A plagiotropic and sympodial branch with flattened internodes (e.g. *Drypetes*, *Savia*). Death of apical meristems indicated by "+".

j A plagiotropic and sympodial branch, leaf with pinnate venation, its petiole pulvinate. Stipules are inconspicuous (e.g. *Antidesma*).

19B Acalyphoideae, Crotonoideae and Euphorbioideae

a A common habit of the Acalyphoideae / Crotonoideae: Pulvinate and long petiolate leaves, rhythmic growth noticeable due to varying size of leaves and internodes (e.g. *Alchornea*, *Ptychopyxis*, *Sagotia*).

b Palmately lobate leaves and LEEUWENBERG's model (e.g. *Jatropha*, *Manihot*).

c Rhythmic growth, leaves glandular and conduplicate when young (a typical character in the Euphorbiaceae), (e.g. *Elateriospermum*). Latex (white) is very common in the Euphorbioideae, frequent (white or coloured) in the Crotonoideae, but absent in the Acalyphoideae and the Phyllanthoideae. gl = gland.

d *Sapium*: apically glandular petiole. Leaves entire, inserted spirally on an orthotropic branch. gl = gland.

e Typical, not densely reticulate, venation of the Euphorbiaceae; leaf bearing minute teeth.

f Long petiolate and faintly serrulate leaf with camptodromous venation, the petiole without pulvinus (e.g. *Hura*, *Macaranga* spp.).

g Trifoliate leaf, latex abundant (e.g. *Hevea*).

h Trifoliate leaves and foliaceous stipules (a few species of *Macaranga*). st = stipule; l = young leaf.

Euphorbiaceae *s.l.* **(continued)**

Architectural models of the important genera of the Euphorbiaceae, Pandaceae and Putranjivaceae
Systematic treatment of the Euphorbiaceae according to Webster [42]:

CORNER (Acalyphoideae: *Agrostistachys*).
LEEUWENBERG (Acalyphoideae: *Acalypha, Alchornea, Mallotus*; Crotonoideae: *Croton, Jatropha, Manihot, Ptychopyxis*?).
KORIBA (Acalyphoideae: *Ptychopyxis*?; Crotonoideae: *Trigonostemon*?; Euphorbioideae: *Glycydendron, Hura, Hippomane, Stomatocalyx*?).
PREVOST (Phyllanthoideae: *Aporusa*; Euphorbioideae: *Excoecaria, Stomatocalyx*?).
NOZERAN (Euphorbioideae: *Anthostema, Dichostemma, Mabea*).
FAGERLIND (Euphorbioideae: *Senefeldera*).
PETIT (Crotonoideae: AF *Tetrorchidium*).
AUBREVILLE (Phyllanthoideae: *Baccaurea, Richeria*; Crotonoideae: *Endospermum*).
RAUH (Phyllanthoideae: *Hieronyma, Uapaca*; Acalyphoideae: *Claoxylon, Macaranga*; Crotonoideae: *Codiaeum, Croton, Elateriospermum, Glycydendron*?, *Hevea, Ricinodendron*?, AM *Tetrorchidium*?; Euphorbioideae: *Actinostemon, Sapium, Synadenium*).
SCARRONE (Crotonoideae: *Croton*).
MASSART (Phyllanthoideae: *Amanoa*).
ROUX (PUTRANJIVACEAE: *Drypetes*, Phyllanthoideae: *Phyllanthus*; PANDACEAE: *Microdesmis*).
COOK (Phyllanthoideae: *Phyllanthus*; PANDACEAE: *Panda*).
TROLL (Phyllanthoideae: *Antidesma, Aporusa, Bridelia, Cleistanthus, Drypetes*?, *Phyllanthus, Savia*; Acalyphoideae: *Chaetocarpus*; Crotonoideae: *Crotonogyne*?; Euphorbioideae: *Maprounea, Pedilanthus*).
CHAMPAGNAT (Phyllanthoideae: *Glochidion, Phyllanthus*; Acalyphoideae: opposite-leaved *Mallotus, Claoxylon*; Crotonoideae: *Croton*).

N.B. The polymorphic genus *Euphorbia* conforms to at least ten different architectural models [36], but they have in common spiral phyllotaxy on all their axes (except for *E. alcicornis*, a species bearing phyllodes [37]). **Twining lianas** in the Acalyphoideae: *Cnesmone*?, *Dalechampia, Omphalea, Plukenetia*; *Tragia* and in the Crotonoideae: *Manniophyton*).
Branches with distichous phyllotaxy are very common in the Phyllanthoideae, but atypical for the Acalyphoideae / Crotonoideae: *Chaetocarpus, Crotonogyne*?, *Crotonoides*?).
Latex is abundant in the Euphorbioideae and almost always white. Crotonoideae commonly have latex, or latex-like exudate, of different colorations, but latex can be opalescent (e.g. *Hura*). Acalyphoideae and Phyllanthoideae do not produce latex. Translucent dots are uncommon (e.g. *Mareya*: AF Acalyphoideae; *Grossera*: AF Crotonoideae).

P19A

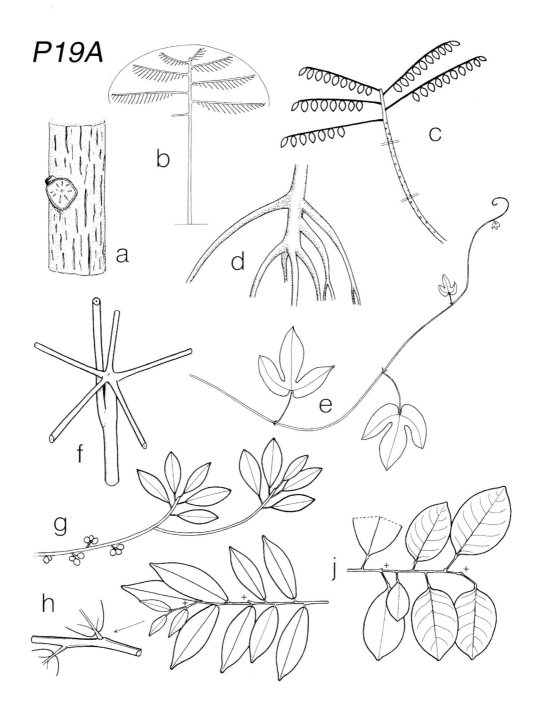

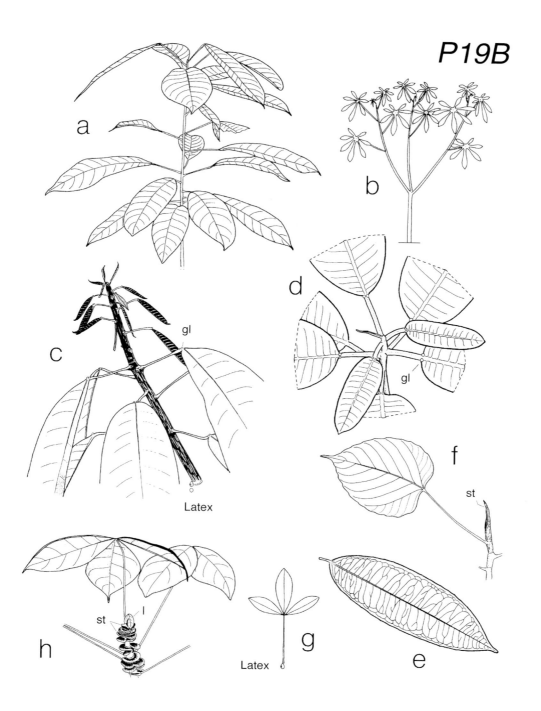

Legends to plates 20A/20B (Leguminosae)

Main tribes of woody Leguminosae in tropical forests; systematic treatment according to Polhill & Raven [44], and Herendeen & Brunneau [45], except for the Dimorphandreae and the Sclerolobieae which are kept as separate tribes:
Papilionoideae: Swartzieae, Sophoreae, Dalbergieae, Tephrosieae, Robinieae, Phaseoleae.
Caesalpinioideae: Caesalpinieae, Dimorphandreae, Sclerolobieae, Cassieae, Cercideae, Detarieae, Amherstieae.
Mimosoideae: Parkieae, Mimoseae, Acacieae, Ingeae.

20A Habit, bark, buds, shape of axes

a TROLL's model: Sympodial trunk consisting of the stacking of modules, each module firstly plagiotropic, secondarily becoming erect, most often after leaf-fall. Tree flat-topped. This model occures in the Caesalpinioideae (Amherstieae and Detarieae), the Mimosoideae (Ingeae: *Calliandra*, *Inga*; Acacieae: *Acacia*), and in a few Papilionoideae (Dalbergieae pp., Swartzieae: *Swartzia*).

b Tree not conforming to TROLL's model, because its branches are orthotropic (Mimosoideae: *Pentaclethra*, *Adenanthera*, *Pithecellobium*; Caesalpinioideae: Dimorphandreae, Sclerolobieae, Caesalpinieae; most Papilionoideae except a few genera, e.g. *Swartzia* and *Lonchocarpus* pp.).

c *Inocarpus*, *Pterocarpus* (Dalbergieae), trunk with large and flat buttresses. Buttressed trees also exist in other tribes (e.g. Swartzieae: *Swartzia*, Cassieae: *Koompassia*, Sclerolobieae: *Vouacapoua*; Detarieae: *Intsia*; Parkieae: *Parkia*).

d Red aqueous exudate flowing out of cut bark (Dalbergieae: *Inocarpus*, *Machaerium*; Phaseoleae: *Mucuna*).

e Spiny, thigmonastic twig of a climber, leaf bipinnate (Mimosoideae: *Acacia* spp.).

f Liana with flattened trunk (Cercideae: *Bauhinia*). Lenticels becoming transversely elongated (typical for the Leguminosae).

g Lianescent *Bauhinia*: tendril subtended by a scale leaf (sl).

h Leaf stipulate with its petiole pulvinate at the base (typical for the Leguminosae).

j A very atypical kind of leaf for the Leguminosae: simple. serrulate and without pulvinate petiole (*Zollernia*).

k **A**. Compound leaf with glands visible only in a very young stage (e.g. Caesalpinioideae: *Brownea*). **B**. detail: glands (gl) are disposed at the base of leaflets.

m **A**. and **B**. Buds forming a condensed branch system (typical for the Mimosoideae and some Papilionoideae, e.g. climbing *Dalbergia*). **C**. and **D**. Buds extended in a vertical series (**C**. *Swartzia*; **D**. Caesalpinieae: *Bussea*, *Caesalpinia*; *Pterolobium*; Sclerolobieae: *Campsiandra*, *Sclerolobium*).

n Rachis section. **A**. grooved (frequent, e.g. *Acacia*, *Pithecellobium*, *Tachigali*, etc.); **B**. raised (uncomon, e.g. *Dialium*, *Swartzia*).

p Young internodes spiny, 5-ribbed (Mimosoideae: *Piptadenia*), or 10-ribbed (Mimosoideae: *Acacia*).

N.B. Leguminosae have always delayed branching in trunk.

Leguminosae (continued)

20B Leaf

Leaf with rachis not ending in a terminal leaflet (ending in a mucro), leaflets opposite or pseudoterminal leaflet

a   Distichous phyllotaxy. Compound leaf with pairs of opposite leaflets and its rachis ending in a mucro (numerous Detarieae and Amherstieae).
b   Distichous phyllotaxy. Compound leaf with a pseudoterminal leaflet, i.e. the rachis ends in a mucro. Leaflets more or less alternate (many Detarieae and Amherstieae).
c   Bifoliolate leaf (Detarieae: *Cynometra*, *Peltogyne*; Cercideae: *Bauhinia*; Detarieae: *Gillietodendron*). Illustration: *Bauhinia*, leaflets with ascending basal secondary veins.
d   Paripinnate compound leaf (Cassieae: *Cassia*; Detarieae: *Leonardoxa*, *Macrolobium*, *Saraca*, *Sindora*).
e   Bipinnate leaf (Mimosoideae, Caesalpinieae), pinnae opposite, and leaflets opposite and asymmetrical (Parkieae: *Pentaclethra*; Ingeae: *Pithecellobium*).
f   Bipinnate leaf with pinnae alternate, but leaflets opposite; leaflets symmetrical (Dimorphandreae: *Burkea*). The five basal pinnae are schematically illustrated.

Leaf with a rachis ending in a terminal leaflet

g   Imparipinnate leaf with pairs of opposite leaflets, leaflets stipellate (many Tephrosieae and Sophoreae).
k   Compound leaf with a terminal leaflet, the other leaflets more or less alternate (Dalbergieae: *Pterocarpus*; Cassieae: *Dialium*).
h   Imparipinnate leaf, rachis winged (Swartzieae: *Swartzia*).
j   Trifoliolate and stipellate leaf (**A**. Phaseoleae: *Erythrina*, *Dioclea*, *Mucuna*). st = stipel. **B**. Detail of glands and stipels disposition (e.g. several species of *Erythrina*)

Raquis, glands and translucent dots

m   Compound leaf with its rachilla baally pulvinate (typical for the Leguminosae). Rachis with cupular glands (gl), **A**. e.g. *Pithecellobium*. **B** and **C**. Convex, more or less stipitate glands (Caesalpinioideae: *Senna*, *Chamaecrista*).
n   Base of rachilla stipellate (st), e.g. *Acacia*.

N.B. Translucent dots do exist in some Caesalpinioideae-Detarieae (*Copaifera*, *Hymenaea*) and Papilionoideae-Sophoreae (*Myrocarpus*, *Myrospermum*, *Myroxylon*).

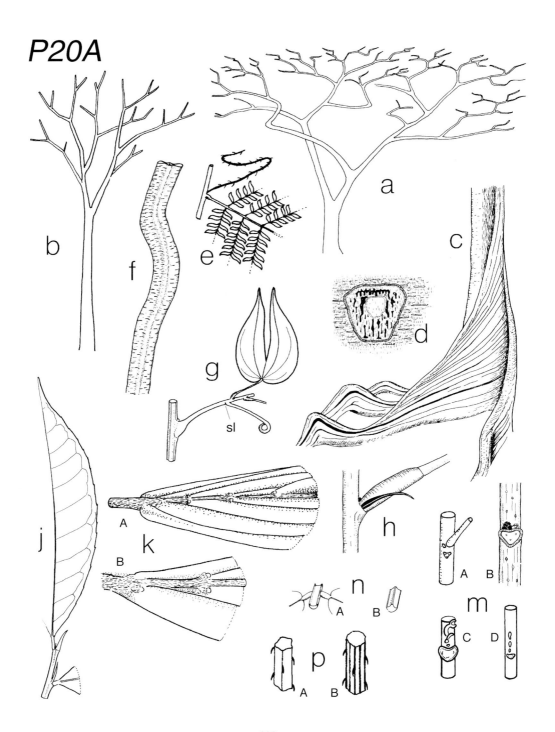

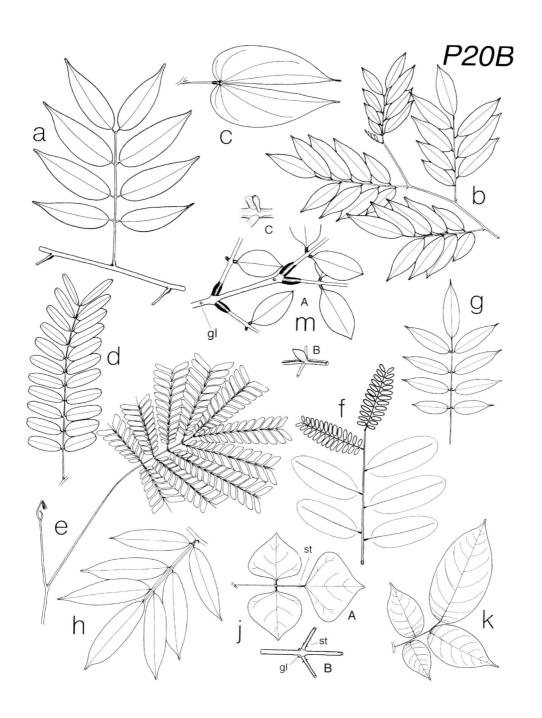

Legends to plates 21A/21B

21A Chrysobalanaceae and Rosaceae (*Prunus*)

a   Crown-shyness in *Parinari*: a situation where the crownlets cease to grow towards each other before they actually touch.
b   Lenticels becoming transversely elongated on trunk (also typical for *Prunus*, Rosaceae).
c   TROLL's model: Sympodial trunk consisting of the stacking of modules, each module firstly plagiotropic, secondarily becoming erect, most often after leaf-fall (e.g. *Chrysobalanus, Couepia, Hirtella, Licania, Parinari, Prunus* pp.). Death of apical meristems indicated by "+".
d   Monopodial and delayed ramification in a branch of *Prunus* (e.g. *P. javanica*). c = cataphylls.
e   Plagiotropic and sympodial branch with distichous phyllotaxy. Branching is delayed (most Chrysobalanaceae, and some tropical species of *Prunus*). sl = scars left by scale-leaves or shed leaves. + = death of an apical meristem.
f   Sympodial ramification in trunk apex. New modules bear numerous scale-leaves at base (e.g. *Couepia*).
g   Immediate ramification occurs very unfrequently in Chrysobalanaceae (e.g. *Chrysobalanus icaco*, a seashore srub). st = stipules, gl = glands, a, b = scaly prophylls.
h   Leaf detail of *Chrysobalanus icaco*; gl =2-4 gland disposed on the underside of the blade.
j   Intrapetiolar stipule (e.g. *Licania*), sometimes inserted on petiole in this genus.
k   Camptodromous (and scalariform) venation; base of blade (or petiole apex) bearing two glands (gl) (e.g. *Parinari*).
m   Leaf of *Licania, Prunus*; glands (gl) are here disposed at the very base of the blade.
n   Glandular stipules (a few species of *Hirtella*).

21B Oxalidaceae and Connaraceae

a   *Averrhoa bilimbi*: KORIBA's model, i.e. trunk and branches modular, modules initially equal, all apparently branches, but later unequal, one becoming a trunk.
b   *Averrhoa carambola*: TROLL's model (see above for definition).
c   *Sarcotheca*: Leaves trifoliolate, leaflets with pulvinate petiolules and two basal ascending secondary veins. Phyllotaxy is distichous in plagiotropic stems (TROLL's model). Death of apical meristems indicated by "+".
d   CHAMBERLAIN's model: Monocaulous tree with a modular and orthotropic trunk (an as yet non-lianescent young plant, e.g. *Connarus, Roureopsis*).
e   Sympodial trunk of a young plant showing transition from simple juvenile leaves to trifoliolate adult leaves (e.g. *Agelaea*). Leaflets with two ascending main secondary veins (typical for some species of the Connaraceae).
f   Petiole with pulvinate base allowing considerable displacement of the leaf (e.g. *Roureopsis*). sc = scale-leaf; l = leaf.
g   Liana with sympodial trunk, apical part of its modules twining (typical for the Connaraceae).

N.B. Delayed branching in trunk for these families (except *Chrysobalanus, Prunus* pp.).

Legends to plate 22A/22B

22A Celastraceae (incl. Hippocrateoideae) and Irvingiaceae

a Hippocrateoideae: Arborescent form, trunk and branches sympodial, phyllotaxy opposite (e.g. *Salacia*).
b Hippocrateoideae: Lianescent form, branches prehensile, phyllotaxy opposite (e.g. *Salacia*).
c Leaves opposite, minutely stipulate, showing laticiferous threads when torn (various Celastraceae); st = stipule.
d Branch orthotropic, spiral phyllotaxy, leaves faintly trinerved (tri), (Celastroideae: *Bhesa*).
e Leaf with scalariform venation and petiole distally pulvinate (*Bhesa*); p = pulvinus.
f *Maytenus*: An American treelet with sympodial architecture (death of apical meristems indicated by "+"), its branches plagiotropic with alternate and distichous phyllotaxy (the branches of the Asiatic *Maytenus* maintain a spiral phyllotaxy). Branching is delayed.
g Young leaf folding involute (typical for the Celastraceae).
h Irvingiaceae: Branches with distichous phyllotaxy (TROLL's model?) and entire leaves. Detail: stipules hood-like, narrow and elongated (e.g. *Desbordesia*, *Irvingia*).

22B Humiriaceae, Linaceae, Ixonanthaceae and Erythroxylaceae

a *Humiria*: Branch orthotropic, rhythmic growth, phyllotaxy spiral (RAUH's model, also for Linaceae, e.g. *Hebepetalum*, *Roucheria*).
b MANGENOT's model (e.g. *Vantanea*, American *Sacoglottis*?, but TROLL's model for *Sacoglottis gabonensis*, the unique African Humiriaceae).
c Typical leaf of the Humiriaceae: the petiole is short and enlarged at its base. Internodes are angular.
d Young leaf folding involute (typical for Humiriaceae, Linaceae, Ixonanthaceae, Irvingiaceae and Erythroxylaceae).
e *Hugonia*: Liana climbing by means of hooks. Phyllotaxy mainly spiral or becoming opposite in small twigs. Detail: leaf serrulate with camptodromous venation.
f Shoot apex showing small stipules (s) and involute leaf folding (typical for the Linaceae, e.g. *Hebepetalum*).
g *Ixonanthes*: Rhythmic growth marked by well-developed scale-leaves (sl). Branch phyllotaxy is spiral (RAUH's model?).
h *Erythroxylum*: Rhythmic growth marked by scale-leaves (MASSART's or MANGENOT's model); sl = scale-leaf; ss = scaly stipule.
j A short twig with distichous phyllotaxy and intrapetiolar, scaly stipules (ss), (*Erythroxylum*).

N.B. All these families have delayed branching in trunk.

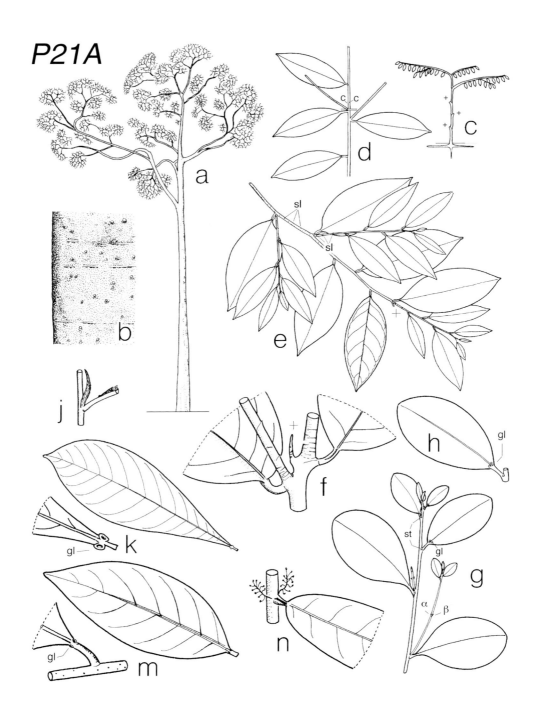

P21B

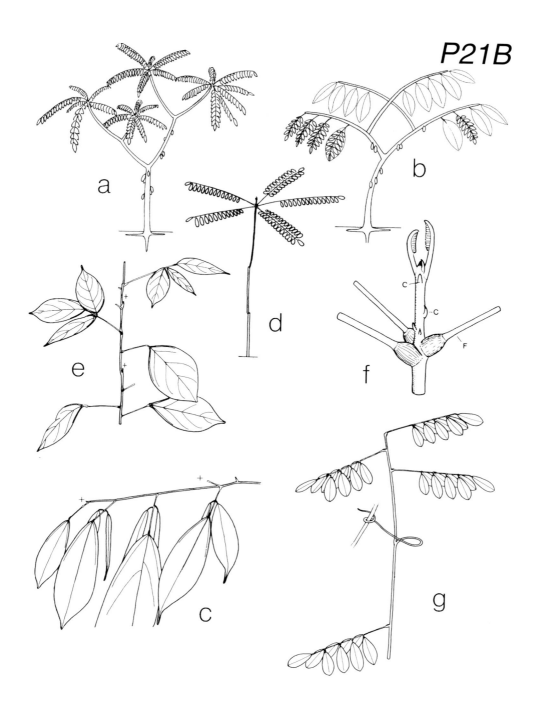

P22A

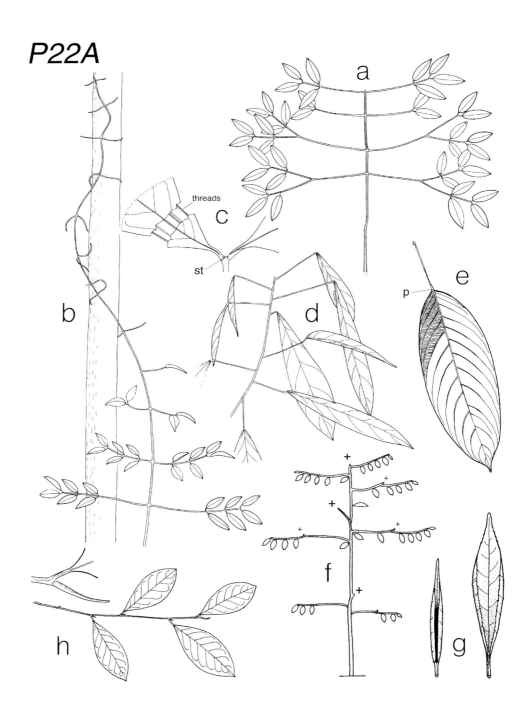

P22B

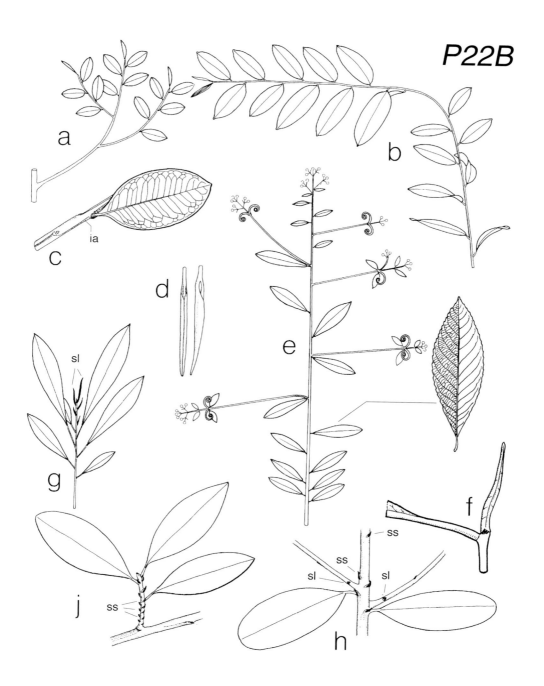

Legends to plate 23A/23B

23A Sapindaceae

a Tree with sympodial architecture (frequent for Sapindaceae).
b CHAMBERLAIN's model (e.g. *Talisia*).
c Young internodes grooved (typical for Sapindaceae).
d Twig apex angular (typical for Sapindaceae).
e Extremity of a sympodial trunk, the older (and basal) module has its periderm and leaf rachis already suberised (heavily stippled), the grooved internodes are noticeable on the upper young module (e.g. *Matayba, Toulicia*).
f Pinnae of a twice compound leaf, leaflets are serrulate (e.g. *Dilodendron, Macphersonia*).
g Leaflet with camptodromous venation (e.g. *Nephelium*).
h Leaf compound, rachis not ending in a leaflet (ending in a mucro), leaflets more or less opposite, (e.g. *Blighia, Deinbollia, Nephelium, Xerospermum*, etc.), if leaflets more or less alternate (e.g. *Cupania, Schleichera, Toechima, Toulicia, Zollingeria*, etc.).
j Leaf with its rachis winged and alternate leaflets (e.g. *Filicium*).
k Leaf of a liana bearing an axillary, tightly coiled (or circinate) tendril, its leaflets serrulate (*Paullinia, Serjania*: leaf twice compound).
m Basal leaflets stipule-like and leaflets serrulate (e.g. *Lepisanthes, Pometia*).
n Trifoliolate leaf with winged petiole (e.g. *Allophylus*).

23B Meliaceae and Simaroubaceae

a Meliaceae-Swietenioideae: an erect branch with marked rhythmic growth (Swietenioideae: e.g. *Toona*, but also in some species of *Trichilia*). Double bars: UE. Leaves paripinnate (e.g. *Carapa, Cedrela, Toona*). Trunk and branches of the Trichilioideae do exhibit generally a weak rhythmic growth (e.g. *Aglaia, Lansium, Trichilia*).
b Camptodromous venation (e.g. *Toona, Swietenia*). Simple leaves in *Turraea, Vavaea*.
c Meliaceae-Swietenioideae: Apex of an erect branch or trunk with coriaceous scale-leaves (rhythmic growth). **A**. Top; **B**. Side (e.g. *Carapa, Entandophragma, Swietenia*).
d Shield-like petiolar scar (typical for Meliaceae).
e Densely lenticellate branch (e.g. *Dysoxylum*).
f Apical part of a rachis showing a bud of leaflets, leaf growth remains active for a very long time (*Chisocheton, Guarea*).
g A very young leaf of *Carapa procera*, its leaflet apices glandular (gl).
h Leaflets more or less alternate, rachis ending in a terminal leaflet (e.g. *Aglaia, Lansium, Trichilia* pp.,), or (not illustrated) rachis ending in a mucro (e.g. *Chukrasia, Dysoxylum*)..
j Leaflets opposite, rachis ending in a terminal leaflet (e.g. *Aglaia, Aphanamixis, Cipadessa, Dysoxylum, Sandoricum, Trichilia* pp., *Walsura*). Petiolules are distally enlarged in *Walsura*.
k *Simaba*: evenly disposed compound L., scaly rhytidome. m. axillary bud (bax) shifted.
n **A**. Venation poorly reticulate (e.g. *Eurycoma, Odyendea, Simarouba amara*). **B**. Venation distinctly reticulate (e.g. *Picrasma, Simaba, Quassia*).
p *Quassia amara*: trifoliolate leaf with winged petiole.
q *Samadera indica*, a very atypical Simaroubaceae: rhythmic growth with cataphylls (c), distichous phyllotaxy, petioles distally enlarged (p) and blade possessing translucent dots.

Legends to plate 24A/24B

24A Anacardiaceae and Burseraceae

a   Viscous latex flowing out of cut bark (e.g. *Gluta, Rhus*).
b   Tree with sympodial architecture and leaves grouped at the end of the stems (e.g. *Dracontomelon*). Imparipinnate leaves are typical for the Anacardiaceae.
c   Typical venation of the Mangifereae (e.g. *Anacardium, Mangifera*), with the secondary veins bending abruptly near the margin.
d   *Bouea*: Rhythmic growth and opposite phyllotaxy! c = cataphylls.
e   Venation secant on the margin, leaves simple: *Ozoroa*, leaves compound: *Euroschinus*.
f   *Spondias*: leaflets bear a submarginal vein.
g   Stout trunk apex and petioles convexe in section are typical for the Anacardiaceae.
h   Auriculate leaf in a young specimen of *Campnosperma*.
j   *Canarium*: an individual of the Burseraceae with stilt buttresses.
k   Rhythmic growth of an erect stem indicated by short internodes (e.g. *Bursera, Trattinickia*). A double-bar, is placed between two adjacent units of extension. Petiolar base are grooved (e.g. *Aucoumea, Canarium, Protium, Santiria*). cs = cataphyll scar.
m   Compound leaf with pulvinate petiolules (e.g. *Canarium, Aucoumea, Protium, Santiria, Trattinickia*). Imparipinnate leaves are typical for the Burseraceae.
n   Compound leaf with lateral petiolules not pulvinate at apex (e.g *Dacryodes, Pachylobus*), however the terminal leaflet can be pulvinate in Dacryodes.
p   *Bursera*: a specimen with serrulate leaflets and winged rachis.
q   *Bursera*: peeling rhytidome (prh).

24B Rutaceae

a   Trunk with rhythmic growth, leaves palmately compound (e.g. *Vepris*), double bars indicate UE delimitation.
b   Hapaxanthic treelet (HOLTTUM's model), large compoun leaves, e.g. *Spathelia*.
c   Shrub conforming to LEEUWENBERG's model, leaves trifoliolate: *Moniera*. The simple leaves disposed terminally are in fact bracts.
d   Spiny trunk (e.g. *Fagara*, possibly RAUH's model).
e   Common leaf form in Rutaceae: compound with leaflets more or less alternate (e.g. *Clausena, Murraya*).
f   Spines axillary, disposed in pair, rachis winged (e.g. *Triphasia, Fagara*).
g   Branch plagiotropic by apposition (AUBREVILLE's model), leaves long persisting with pulvinate (p) petiole: *Esenbeckia*.
h   *Esenbeckia*: detail of twig apex, petioles are suberised at both ends, p = pulvinus.
j   Leaves crenelate with winged petiole and articulation (unifoliolate leaf), short shoots modified into spines, (e.g. *Citrus*). sp = spine.
k   Opposite-decussate phyllotaxy, immediate ramification (leaves simple or compound), ATTIMS's model: *Evodia*.
m   Opposite-decussate leaves, delayed ramification, p = pulvinus (e.g. *Tetractomia*).

N.B. Trees of the Sapindales have almost always delayed branching in trunk (an exception: *Evodia*).

P23A

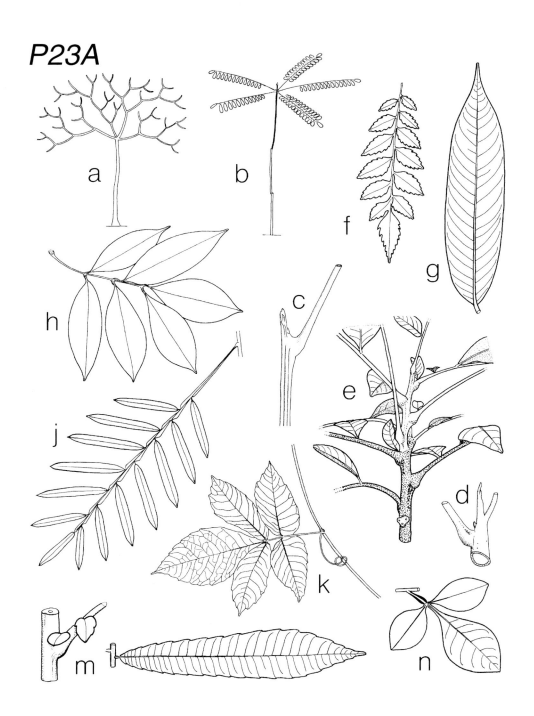

P23B

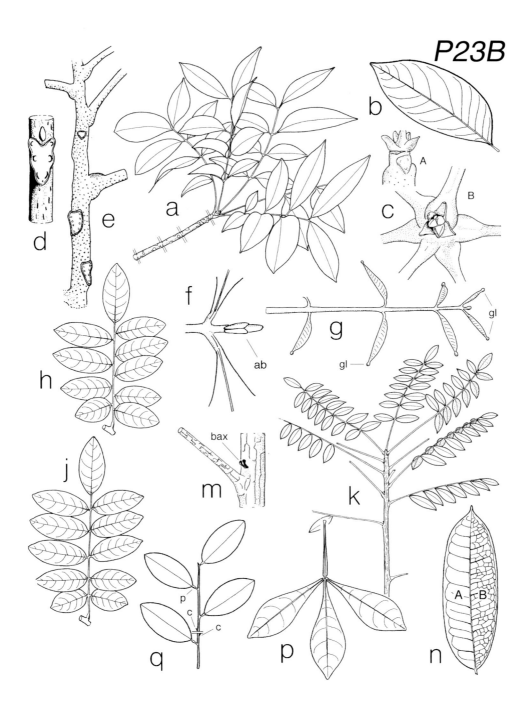

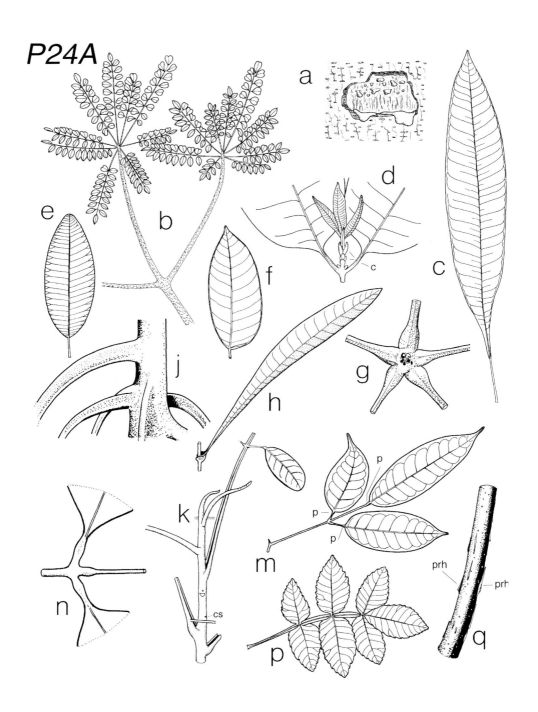

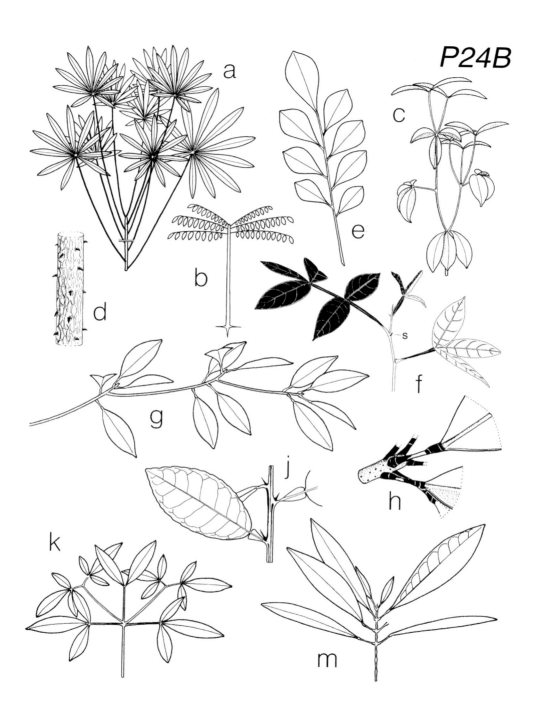

Legends to plates 25A/25B

25A Thymelaeaceae

a   Fibrous inner bark rendering the twigs difficult to break (a general trait of the Thymeleaceae).
b   Orthotropic, sympodial trunk and branches with opposite phyllotaxy (KORIBA's model), (e.g. *Phaleria*). Detail: leaf entire, venation brochidodromous. The trunk bears inflorescences (infl).
c   MANGENOT's model: Trunk formed by the stacking of modules, each module firstly orthotropic, then becoming abruptly plagiotropic (Thymelaeoideae: *Dicranolepis*). Detail: secondary veins ending in a fimbrial vein. Phyllotaxy is truely distichous in branches.
d   Monopodial, rhythmic growth and branching (MASSART's or RAUH's model); leaves pellucid-dotted (Gonystyloideae, e.g. *Gonystylus*). Detail: leaf entire and pellucid-dotted, brochidodromous venation forming an intramarginal vein.
e   LEEUWENBERG's model in *Lasiosiphon*, inflorescences (infl) are axillary.
f   ATTIM's model in *Dais* (continuous branching in trunk), phyllotaxy is opposite.
g   A peculiar variant of a LEEUWENBERG's model in *Edgeworthia*, where branching is immedaite, but occurs somewhat above the initiation of the UE.
h   MASSART's model in *Daphnopsis*, phyllotaxy is pseudodistichous in plagiotropic branches (Double bars indicate adjacent UE). Detail: camptodromous venation of a leaf.
j   *Enkleia*: a liana with opposite phyllotaxy and small prehensile twigs bearing reduced leaves.
k   Apical bud covered with appressed hairs (a general trait of the Thymelaeaceae, however absent in *Phaleria*).

25B Dichapetalaceae and Malpighiaceae

a   *Dichapetalum*: Section of trunk showing interxylary phloem (in black).
b   Lenticels in longitudinal rows (*Dichapetalum*).
c   ROUX's or MANGENOT's model, i.e. trunk monopodial or, if sympodial, the sympodial units firstly orthotropic, than becoming abruptly plagiotropic; branching more or less continuous. Branches sympodial (but not modular), with distichous phyllotaxy (*Dichapetalum*).
d   *Tapura*: inflorescence with its peduncle adnate to the petiole.
e   Variation in stipular shape for the Dichapetalaceae: from large and divided to narrow.
f   Twining liana with glandular leaves (typical for the lianescent Malpighiaceae).
g   Cordate leaves with two basal glands (e.g. *Stigmaphyllon*).
h   Branches somewhat modular or plagiotropic by apposition growth (FAGERLIND's or AUBREVILLE's model, e.g. *Byrsonima*), gl = gland.
j   Intrapetiolar stipules (e.g. *Byrsonima*).
k   Leaves lanceolate with long appressed hairs on the underside (e.g. *Malpighia*).
m   *Spachea*: margin of leaf glandulate on the underside.

N.B. Thymelaeaceae and Dichapetalaceae have in general immediate branching in trunk,.

Legends to plates 26A/26B

26A Polygalaceae and Xanthophyllaceae

a   A liana with prehensile twigs (e.g., *Securidaca*). Detail: gland (gl) in stipular position.
b   *Securidaca*: plagiotropic branches with distichous phyllotaxy.
c   *Bredemeyera*: imbricate aesivation of leaves (also in *Securidaca*).
d   Twining liana, Leaf with glands situated on very thin secondary veins, veins of higher order not visible (*Moutabea*).
e   Scandent shrub with, branches thigmonastic, spiral phyllotaxy (*Bredemeyera*).
f   *Diclidanthera*: erect branches with short twigs bearing terminal inflorescence. Phyllotaxy is spiral.
g   *Xanthophyllum*: trunk rhytidome irregularly embossed.
h   *Xanthophyllum*: Extremity of a plagiotropic and sympodial branch, phyllotaxy is distichous (ROUX's or MANGENOT's model?). Sympodial construction noticeable by the presence of dead apical meristems (+) and scale-leaves (c) at the base of the sympodial unit. Leaves entire, not stipulate, glands situated at the junction of the veins (twig seen from the underside).

26B Brunelliaceae, Cunoniaceae and Rhizophoraceae

a   *Brunellia*: Comound leaves visible in apical bud, interpetiolar ridge (ir), and grooved stem.
b   ATTIMS's model: Trunk monopodial, continuous branching, branches orthotropic and monopodial (e.g. *Pancheria*, *Weinmannia*, this last genus also conforming to RAUH's model).
c   Leaves in whorls of three (e.g. *Pancheria*). Interpetiolar stipules leaving large scars (sc) after leaf-fall.
d   *Weinmannia*: Leaves compound and opposite, leaflets serrulate. s = interpetiolar stipules.
e   *Geisssois*: Trifoliolate leaves and large **intra**petiolar stipules (i st).
f   In terrestrial forests: ROUX's model (*Cassipourea*, *Pellacalyx*), more rarely AUBREVILLE's model (e.g. *Sterigmapetalum*).
g   Bijugate phyllotaxy in a erect stem: leaves are not decussate, i.e. the succesive pairs, (1, 2, 3, 4 ...), are less than 90° apart (e.g. *Cassipourea*, *Rhizophora*).
h   Interpetiolar stipules (s) forming a hood (e.g. *Ceriops*, *Rhizophora*).
j   Mangrove tree with stilt-roots (*Rhizophora*).
k   In mangroves: AUBREVILLE's model, (indicated by leaves grouped at the end of the modules), (*Bruguiera*, *Ceriops*) or ATTIM's model (*Rhizophora*).

N.B. Polygalaceae, Xanthophyllaceae, and Brunelliaceae have delayed branching in trunk, Malpighiaceae are variable in this respect. Cunoniaceae and Rhizophoraceae have in general immediate branching in trunk.

P25A

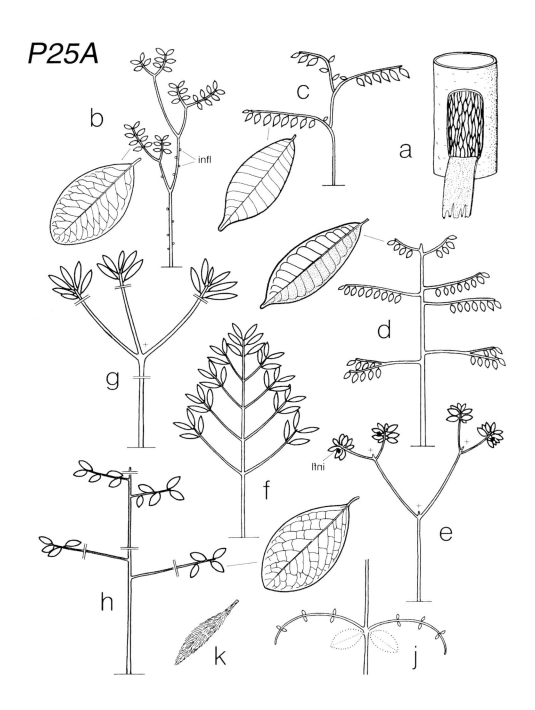

P25B

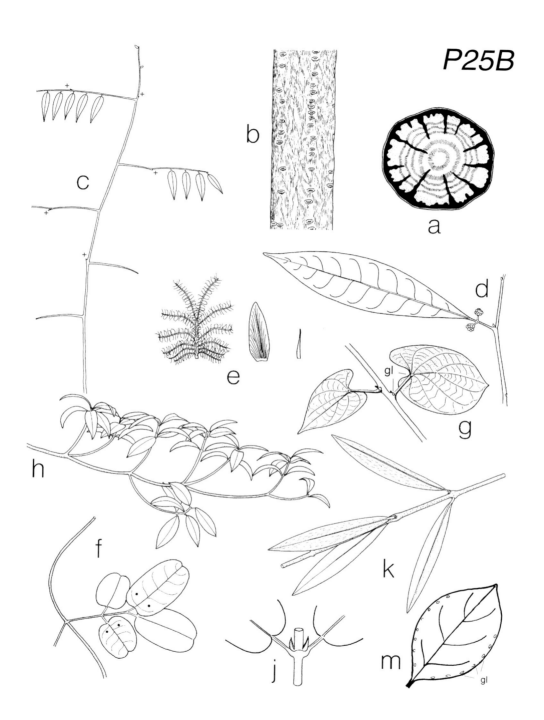

249

P26A

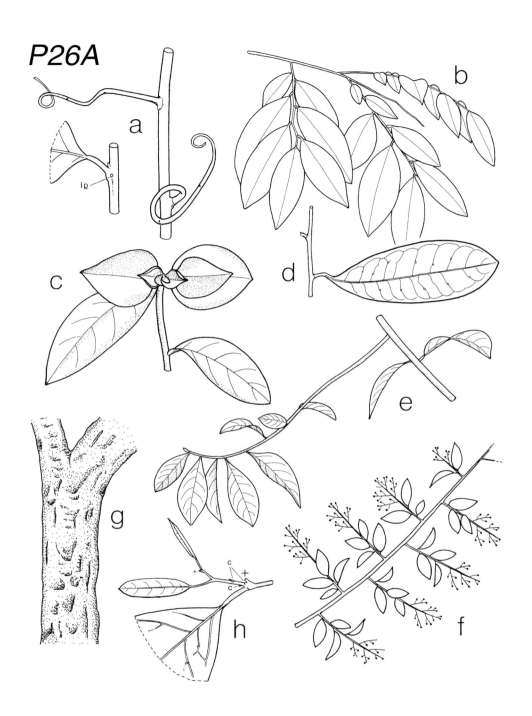

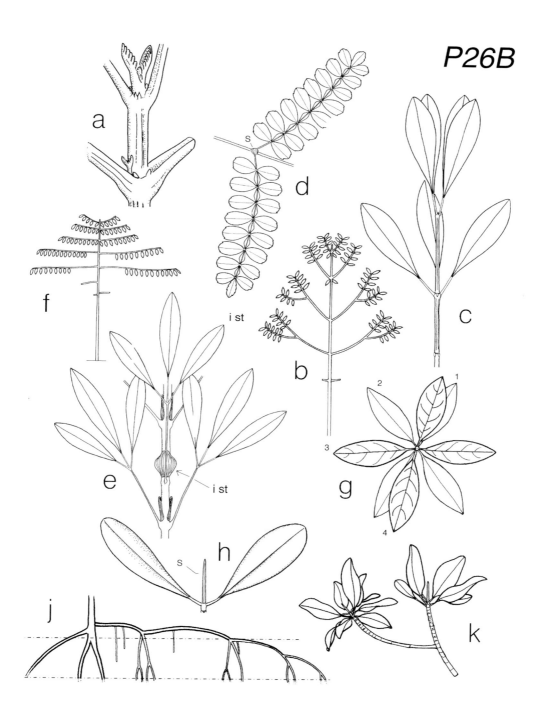

Legends to plates 27A/27B

27A Pittosporaceae, Aquifoliaceae and Araliaceae

a *Pittosporum*: Leaves entire, grouped at the apex of the UE. Leaves acuminate (ac) for some species.
b *Pittosporum* species have a modular architecture (KORIBA's model). Modules consist of one or several UE, depending of the species. Leaves always grouped at distal end of UE.
c *Ilex*: RAUH's or MASSART's model, i.e. monopodial trunk, rhythmic branching, branches monopodial, orthotropic or plagiotropic.
d Old units of extension of branch with shed leaves (somewhat typical for *Ilex*), phyllotaxy is spiral, leaves are shortly petiolate.
e *Ilex*: Entire leaf, coriaceous and with its margin slightly revolute (mr), (quite common for the species growing in open places).
f *Ilex*: Serrulate leaf, its secondary and tertiary veins visible, but veins of higher order not visible (typical).
g Araliaceae usually do conform to LEEUWENBERG's model: Stems orthotropic and sympodial, not distinctly differentiated into trunk and branches (all stems equivalent), (*Cussonia*, *Polyscias*, *Schefflera*, or *Eryngium* in Apiaceae).
h CHAMBERLAIN's model: Monocaulous tree with a modular and orthotropic trunk (*Brassaiopsis*, *Trevesia*).
j Peculiar leaf forms: **A**. palmately lobate with pinnatifid divisions (e.g., *Trevesia*, juvenile leaf). **B**. compound and pinnate (e.g., *Aralia*).
k Modular branches and very rythmic growth, leaves entire (*Dendropanax*).
l *Hedera*: Adventicious roots (ar), thin rhytidome with longitudinal slits (ls), leaf lobate.
m Different forms of stipular expansions in the Araliaceae. **A**. fused; **B**. and **C**. free.

27B Cornaceae, Alangiaceae (*Alangium*) and Anisophylleaceae (*Anisophyllea*)

a *Cornus*: opposite and decussate phyllotaxy, petiolar bases decurrent to the stem and somewhat jointed.
b Leaf with camptodromous venation (*Cornus*).
c *Mastixia*: spiral phyllotaxy in an orthotropic stem. Flowering is terminal.
d *Alangium*: Rhythmic ramification in a monopodial and plagiotropic branch (MASSART's model, not all leaves are represented).
e *Anisophyllea*: Asymmetric trinerved leaf, venation camptodromous and scalariform.
f MASSART's model: Monopodial trunk, rhythmic branching, monopodial and plagiotropic branches (*Anisophyllea*).
g Extreme dorsiventrality in a plagiotropic branch expressed by strong anisophylly: the smaller leaves are inserted on the upper surface, the larger leaves are asymmetrical and inserted in a distichous position.

N.B. Pittosporaceae, Aquifoliaceae and Araliaceae have in general delayed branching in trunk. Alangiaceae and Anisophylleaceae have in general immediate branching in trunk Cornaceae are variable in this respect

Legends to plates 28A/28B

28A Icacinaceae

a   Plagiotropic and monopodial branch, branching continuous (ROUX's model, e.g. *Dendrobangia, Discophora, Medusanthera*).
b   Variant of PETIT's model: Trunk sympodial (monopodial in the definition of the model sensu stricto), continuous branching, branches plagiotropic and modular (e.g. *Gomphandra, Leptaulus*). Death of apical meristems indicated by "+".
c   Surnumerary buds (*Discophora*). A1: main stem; A2: first order branch somewhat displaced from leaf axil; A3: second order branch.
d   Twining liana with cordate leaves (*Phytocrene*).
e   MANGENOT's model: Sympodial trunk consisting of the stacking of modules, each module firstly orthotropic, then becoming abruptly plagiotropic (e.g. *Rhyticaryum*).
f   *Iodes*: Sympodial trunk, the sympodial units twining only at their extremity. Leaves are opposite. One leaf subtends an axillary bud while the new module of the trunk and an inflorescence are subtended by the other leaf.
g   *Emmotum*: Fimbrial vein, and quaternary veins more or less parallel to the tertiary.
h   Sclerenchymatous yellow inclusions in bark (e.g. *Gonocaryum, Iodes*).
j   *Citronella brassii*: trunk monopodial (**A**) or sympodial (**B**) on different individuals of a same species. Leaves pseudowhorled with branches displaced from their axil.

28B Asteraceae

a   *Mikania*, a twining genus. **A**. External wood furrowed (as for many other Asteraceae). **B**. Opposite phyllotaxy, leaves long petiolate. **C**. Leaf bases somewhat enlarged.
b   Sympodial and modular architecture in *Vernonia*. Sessile capitula (cap).
c   Leaves bearing glandular trichomes (*Vernonia*).
d   *Baccharis* (Senecioneae), distinct decurrent ridge in stem (also in other Senecioneae).
e   LEEUWENBERG's model (e.g. *Centauropsis*, Vernonieae).
f   SCARRONE-like model of many Eupatorieae and Heliantheae: main stem is at first monopodial, but bears thereafter a terminal inflorescence.
g   A climbing species of *Dasyphyllum* (Barnadesieae). Stem spiny. Leaf coriaceous, its pair of secondary veins run parallel to the margin. External wood furrowed.
h   *Gochnatia*: Leaf and bud estivation, internode with protruding decurrent ridges (dr).
j   Mutisieae: **A**. Young rhitidome exhibiting longitudinal slits (Mutisieae: *Stifftia*, *Gochnatia*). **B**. *Stifftia*: Leaf with a hyaline margin (hy m).
k   Sessile capitula in *Vernonia, Piptocarpha* and other Vernonieae.
m   *Baccharis platypoda*: leaves coriaceous with numerous oblicuous secondary veins.
n   Scabrous leaf (Heliantheae: *Verbesina*). **A**. Leaf. **B**. Long and rigidous hairs on underside of blade. **C** Hispidous main vein.
p   Deeply pinnatifid leaf (Heliantheae: *Dahlia*). (Leaves can be compound in Heliantheae).
q   Two basal ascending secondary veins in *Baccharis* sp.
r   *Porophyllum* (Helenieae): leaf bearing large marginal translucent pouches (tp).

N.B. Icacinaceae have in general immediate branching in trunk. Woody Asteraceae have delayed branching in trunk and their branches are never truely plagiotropic.

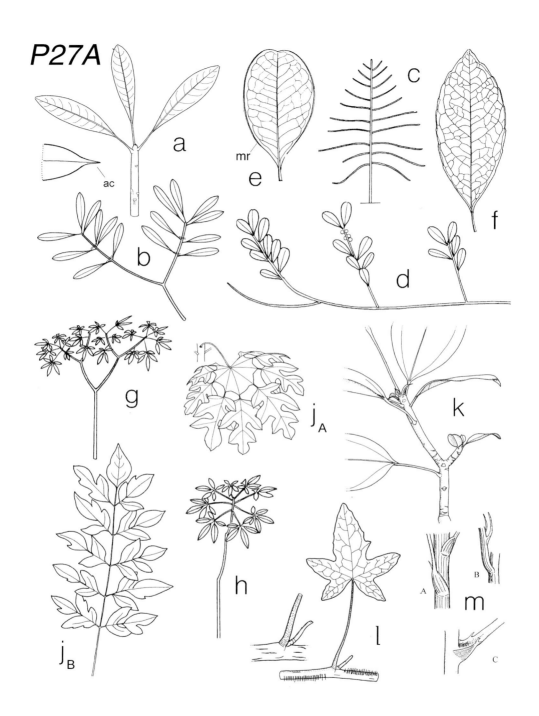

P27B

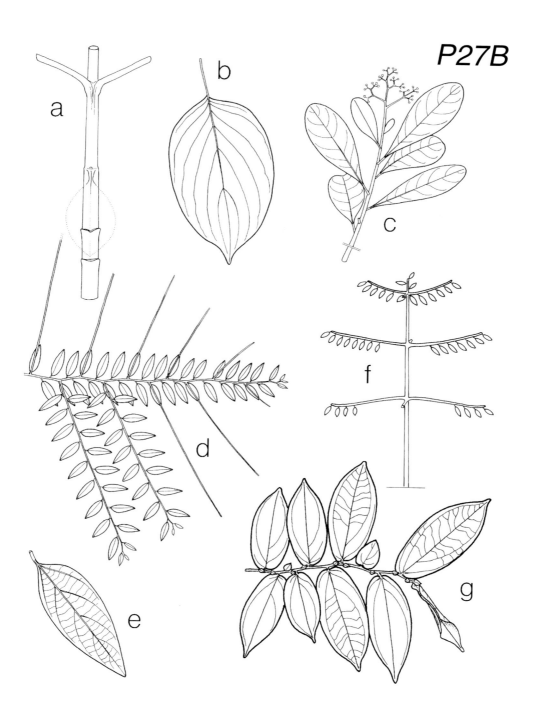

P28A

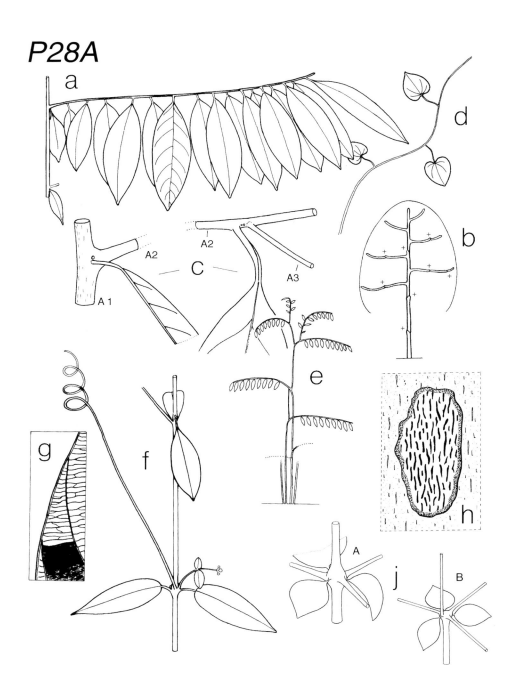

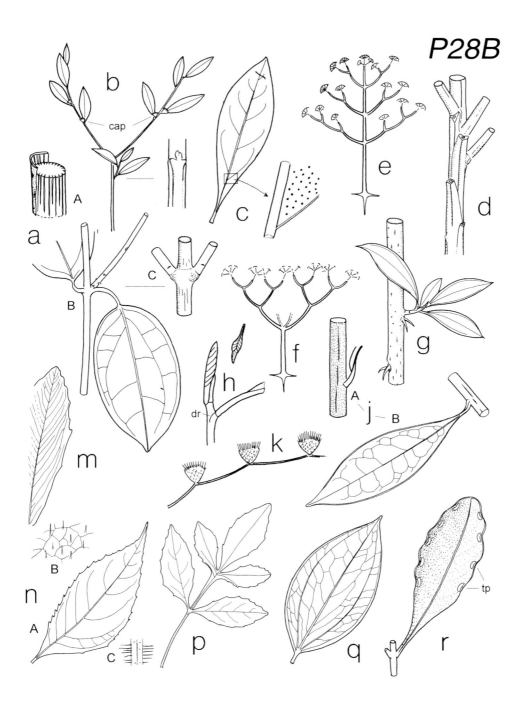

Legends to plates 29A/29B

29A Convolvulaceae and Boraginaceae

a   Liana, stem twining for all its length. Leaves simple, entire with pinnate venation (e.g. *Erycibe, Dicranostyles, Neuropeltis*).
b   Young stem with broad pith cylinder, stem easy to squeeze (typical for the Convolvulaceae).
c   Cordate leaf with palmate venation (e.g. *Bonamia, Ipomoea*). Latex is typical for the herbaceous species of the Convolvulaceae, but rather uncommon in their woody species.
d   Cordate leaf with pinnate venation, the basal secondary veins grouped (e.g. *Ipomoea*).
e   Various kinds of hairs: **A**. erect and simple (e.g. *Bonamia* pp., *Erycibe*); **B**. stellate (e.g. Jacquemontia); **C**. simple and appressed (e.g. *Bonamia* pp.); **D**. sessile and spherical (e.g. *Maripa*).
f   PREVOST's model: Sympodial, modular trunk and branches, branches plagiotropic and sympodial (frequent for the American *Cordia*). **A**. side view; **B**. top view.
g   CHAMPAGNAT's model: Trunk and branches firstly orthotropic, than becoming decumbent (branches bearing short shoots, e.g. *Bourreria, Cordia*). Many Solanaceae also conform to this model (e.g., *Cestrum, Markea, Solanum*).
h   Leaves of Boraginaceae. **A**. entire, venation pinnate and secant; **B**. trinerved and subentire.

29B Solanaceae

a   Concaulescence in Solanaceae (e.g. *Datura*), lateral axes being partly fused with the petiole (P) of axillary leaf. Fl = terminal flowe; α, β the two unequal leafy prophylls.
b,c A climbing species of *Lycianthes*. **b**: the young trunk exhibiting a monopodial phasis, while canopy twigs (**c**) are sympodial. Leaves hap to shed when the liana is fruiting.
d   Lenticellate twig, leaf axil bearing a twig and a small bud (e.g. *Datura*). Ls = proeminent leaf scar.
e   LEEUWENBERG or KORIBA model in *Solanum*, sympodial units alternating in black or white. Flowering is terminal.
f   Sarmentous climbing individual bearing relatively short twigs (e.g. *Markea*). *Cestrum* individuals can also conform to CHAMPAGNAT model, but these are not climbing.
g   *Dunalia, Witheringia* have plagiotropic modular branches. In *Witheringia* each plagiotropic module bearing two leaves (a small one and a large one). Flowering is terminal.
h   A top sided view of the previous individual showing alternating small and large leaves.
j   Secant venation in a pinnatilobate leaf (e.g. *Solanum*).
k   Spiny twigs in some species of *Solanum*.
m   Cordate-sagittate leaf of *Cyphomandra betacea*.
n   Lanceolate leaf with camptodromous venation (e.g. *Cestrum*).

N.B. Boraginaceae and Convolvulaceae have in general delayed branching in trunk. Solanaceae is variable in this respect.

Legends to plates 30A/30B

30A Bignoniaceae, Oleaceae and Verbenaceae

Bignoniaceae

a   KORIBA's model: Trunk and branches modular, modules initially equal, all apparently branches, but later unequal, one becoming a trunk (e.g., *Godmania, Tabebuia*).
b   Liana climbing by means of foliar tendrils. Leaves of simpler form in the young stage, then acquiring more leaflets and, ultimately, their terminal tendril (heterophylly).
c   Leaves bipinnate (e.g. *Jacaranda*).

Bignoniaceae, Oleaceae and Verbenaceae

d   A. stem bearing opposite leaves; **B.** stem more or less quadrangular in section, without protruding lenticels, but with interpetiolar ridges (Verbenaceae).
e   A. stem bearing opposite leaves; **B.** stem with protruding lenticels, but without interpetiolar ridges (Oleaceae).
f   Leaves palmately compound, its petiolules distally pulvinate (Bignoniaceae: *Tabebuia*) or not distally pulvinate (Verbenaceae: *Vitex*).
g   Leaves trifoliolate (Oleaceae: e.g. *Jasminum*).
h   CHAMPAGNAT's model: Trunk and branches firstly orthotropic, than becoming decumbent (Bignoniaceae: *Crescentia*, Oleaceae: *Jasminum*, Verbenaceae: *Petrea* pp.).
j   Short shoots bearing pseudowhorls of simple leaves (e.g., *Crescentia*).
k   Leaf simply pinnate (Bignoniaceae: *Spathodea*; Oleaceae: *Fraxinus*; Verbenaceae: *Peronema*); the winged rachis is for *Peronema*.
m   *Avicennia* (Avicenniaceae): A mangrove tree with pneumatophores, continuous branching, the branches all orthotropic and monopodial (ATTIMS's model).

30B Acanthaceae

a   Quadrangular internode, typical for the Acanthaceae.
b   Quadrangular twig, constricted at internode basis (e.g. *Ruellia*).
c   Lanceolate leaf with camptodromous venation (frequent for the family).
d   As in previous illustration. but leaf broader and basal secondary veins grouped.
e   Spiny leaf (e.g. mangrove-dwelled species of *Acanthus*).
f   A typical Acanthaceae with opposite-decussate phyllotaxy, internodes swollen at base, subcylindrical in section, forming four weak marked ribs.
g   A schematical representation of the sectorial anisoclady prevailing in the family.
h   Twining vine with sagittate leaves (*Thunbergia*, in part).
j   Cordate leaf (*Thunbergia*, in part).

N.B. All these families have in general delayed branching in trunk.

P29A

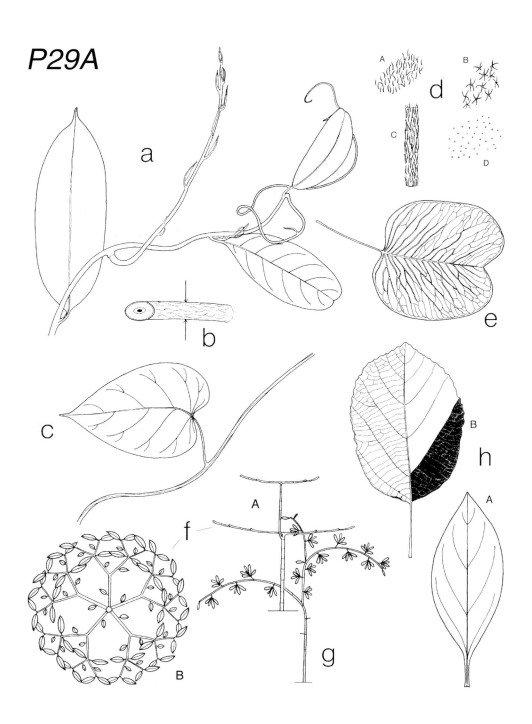

P29B

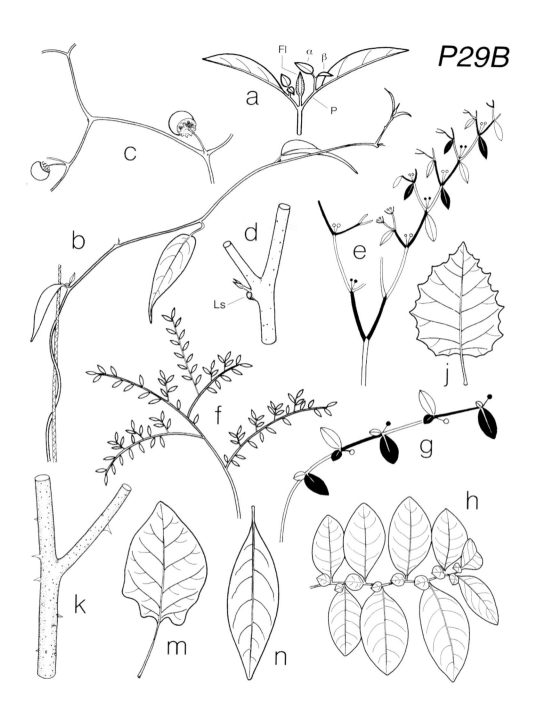

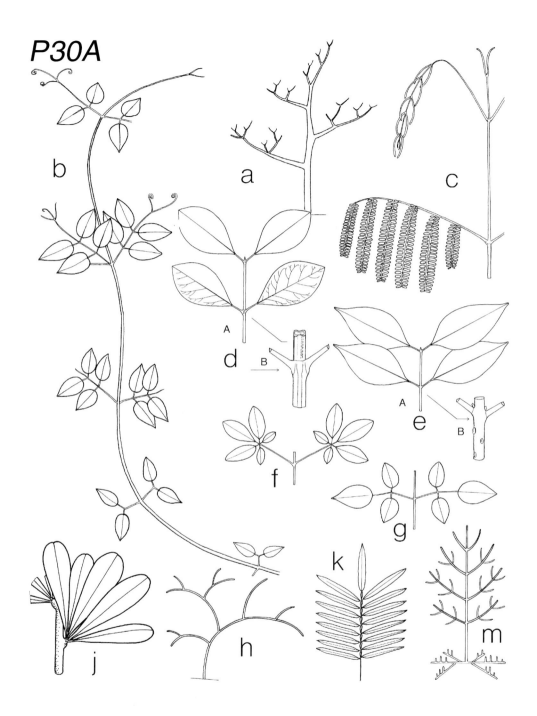

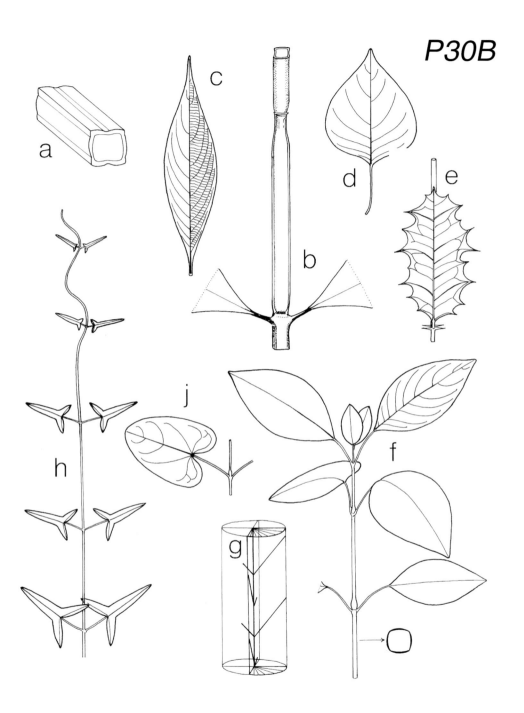

P30B

Legends to plates 31A/31B

31A Loganiaceae and Gentianaceae

a   KORIBA or LEEUWENBERG's model: Stems orthotropic and sympodial, more or less differentiated into trunk and branches (e.g. *Anthocleista*, Gentianaceae).

b   Large, cuneate leaves, with an stipuliform appendage (sa), (e.g. *Anthocleista*).

c   MANGENOT's model: Trunk formed by the stacking of modules, each module firstly orthotropic, then becoming abruptly plagiotropic (almost all species of *Strychnos*).

d   *Strychnos*: Branch with hooks subtended by scale-leaves. Typical, supratrinerved, leaf venation. Interpetiolar ridge (all these characters typical for the Loganiaceae).

e   FAGERLIND's model: Monopodial trunk; rhythmic branching, branches plagiotropic by apposition growth, flowering terminal or, if lateral, death of the apical meristem (e.g. *Fagraea*, Loganiaceae).

f   *Fagraea*: Sylleptic branching; petiolar scars large and jointed. inf = inflorescence.

g   Various kinds of stipules or foliar appendages: **A**. petiolar spur (e.g. *Fagraea*); **B**. annular appendage (e.g. *Fagraea*); **C**. auriculate appendage (e.g. *Anthocleista*, *Fagraea*); **D**. stipular collarette (e.g. *Labordia*, *Mostuea*); **E**. stipules adnate to the petiole (e.g. *Neuburgia*). **F**. interpetiolar stipules (e.g. *Logania*).

h   *Tachia* (Gentianaceae): small tree conforming to ROUX's model, branch internodes alternatively twisting (see F9 k), leaves with two basal ascending secondary veins.

31B Apocynaceae

a   Three, structurally related, architectural models: **A**. LEEUWENBERG; **B**. KORIBA; **C**. PREVOST. Branches are always modular. LEEUWENBERG's model is very common in the Apocynaceae (e.g. *Alstonia, Kopsia, Plumeria, Rauvolfia, Tabernaemonta*); KORIBA (e.g. *Alstonia, Cerbera, Himatanthus, Stemmadenia*); PREVOST (e.g. *Alstonia, Geissospermum*). RAUH models are uncommon: e.g. *Couma*.

b   Twining liana with opposite phyllotaxy. Detail: trunk bearing many lenticels (typical in the climbing Apocynaceae). Not twinig trunks but thigmotropic twigs do exist in tribe Willughbeeae (e.g. *Pacouria* in AM, *Landolphia* in AF and MA, *Willughbea* in AS).

c   A young tree bearing plagiotropic branches with distichous phyllotaxy (*Aspidosperma*). However the branch phyllotaxy is quite variable for this genus: from distichous (or pseudodistichous) to opposite. (MASSART's or RAUH's model).

d   LEEUWENBERG's model and whorled leaves (e.g. *Rauvolfia, Tabernaemontana*).

e   Spiral phyllotaxy in an erect, somewhat fleshy stem (e.g. *Plumeria*). Intramarginal veins are frequent in the family.

f   *Carissa*: A spiny shrub with modular architecture. Along a given branch the modules progressively acquire plagiotropic growth. The last pair of shoots produced by the module is modified into two opposite spines.

g   Apical bud protected by the petiolar bases (typical for the Apocynaceae, e.g. *Couma, Tabernaemontana*).

h   Stipuliform annular expansions (e.g. *Odontadenia*).

N.B. Loganiaceae and Apocynaceae have in general immediate branching in trunk.

Legends to plate 32A/32B

32A Rubiaceae

Taxonomic treatment according to Robbrecht [52]. ANT: Anthospermeae, CHI: Chioccoceae, CIN: Cinchonoideae, COF: Coffeeae, CON: Condamineeae, COU: Coussareeae, GAR: Gardenieae, GUE: Guettardeae, HEN: Henriquezieae, ISE: Isertieae, NAU: Naucleeae, PAV: Pavetteae, PSY: Psychotrieae, RON: Rondeletieae, SPE: Spermacoceae, VAN: Vanguerieae.

- **a** Modular branch of an individual conforming to LEEUWENBERG's or KORIBA's model; PSY: *Cephaelis*, *Psychotria*.
- **b** *Morinda citrifolia* (CIN): a modular branch, its modules consisting of only two internodes, the first internode bearing a pair of leaves distally, the second internode bearing only one developed leaf (PETIT's model); GAR: *Aidia*, *Rothmannia* or *Schumanniophyton*; CIN?: *Bertiera* conform to a less stereotyped PETIT's model.
- **c** Branch plagiotropic by apposition growth with terminal flowering, individuals conforming to FAGERLIND's model; GAR: *Genipa*, *Randia*; PSY: *Rudgea*.
- **d** Monopodial and plagiotropic branch in trees conforming to ROUX-COOK's model; CIN?: *Bertiera*, COF: *Coffea*, NAU: Naucleae, MOR: *Lasianthus*, VAN: *Canthium*.
- **e** Branch firstly orthotropic, than becoming decumbent; SPE: *Ernodea*.
- **f** Unbranched monopodial treelet, CORNER's model, GAR: *Duroia*, *Gardenia*.
- **g** Liana with twigs modified into hooks: Isertieae: *Mussaenda*, CIN: *Uncaria*.
- **h-t** Various kinds of stipules (**h** to **s**: interpetiolar). **h**. Curvilinear, ISE: *Mussaenda*. **j**. Bilobate, ISE: *Isertia*, PSY: *Psychotria* (stipules are completely free in *Henriquezia*, HEN). **k**. Notched, PSY: *Palicourea*. **m**. Forming a collarette, PSY: *Pagamea*. **n**. More or less triangular, COF: *Coffea*, PSY: *Gaertnera*, GAR: *Randia*, RON: *Rondeletia*, CIN: *Uncaria*. **p**. Collarette with a linear appendage, GAR: *Argocoffeopsis*. **q**. Fringed, SPE: *Borreria*. **r**. Spatulate: CIN: *Cinchona*, NAU: *Nauclea* (forming a hood in *Platycarpum*, HEN). **s**. Acuminate: COU: *Faramea*. **t**. Intrapetiolar stipules (CIN: *Capirona*).

32B Rubiaceae

- **a** Buttressed tree; CON: *Chimarrhis*. CIN-Cinchoneae do have bitter bark.
- **b** Rhytidome peeling in large sheaths; CIN: *Calycophyllum*.
- **c** Fibrous rhitidome; GAR: *Duroia*.
- **d** Rhytidome not peeling, trunk lenticellate (uncommon), e.g. GAR: American *Randia*.
- **e** Suberization forming well-anchored scales; PAV: *Ixora*.
- **f** Rhytidome peeling is typical for Rubiaceae; e.g. COF: *Coffea*.
- **g** Progressive reduction of high order axis bearing terminal inflorescence into axillary (or pseudoaxillary ?) inflorescences in the family. **A**. CON: *Dioicodendron*. **B**. CIN: *Cinchona*, CON: *Pogonopus*. **C**. COF: *Coffea*.
- **h** Helicoidal anisoclady is quite common in Rubiaceae conforming to ROUX's model; MOR: *Lasianthus*.
- **j** Furrowed internode (f i), (COU: *Coussarea*). Venation brochidodromous.
- **k** Short twigs modified into spines; GAR: American *Randia*. Venation camptodromous.
- **m** Cupular domatia (e.g. ANT: *Coprosma*). Hairy domatia do exist also (*Calycophyllum*).

P31A

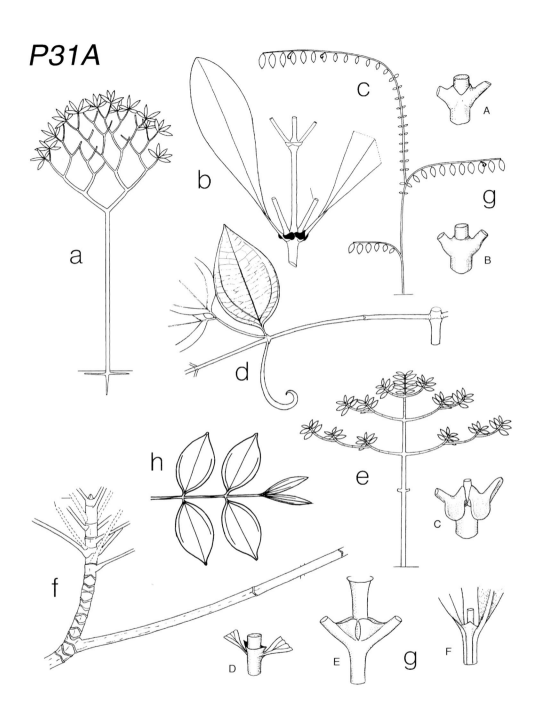

P31B

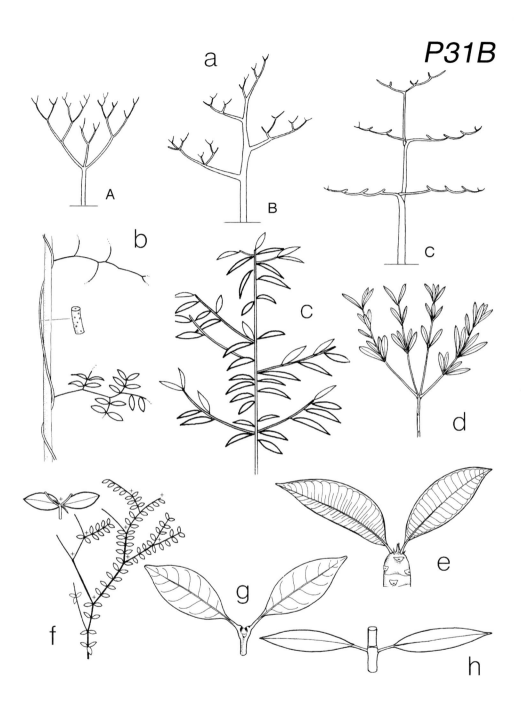

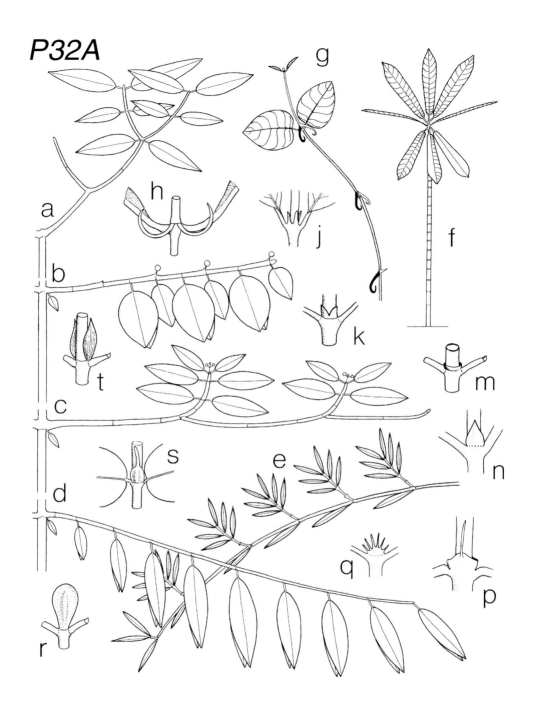

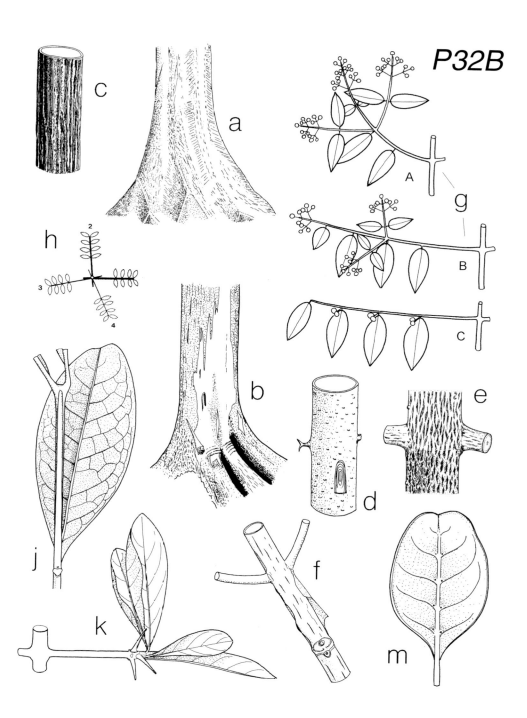

Bibliography

Morphology, anatomy and architecture

[1] **Tomlinson PB. 1991.** *The Structural Biology of Palms.* Oxford University Press, New York.

[2] **Rejmánek M, Brewer SW. 2001.** Vegetative identification of tropical woody plants: State of the art and annotated bibliography. *Biotropica* **33**: 214-228.

[3] **Little RJ, Jones CE. 1980.** *A Dictionary of Botany.* Van Nostrand Reinhold Company, New York.

[4] **Romberger JA, Hejnowicz Z, Hill JF. 1993.** *Plant Stucture: Function and Development - A Treatise on Anatomy and Vegetative Development, with Special Reference to Woody Plants.* Springer, Berlin.

[5] **Roth I. 1981.** *Structural patterns of tropical barks.* Borntraeger, Berlin.

[6] **Rollet B. 1980.** Intérêt de l'étude des écorces dans la détermination des arbres tropicaux sur pied. *Revue Bois et Forêts des Tropiques* **194**: 3-28.

[7] **Rollet B. 1982.** Intérêt de l'étude des écorces dans la détermination des arbres tropicaux sur pied. *Revue Bois et Forêts des Tropiques* **195**: 31-50.

[8] **Fahn A. 1985.** *Plant Anatomy. 2d ed.* Pergamon Press, Oxford.

[9] **Junikka L. 1994.** Survey of English macroscopic bark terminology. *IAWA Journal* **15**: 3-45.

[10] **Letouzey R. 1972.** *Manuel de Botanique forestière d'Afrique tropicale. Tomes I, II et III.* CTFT, Nogent-sur-Marne.

[11] **Whitmore TC. 1966.** Studies in systematic bark morphology. I. Bark morphology in Dipterocarpaceae. II. General features of bark construction in Dipterocarpaceae. *The New Phytologist* **61**: 191-220.

[12] **Wyatt-Smith J. 1954.** Suggested definitions of field characters (for use in the identification of tropical forest trees in Malaya). *Malayan Forester* **17**: 170-183.

[13] **Serier JB. 1986.** Les sécrétions d'arbres. *Revue Bois et Forêts des tropiques* **213**: 33-39.

[14] **Carlquist S. 1991.** Anatomy of vine and liana stems: a review and synthesis, In F.E. Putz and H.A. Mooney, eds. *The Biology of Vines.* Cambridge University Press, Cambridge, 53-71.

[15] **Caballé G. 1993.** Liana structure, function and selection: a comparative study of xylem cylinders of tropical rainforest species in Africa and America. *Botanical Journal of the Linnean Society* **113**: 41-60.

[16] **Rawitscher R. 1932.** *Der Geotropismus der Pflanzen.* Gustav Fischer, Jena.

[17] **Tronchet A. 1974.** *La sensibilité des plantes.* Masson, Paris.

[18] **Hallé F, Martin R. 1968.** Etude de la croissance rythmique chez l'Hévéa (*Hevea brasiliensis* Muell. Arg. Euphorbiacées-Crotonoidées). *Adansonia* **7**: 475-503.

[19] **Prévost MF. 1967.** Architecture de quelques Apocynaceae ligneuses. *Mémoires de la Société Botanique de France. Colloque sur la physiologie de l'arbre*, Paris.

[20] **Hallé F, Oldeman RAA, Tomlinson PB. 1978.** *Tropical trees and forests - An architectural analysis.* Springer, Berlin.

[21] **Hallé F, Oldeman RAA. 1970.** *Essai sur l'architecture et la dynamique de croissance des arbres tropicaux.* Masson, Paris.

[22] **Keller R. 1994**. Neglected vegetative characters in field identification at the supraspecific level in woody plants: phyllotaxy, serial buds, syllepsis and architecture. *Botanical Journal of the Linnean Society* **116**: 33-51.

[23] **Ducke JA. 1969**. On tropical tree seedlings – Seeds, seedlings, systems and systematics. *Annals of the Missouri Botanical garden* **56**: 125-161.

[24] **Bell AD. 1991**. *Plant Form. An illustrated guide to flowering plant morphology*. Oxford University Press, London.

[25] **Champagnat P. 1948.** Ramification à régime rythmique et anisophyllie chez les végétaux ligneux supérieurs. *Lilloa* **16**: 161-191.

[26] **Sandt W. 1925**. Zur Kenntnis der Beiknospen. *Botanische Abhandlungen, herausgeben von K. Goebel, Heft* 7. Gustav Fischer, Jena.

[27] **Hoogland R.D. 1952**. A revision of the genus Dillenia. *Blumea* **7**: 1-145.

[28] **Weberling F, Leenhouts PW. 1965**. Systematisch-morphologische Studien an Therebinthales-Familien. *Abhandlungen der Mathematisch - Naturwissenschaftlichen Klasse, Akademie der Wissenschaften und der Literatur, Mainz* **10**: 495-584.

[29] **Cullen J. 1978**. A preliminary survey of ptyxis (vernation) in the Angiosperms. *Notes of the Royal Botanic Garden Edinburgh* **37**: 161-214.

[30] **Hickey LJ. 1973**. Classification of the architecture of Dicotyledons leaves. *American Journal of Botany* **60**: 17-33.

[31] **Gifford EM, Foster AS. 1989**. *Morphology and Evolution of Vascular Plants. 3d ed.* W.H. Freeman and Co., New York.

[32] **Belin-Depoux M. 1989**. Des hydathodes aux nectaires foliaires chez les plantes tropicales. *Bulletin de la Société Botanique de France, Actualités Botaniques* **136**: 151-168.

[33] **Schnell R, Cusset G, Quenum M. 1963**. Contributions à l'étude des glandes extraflorales chez quelques groupes de plantes tropicales. *Revue Générale de Botanique* **70**: 269-342.

[34] **Hickey L.J., Wolfe J.A. 1975**. The bases of Angiosperm phylogeny: vegetative morphology. *Annals of the Missouri Botanical Garden* **62**: 538-589.

[35] **Hallé F. In press**. *Architectures de Plantes*. JPC EDITION, New York.

[36] **Cremers G. 1975**. Sur la présence de dix modèles architecturaux chez les Euphorbes malgaches. *Comptes Rendus de l'Académie des Sciences, Paris, série D* **281**: 1575-1578.

[37] **Troll W. 1937**. *Vergleichende Morphologie der höheren Pflanzen. Erster Band*. Berlin: Borntraeger.

Systematics

[38] **Mabberley DJ. 1997**. *The Plant Book, 2nd ed.* University Press, Cambridge.

[39] **Magallón S, Crane PR, Herendeen PS. 1999**. Phylogenic pattern, diversity, and diversification of Eudicots. *Annals of the Missouri Botanical Garden* **86**: 297-372.

[40] **Chase M. *et al*. 2003**. An update of the Angiosperm phylogeny group classification for the orders and families of flowering plants: APG II. *Botanical Journal of the Linnean Society* **141**: 399-436.

[41] **Uhl NW, Dransfield J. 1987. Genera Palmarum** – *A classification of Palms Based on the Work of Harold E. Moore, Jr.* The L.H. Bailey Hortorium and the International Palm Society, Allen Press, Lawrence.

[42] **Hoot SB, Magallón S, Crane P.R. 1999**. Phylogeny of basal Eudicots based on three molecular data sets: atpB, rbcL, and 18S nuclear ribosomal DNA sequences. *Annals of the Missouri Botanical garden* **86**: 1-32.

[43] **Cronquist A. 1981.** *An integrated system of classification of flowering plants.* Columbia University Press, New York.

[44] **Graham SA, Crisci JV, Hoch PC. 1993.** Cladistic analysis of the Lythraceae *sensu lato* based on morphological characters. *Botanical Journal of the Linnean Society.* **113**: 1-33.

[45] **Simmons MP, Hedin JP. 1999.** Relationships and morphological character change among genera of Celastraceae *sensu lato* (including Hippocrateaceae). *Annals of the Missouri Botanical garden* **86**: 723-751.

[46] **Juncosa AM, Tomlinson PB. 1988.** Systematic comparison and some biological characteristics of Rhizophoraceae and Anisophylleaceae. *Annals of the Missouri Botanical garden* **75**: 1296-1318.

[47] **Swensen SM, Luthi JN, Rieseberg LH. 1998.** Datiscaceae revisited: Monophyly and the sequence of breeding system evolution. *Systematic Botany* **23**: 157-169.

[48] **Polhill RM, Raven PH. 1981.** *Advances in Legumes systematics, Vol. 1.* Royal Botanic Gardens, Kew.

[49] **Herendeen PS, Brunneau A. 2000.** *Advances in Legume Systematics.* Vol. **9**. Royal Botanic Gardens, Kew.

[50] **Anderberg AA, Stohl B, Källersjo M. 2000.** Maesaceae, a new primuloid family in the order Ericales, s.l. *Taxon*: **49**: 183-187.

[51] **Alverson WS, Whitlock BA, Nyffeler R, Byer C Baum DA. 1999.** Phylogeny of the core Malvales: Evidence from NDHF sequence data. *American Journal of Botany* **86**: 1474-1486.

[52] **Lemke DE. 1988.** A synopsis of Flacourtiaceae. *Aliso* **12**: 29-43.

[53] **Webster GL. 1994.** Synopsis of the genera and suprageneric taxa of Euphorbiaceae. *Annals of the Missouri Botanical garden* **81**:33-144.

[54] **Endress ME, Bruyns PV. 2000.** A revised classification of the Apocynaceae *s.l.*. The Botanical Review 66 (1).

[55] **Struwe AA, Bremer B. 1994.** Cladistics and family level classification of the Gentianales. *Cladistics* **10**: 175-206.

[56] **Robbrecht E. 1988.** Tropical woody Rubiaceae. *Opera Botanica Belgica* **1**. National Plantentuin van België.

[57] **Schwarzbach AE, MacDade LA. 2002.** Phylogenetic relationships of the mangrove family Avicenniaceae based on chloroplast and nuclear ribosomal DNA sequences. *Systematic Botany* **27**: 84-98.

Field guides consulted for vegetative characters

Pantropical

Keller R. 1996. Identification of tropical woody plants in the absence of flowers – A Field Guide. Birkhäuser, Basel, Berlin & Boston.

Africa

Aubréville A. 1959. *Flore forestière de la Côte d'Ivoire. Tomme III*. CTFT, Nogent-sur-Marne.

Breteler FJ. 1973. The African Dichapetalaceae – A taxonomic revision. *Mede. Landbouwhogeschool Wageningen* **73**-13.

Breteler FJ. 1989. The Connaraceae - A taxonomic study with emphasis on Africa. *Agricultural University, Wageningen Papers* **89**-6.

Geerling C. 1987. *Guide de terrain des ligneux Sahéliens et Soudano Guinéens*. Agricultural University, Wageningen, The Netherlands.

Letouzey R. 1972. *Manuel de botanique forestière d'Afrique tropicale. Tomes I, II et III*. CTFT, Nogent-sur-marne, France.

Schatz GE. 2001. *Generic tree flora of madagascar*. Royal Botanic Gardens, Kew.

Americas

Aristiguieta L. 1974. *Familias y generos de los árboles de Venezuela*. Fundacion Instituto Botanico de Venezuela, Caracs, Venezuela.

Gentry AH. 1993. *A field guide to the families and genera of woody plants of northwest South America (Colombia, Ecuador, Peru) with supplementary notes on herbaceous taxa*. Conservation International, Washington DC.

Henderson A, Galleano G, Bernal R. 1995. *Palms of the Americas*. Princeton University Press, Princeton.

Lorenzi H. 1998. *Arvores brasileiras. Vol. 2,3rd ed.* Instituto Plantarum da Estudos da Flora LTDA. Nova Odessa.

Lorenzi H. 2000. *Arvores brasileiras. Vol. 1*. Instituto Plantarum da Estudos da Flora LTDA. Nova Odessa

Australasia

Nair KKN. 2000. *Manual of non-wood forest produce plants of Kerala*. Kerala Forest Research Institute, Peechi, Kerala.

Ng FSP. 1978, 1989. *Tree Flora of Malaya. Vol.3 and 4*. Longman, London.

Whitmore TC. 1962. *Guide to the forets of the British Solomon islands*. Oxford University Press, London.

Whitmore TC. 1972, 1973. *Tree Flora of Malaya. Vol.1 and 2*. Longman, London.

Williams JB, Harden GJ. 1984. *A field guide to the rainforests climbing plants of New South Wales using vegetative characters*. University of New England, Armidale, Australia.

Williams JB, Harden GJ. 1984, Mac Donald WJF. *Trees and shrubs in rainforests of New South Wales and southern Queensland*. University of New England, Armidale.

Index of the genera and families
K: in Part I, keys
F: in Part I, glossary
P: in Part II, families

Abatia, **F** *162*
Abrus, **K** *118*
Abuta, K *34*, **P** *179*
Acacia, **K** *52, 97, 116, 119, 120,* **F** *130, 142, 146, 148,* **P** *230, 231*
Acalypha, **K** *33, 100,* **F** *162,* **P** *226*
ACANTHACEAE, **K** *67, 91, 94,* **F** *125, 134, 140, 144, 148, 154,* **P** *171, 259*
Acanthus, **P** *259*
Acer, **K** *71*
ACERACEAE, see SAPINDACEAE
Aceratium, **K** *66*
Acosmium, **K** *118*
Acridocarpus, **K** *30, 86,* **F** *160*
Acrocomia, **P** *172*
ACTINIDIACEAE, **K** *43, 50,* **F** *125, 156, 162,* **P** *202*
Actinostemon, **K** *16, 104*
Adansonia, **K** *109,* **P** *221*
Adelia, **F** *144*
Adelobotrys, **K** *94,* **P** *209*
Adenanthera, **K** *120,* **P** *230*
Adenia, **K** *83,* **F** *160,* **P** *215*
Adinandra, **K** *40,* **P** *202*
Aegiphila, **K** *92*
Aegle, **K** *57*
Aesculus, **K** *79*
Afrobrunnichia, **K** *87,* **F** *148,* **P** *185*
Afrostyrax, **K** *25,* **F** *162*
Afzelia,, **F** *150*
Agapetes, **K** *89*
Agathis, **K** *60, 62, 74,* **F** *128, 152*
AGAVACEAE, **K** *59,* **F** *138, 152*
Agave, **F** *138*
Agelaea, **K** *56, 57, 95,* **P** *234*
Aglaia, **K** *57, 112, 113, 114,*

P *240*
Agonandra, **K** *47,* **P** *190*
Agrostistachys, **K** *32,* **P** *227*
Aidia, **P** *265*
Ailanthus, **K** *112, 113,* **F** *162*
Aiouea, **F** *126,* **P** *173*
ALANGIACEAE, **K** *35, 84,* **F** *140, 156, 162,* **P** *252*
Alangium, **K** *35, 84,* **F** *136, 156, 158, 162,* **P** *252*
Alchornea, **K** *29, 100, 101,* **F** *160,* **P** *226*
Alchorneopsis, **F** *160*
Aleurites, **K** *103,* **F** *160*
Alexa, **K** *56, 118*
Allamanda, **K** *61,* **F** *132*
Allanblackia, **P** *214*
Allophylus, **K** *113,* **P** *240*
Alphitonia, **P** *197*
Alstonia, **K** *62,* **F** *138,* **P** *264*
Altingia, **K** *31, 36, 51,* **F** *142,* **P** *191*
Amanoa, **K** *98,* **P** *227*
AMBORELLACEAE, **F** *124*
Ambroma, **K** *107*
Amherstia, **K** *116*
Amicia, **K** *119*
Amphirrhox, **K** *31,* **P** *220*
ANACARDIACEAE, **K** *15, 42, 50, 62, 75, 111, 112, 113,* **F** *134, 140, 142, 144, 150, 152, 156, 158,* **P** *241*
Anacardium, **K** *111,* **F** *136, 158,* **P** *241*
Anacolosa, **K** *88*
Anadenathera, **K** *120*
Anaxagorea, **F** *138*
Anchietea, **K** *85*
ANCISTROCLADACEAE, **K** *87,* **F** *125, 130*
Ancistrocladus, **K** *87*
Ancistrophyllum, **K** *96*
Ancistrothyrsus, **K** *87*
Andira, **K** *117*
Anemopaegma, **F** *146*
Angostura, **K** *110*
Aniba, **K** *42, 43, 53,* **F** *130, 136, 144,* **P** *173*
Anisacanthus, **K** *94*

Anisomallon, **K** *52*
Anisophyllea, **K** *35,* **F** *142, 158,* **P** *252*
ANISOPHYLLEACEAE, **K** *35,* **F** *142, 158,* **P** *252*
Anisoptera, **K** *13, 20,* **F** *162,* **P** *221*
Annona, **K** *37,* **F** *138,* **P** *178*
ANNONACEAE, **K** *37, 39, 85,* **F** *128, 130, 132, 138, 140, 146, 152, 154, 156, 158, 162*
Anogeissus, **K** *17*
Anthocleista, **K** *63, 68,* **F** *138,* **P** *171, 264*
Anthodiscus, **K** *55*
Anthonotha, **K** *116*
Anthostema, **K** *105,* **P** *227*
Antidaphne, **K** *90,* **P** *190*
Antidesma, **K** *20, 98,* **P** *226, 227*
Antigonon, **P** *185*
Antonia, **K** *68,* **F** *134*
Apeiba, **K** *107,* **F** *158, 162,* **P** *221*
Aphanamixis, **K** *113,* **P** *240*
Aphelandra, **F** *154*
Aphloia, **K** *25,* **F** *124,* **P** *171, 215*
APHLOIACEAE, **K** *25,* **F** *124*
APOCYNACEAE, **K** *15, 16, 61, 62, 80,* **F** *126, 128, 132, 138, 144, 148, 154, 156, 158, 162,* **P** *171, 264*
Apodytes, **K** *46*
Aporusa, **K** *21, 22, 29, 31, 98,* **F** *158,* **P** *227*
Apuleia, **K**
AQUIFOLIACEAE, **K** *30, 33, 52,* **F** *124, 125, 134,* **P** *252*
Aquilaria, **K** *37,* **F** *156*
ARACEAE, **K** *89,* **F** *132, 146, 152*
Aralia, **K** *97,* **P** *252*
ARALIACEAE, **K** *27, 34, 55, 56, 57, 83, 97,* **F** *125, 148, 152, 156*
ARALIDIACEAE, **K** *45,* **F** *150*
Aralidium, **K** *45,* **F** *150*
Araucaria, **K** *59*

ARAUCARIACEAE, **K** *59, 60, 62, 74,* **F** *125, 152, 156*
Arbutus, **P** *203*
Archytaea, **K** *43,* **P** *202*
Ardisia, **K** *38, 43, 44, 48, 49,* **F** *144,* **P** *202*
Areca, **P** *172*
ARECACEAE, **K** *59, 60, 96,* **F** *138, 154,* **P** *172*
Arenga, **P** *172*
Argocoffeopsis, **P** *265*
Aristolochia, **K** *34, 39, 82,* **F** *158,* **P** *184*
ARISTOLOCHIACEAE, **K** *34, 39, 82,* **F** *144, 146, 154, 158,* **P** *184*
Aristotelia, **K** *66*
Arrabidaea, **K** *95*
Artabotrys, **K** *85,* **F** *132,* **P** *178*
Artocarpus, **K** *13,* **F** *136, 150,* **P** *196*
Arytera, **K** *113,* **F** *162*
Ascarina, **K** *63*
ASPARAGACEAE, **K** *85*
Asparagus, **K** *85*
Aspidosperma, **K** *16,* **P** *264*
Asplundia, **K** *59, 96*
ASTERACEAE, **K** *49, 51, 53, 69, 73, 88, 91,* **F** *128, 132, 162,* **P** *253*
Astronium, **K** *111, 112, 113,* **F** *156*
Ateleia, **K** *118*
Atractogyne, **K** *91*
Atropa, **F** *146*
Attalea, **P** *172*
Atuna, **K** *18*
Aucoumea, **K** *111,* **P** *241*
Aucuba, **K** *75,* **F** *124*
AUCUBACEAE, **K** *75*
AUSTROBAILEYACEAE, **F** *124*
Averrhoa, **K** *56,* **P** *234*
Avicennia, **K** *69,* **F** *136,* **F** *148,* **P** *259*
AVICENNIACEAE, **K** *69,* **F** *148,* **P** *259*
Azadirachta, **K** *113*

Baccaurea, **K** *28, 99,* **F** *126, 136,* **P** *226, 227*
Baccharis, **K** *49,* **P** *253*
Bactris, **K** *60,* **P** *172*
Bagassa, **K** *61*
BALANITACEAE, **K** *56,* **F** *124, 144, 152*
Balanites, **K** *56*
BALANOPACEAE, **F** *124*
Balfourodendron, **K** *110*
Baloghia, **K** *102*
Bambusa, **K** *60*
BAMBUSACEAE, **K** *60,* **F** *146, 156*
Banara, **K** *21,* **F** *160,* **P** *215*
Bandereia, **K** *82, 115*
Banisteriopsis, **K** *92, 93,* **F** *146*
Baphia, **K** *115,* **F** *152*
Barleria, **K** *67*
Barringtonia, **K** *45,* **F** *128, 134,* **P** *208*
Barteria, **K** *22,* **F** *144, 160,* **P** *215*
Bauhinia, **K** *20, 23, 80, 83, 96, 97, 115, 116,* **F** *144, 152,* **P** *230, 231*
Beilschmiedia, **K** *75,* **P** *173*
Bejaria, **K** *47,* **F** *154,* **P** *203*
BERBERIDACEAE, **K** *42, 56*
Berberis, **K** *42*
Berrya, **K** *108,* **F** *158*
Bersama, **K** *54, 110*
Bertholletia, **P** *208*
Bertiera, **P** *265*
BETULACEAE, **F** *144*
Bhesa, **K** *29,* **F** *152,* **P** *171, 235*
BIGNONIACEAE, **K** *42, 57, 67, 71, 74, 79, 94, 95,* **F** *126, 130, 132, 134, 144, 146, 150, 152, 158, 160, 162,* **P** *259*
Billia, **K** *79*
Bischofia, **K** *55*
BISCHOFIACEAE, **K** *55, 98,* **F** *124*
Bixa, **K** *27,* **F** *158, 162,* **P** *220*
BIXACEAE, **K** *14, 27, 55,* **F** *124, 128, 158, 162,* **P** *220*
Blachia, **K** *102*
Blepharistemma, **K** *64*

BLEPHAROCARYACEAE, **F** *124*
Blighia, **K** *113,* **P** *240*
Bocoa, **K** *115,* **F** *152*
Boehmeria, **K** *66,* **F** *158,* **P** *196*
BOMBACACEAE, **F** *128, 144, 150, 152, 158, 162,* **P** *171, 221*
Bombax, **K** *108,* **F** *128, 150,* **P** *221*
Bonamia, **P** *258*
Bonnetia, **K** *50,* **P** *202*
BONNETIACEAE, **K** *43, 50,* **F** *125, 134, 150, 152, 154,* **P** *171, 202*
BORAGINACEAE, **K** *35, 42, 43, 46, 90,* **F** *128, 138, 140, 162,* **P** *258*
Borassus, **K** *60*
Borreria, **P** *265*
Bouea, **K** *62, 75,* **F** *142,* **P** *241*
Bougainvillea, **K** *51,* **F** *144, 146,* **P** *190*
Bourreria, **K** *42,* **P** *258*
Bowdichia, **K** *118,* **F** *138*
Bowringia, **K** *117*
Brassaiopsis, **P** *252*
Bravaisia, **K** *69*
Brazzeia, **K** *40*
Bredemeyera, **K** *22, 47, 86,* **F** *132, 154, 160,* **P** *247*
Brexia, **K** *47,* **F** *156,* **P** *171*
BREXIACEAE, **K** *47,* **F** *124, 156,* **P** *171*
Bridelia, **K** *98,* **P** *227*
Brosimum, **K** *13,* **P** *196*
Brownea, **K** *116,* **P** *230*
Brownlowia, **K** *108*
Bruguiera, **K** *64,* **F** *134,* **P** *247*
Brunellia, **K** *65, 78,* **P** *247*
BRUNELLIACEAE, **K** *65, 78,* **F** *124, 156, 162,* **P** *247*
Brunfelsia, **K** *42*
Buchenavia, **P** *208*
Buddleja, **K** *64, 69,* **P** *171*
BUDDLEJACEAE, **P** *171*
Bulnesia, **K** *78*
Bumelia, see Sideroxylon
Burasaia, **K** *95,* **P** *184*
Burdachia, **K** *63*

Burkea, **P** *231*
Bursera, **P** *241*
BURSERACEAE, **K** *51, 110, 111, 112, 113,* **F** *126, 128, 134, 144, 150, 152,* **P** *241*
Bussea, **K** *121,* **P** *230*
Butia, **P** *172*
BUXACEAE, **K** *71, 75,* **F** *126, 144, 154*
Buxus, **K** *71, 75,* **F** *126, 144, 154*
Byrsonima, **K** *63,* **F** *148,* **P** *246*
Byttneria, **K** *82, 107,* **F** *130, 160*

CACTACEAE, **K** *42, 85,* **F** *124, 148*
Caesalpinia, **K** *117, 121,* **F** *136, 146,* **P** *230*
Calamus, **K** *96,* **P** *172*
Calea, **K** *69*
Calliandra, **K** *117, 144,* **P** *230*
Callicarpa, **K** *76*
Callisthene, **K** *77,* **F** *142*
Caloncoba, **K** *33,* **F** *158,* **P** *215*
Calophyllum, **F** *156*
Calycophyllum, **P** *265*
Calyptranthes, **K** *72, 73,* **F** *158*
Camellia, **F** *154,* **P** *202*
Campnosperma, **K** *42, 57,* **F** *140,* **P** *241*
Campomanesia, **K** *76*
Campsiandra, **K** *118,* **P** *230*
Canarium, **K** *111,* **F** *150,* **P** *241*
Canella, **K** *44*
CANELLACEAE, **K** *39, 44,* **F** *125, 160,* **P** *173*
Canthium, **K** *93*
Capirona, **P** *265*
CAPPARIDACEAE, **K** *25, 29, 45, 48, 49, 55, 88,* **F** *124, 148, 150, 152, 154*
Capparis, **K** *25, 88,* **F** *148,* **P** *215*
CAPRIFOLIACEAE, **K** *67, 69, 73, 79, 91,* **F** *126, 146, 148, 154*
Caraipa, **K** *16, 40,* **P** *214*

Carapa, **F** *150,* **P** *240*
Carica, **K** *35,* **P** *220*
CARICACEAE, **K** *35,* **F** *125, 128,* **P** *220*
Carissa, **F** *144,* **P** *264*
Carludovica, **K** *60*
Caryocar, **K** *78*
CARYOCARACEAE, **K** *55, 78,* **F** *124*
Caryodaphnopsis, **K** *75*
Caryodendron, **K** *100*
Caryota, **F** *138,* **P** *172*
Casearia, **K** *23, 24, 26,* **F** *136,* **P** *215*
Cassia, **K** *116, 119, 150,* **P** *231*
Cassinopsis, **K** *76*
Cassipourea, **K** *64,* **F** *140,* **P** *247*
Castanopsis, **P** *191*
Castilla, **K** *13,* **F** *136,* **P** *196*
Casuarina, **K** *67*
CASUARINACEAE, **K** *67,* **F** *124, 128*
Catha, **K** *65*
Cathedra, **K** *41*
Catostemma, **K** *28, 108, 109*
Cavanillesia, **K** *108*
Cavendishia, **K** *39,* **P** *203*
Cecropia, **K** *13, 27,* **F** *142, 156,* **P** *196*
CECROPIACEAE, **P** *171*
Cedrela, **K** *114,* **P** *240*
Cedrelinga, **K** *120*
Ceiba, **K** *109,* **F** *136, 144,* **P** *221*
CELASTRACEAE, **K** *26, 29, 32, 33, 46, 65, 66, 80, 85, 93,* **F** *124, 130, 132, 140, 146, 148, 152, 154, 158,* **P** *171, 235*
Celastrus, **K** *32, 85,* **F** *132*
Celtis, **K** *19, 23, 84,* **F** *158,* **P** *196*
Centauropsis, **P** *253*
Centrolobium, **K** *117*
Cephaelis, **P** *265*
Cephalaralia, **K** *97*
Cephalomappa, **K** *100*
Cerbera, **K** *16,* **F** *126*
Ceriops, **P** *247*

Cespedesia, **K** *32,* **P** *214*
Cestrum, **K** *53,* **F** *144, 146,* **P** *258*
Chaetachme, **K** *19*
Chaetocarpus, **K** *25, 99,* **P** *227*
Chamaecrista, **K** *116,* **F** *160,* **P** *231*
Chamaecyparis, **F** *134*
Chamaedorea, **P** *172*
Chamaesyce, **K** *61, 105*
Cheiloclinium, **F** *140*
Chimarrhis, **P** *265*
Chionanthus, **K** *77*
Chisocheton, **K** *112, 114,* **F** *150,* **P** *240*
CHLAENACEAE, **F** *124*
CHLORANTHACEAE, **K** *63, 67, 94,* **F** *125, 140, 148, 152,* **P** *179*
Chloroleucon, **K** *120*
Chlorophora, **K** *13*
Chomelia, **F** *162*
Chorisia, **K** *109*
Christiana, **P** *221*
Chrysalidocarpus, **K** *60*
CHRYSOBALANACEAE, **K** *18, 21, 22, 25, 41,* **F** *126, 134, 138, 140, 148, 154, 156, 158, 160,* **P** *171, 234*
Chrysobalanus, **F** *160,* **P** *234*
Chrysochlamys, **P** *214*
Chrysophyllum, **K** *16, 81,* **F** *156,* **P** *203*
Chukrasia, **P** *240*
Cinchona, **K** *61,* **P** *265*
Cinnamodendron, **K** *44,* **F** *160,* **P** *173*
Cinnamomum, **K** *34, 35, 71,* **F** *136, 158*
Cinnamosma, **K** *39,* **P** *173*
Cipadessa, **P** *240*
Cissus, **K** *83, 96,* **F** *144,* **P** *197*
Citharexylum, **K** *67, 68, 69, 72,* **F** *160*
Citronella, **K** *41,* **F** *162,* **P** *253*
Citrus, **F** *144, 150, 152,* **P** *241*
Claoxylon, **K** *33, 100,* **P** *227*
Clappertonia, **K** *108*
Clarisia, **K** *14*

Clathrotropis, **K** *118*
Clausena, **P** *241*
Clavija, **K** *45*, **P** *202*
Cleidion, **K** *99*
Cleistanthus, **K** *98*, **P** *227*
Clematis, **K** *95*, **F** *132*, **P** *184*
Clerodendron, **K** *76, 92, 93*, **F** *132*
Clethra, **K** *50*
CLETHRACEAE, **K** *50*, **F** *125*
Cleyera, **K** *39*
Clitoria, **K** *118*
Clusia, **K** *61, 81*, **F** *154, 156*, **F** *162*, **P** *214*
CLUSIACEAE, **K** *13, 16, 32, 40, 51, 61, 68, 70, 74, 77, 81*, **F** *136, 140, 148, 154, 156, 158, 162*, **P** *214*
Cnesmone, **K** *101*, **P** *227*
Cobaea, **K** *96*, **F** *124, 132*
COBAEACEAE, **F** *124*
Coccoloba, **K** *18, 27, 85, 88*, **F** *130*, **P** *185*
Coccothrinax, **P** *172*
Cocculus, **K** *34*, **P** *184*
Cochlospermum, **K** *14, 28, 55*, **F** *124*, **P** *220*
Cocos, **K** *59*, **P** *172*
Codiaeum, **K** *49, 102*, **P** *227*
Coelostegia, **P** *221*
Coffea, **F** *136, 140, 148*, **P** *265*
Cola, **K** *108, 109*, **F** *150*, **P** *221*
Colea, **K** *79*
Colona, **K** *106*, **P** *221*
Colubrina, **K** *30, 65, 66*, **F** *160*, **P** *197*
COMBRETACEAE, **K** *17, 43, 50, 53, 73, 92, 94*, **F** *126, 130, 134, 140, 156, 158, 160*, **P** *208*
Combretum, **K** *72, 73, 92, 94*, **F** *126, 130, 140, 156*, **P** *208*
Comoranthus, **K** *76*
Conceveiba, **K** *101*, **F** *126*
CONNARACEAE, **K** *39, 56, 57, 80, 95*, **F** *126, 130, 132, 142, 146, 150, 152, 158*, **P** *171, 234*
Connarus, **K** *56, 80, 95*, **F** *146*, **P** *234*

Conocarpus, **K** *50*
CONVOLVULACEAE, **K** *16, 80, 82, 86, 95*, **F** *130, 144, 154, 158*, **P** *258*
Copaifera, **P** *231*
Oprosma, **P** *265*
Cordia, **K** *35, 42, 43, 46*, **F** *128, 138, 140, 162*, **P** *258*
Cordyla, **K** *115*
Cordyline, **K** *59*
CORNACEAE, **K** *68, 76*, **F** *130, 152, 156*, **P** *252*
Cornus, **K** *68, 76*, **F** *130, 152, 156*, **P** *252*
CORYNOCARPACEAE, **K** *43*
Corynocarpus, **K** *43*
Corynostylis, **K** *85*
Corypha, **F** *138*
Cossinia, **K** *113*
Cotylelobium, **K** *20*, **F** *156*, **P** *221*
Couepia, **K** *25*, **F** *156*, **P** *234*
Coula, **F** *152*, **P** *190*
Couma, **K** *61*, **F** *148, 162*, **P** *264*
Coumarouna, **K** *119*
Couratari, **K** *37, 40*, **F** *144*, **P** *208*
Couroupita, **K** *31*, **P** *208*
Coursetia, **K** *116*
Coussapoa, **K** *13, 81*
Coussarea, **P** *265*
Coutoubea, **K** *68*, **P** *264*
Craterosiphon, **K** *75, 93*
Crateva, **K** *55*, **P** *215*
Cratoxylum, **K** *62*
Crescentia, **K** *42, 57*, **F** *136, 144, 150*, **P** *259*
Croton, **K** *13, 14, 29, 34, 35, 45, 83, 100, 101, 104*, **F** *150, 158, 160, 162*, **P** *227*
Crotonogyne, **K** *102*, **P** *227*
Crotonoides, **K** *99*
Crudia, **K** *116*, **F** *150*
Crypteronia, **K** *69*
CRYPTERONIACEAE, **K** *69*, **F** *124*
Cryptocarya, **K** *34, 39, 40*, **F** *140*, **P** *173*

Ctenolophon, **K** *64*
CTENOLOPHONACEAE, **K** *64*, **F** *125*
CUCURBITACEAE, **K** *83*
CUNONIACEAE, **K** *63, 64, 78*, **F** *125, 148, 150, 152*, **P** *247*
Cupania, **K** *113*, **P** *240*
CUPRESSACEAE, **F** *134*
Cupressus, **F** *134*
Curatella, **K** *50*, **F** *148, 156*, **P** *185*
Cusparia, **K** *57, 150*
Cussonia, **K** *57*, **P** *252*
Cyathocalyx, **F** *162*
Cybianthus, **P** *202*
CYCLANTHACEAE, **K** *59, 60, 96*, **F** *124, 152, 154*
Cyclea, **P** *184*
Cylicomorpha, **P** *220*
Cynometra, **K** *116*, **F** *138, 146, 150*, **P** *231*
Cyphomandra, **K** *57*, **F** *146*, **P** *258*
Cyrilla, **K** *50*, **P** *203*
CYRILLACEAE, **K** *50*, **F** *124*, **P** *203*

Dacryodes, **K** *111, 113*, **P** *241*
Dahlia, **P** *253*
Dais, **P** *246*
Dalbergia, **K** *88, 115*, **F** *150*, **P** *230*
Dalechampia, **K** *101*, **P** *226, 227*
Dapania, **K** *89*
Daphnopsis, **K** *48*, **P** *246*
DAPHNIPHYLLACEAE, **K** *51*, **F** *124*
Daphniphyllum, **K** *51*
Dasyphyllum, **K** *88*, **P** *253*
Datisca, **P** *171*
DATISCACEAE, **F** *124*, **P** *171*
Datura, **F** *146*, **P** *258*
DAVIDSONIACEAE, **F** *124*
Davilla, **K** *53, 85*
DEGENERIACEAE, **F** *124*
Deinbollia, **P** *240*
Delonix, **K** *121*
Dendrobangia, **K** *37*, **P** *253*

Dendrocalamus, **K** *60*
Dendrocnide, **K** *28*, **F** *128*, **P** *196*
Dendrohyptis, **K** *67*
Dendropanax, **K** *34*, **P** *252*
Dendrophthoe, **K** *94*
Dendrophthora, **K** *94*
Denhamia, **K** *46*
Deplanchea, **K** *71*, **F** *158*
Derris, **K** *95*, *117*
Desbordesia, **P** *235*
Desmoncus, **K** *96*
Desmostachys, **K** *86*, *87*, **F** *132*
Desplatsia, **K** *106*
Dialium, **P** *231*
DIALYPETALANTHACEAE, **K** *63*, **F** *124*, *148*
Dialypetalanthus, **K** *63*
DICHAPETALACEAE, **K** *21*, *25*, *85*, *88*, **F** *126*, *140*, *144*, *158*, *160*, *162*, **P** *171*, *246*
Dichapetalum, **K** *21*, *25*, *85*, *88*, **F** *144*, *160*, **P** *246*
Dichostemma, **K** *105*, **F** *138*, **P** *226*, *227*
Dichrostachys, **K** *117*
Diclidanthera, **K** *51*, **P** *247*
Dicorynia, **K** *115*, *118*
Dicoryphe, **K** *33*
Dicraeopetalum, **K**
Dicranolepis, **K** *37*, *89*, **F** *132*, *138*, **P** *246*
Dicranostyles, **K** *86*, **F** *158*, **P** *258*
DIDYMELACEAE, **K** *47*
Didymeles, **K** *47*
Didymopanax, **F** *156*
DIEGODENDRACEAE, **K** *18*, **F** *124*
Diegodendron, **K** *18*
Dilkea, **F** *160*
Dillenia, **K** *27*, *50*, *85*, *148*, **F** *156*, **P** *185*
DILLENIACEAE, **K** *27*, *50*, *53*, *85*, **F** *130*, *148*, *152*, *154*, *156*, **P** *185*
Dilobeia, **F** *142*
Dilodendron, **K** *11*, **P** *240*
Dimorphandra, **K** *121*

Dioclea, **K** *95*, *117*, **P** *231*
Dioicodendron, **P** *265*
DIONCOPHYLLACEAE, **K** *87*, **F** *124*, *132*
Dioncophyllum, **K** *87*
Dioscorea, **K** *82*, *92*
DIOSCOREACEAE, **K** *82*, *92*, **F** *156*
Diospyros, **K** *38*, *46*, **F** *134*, *136*, *140*, *142*, **P** *203*
Diplodiscus, **P** *221*
Diplotropis, **K** *118*
DIPTEROCARPACEAE, **K** *13*, *19*, *20*, *21*, *22*, *23*, *28*, *30*, *32*, **F** *124*, *128*, *148*, *152*, *156*, *160*, *162*, **P** *171*, *221*
Dipterocarpus, **K** *13*, *19*, **F** *134*, *148*, *156*, **P** *221*
Dipteryx, **K** *56*, *119*
Disciphania, **K** *95*
Discoglypremna, **K** *10*
Discophora, **F** *140*, **P** *253*
Distylium, **K** *24*
Dodonaea, **K** *49*
Doliocarpus, **K** *85*, **P** *185*
Dombeya, **K** *107*, **F** *150*, *162*, **P** *221*
Domohinea, **K** *103*
Dovyalis, **P** *215*
Dracaena, **K** *59*
Dracontomelon, **P** *241*
Drimys, **K** *44*, **F** *125*
Dryobalanops, **K** *20*, **F** *156*, **P** *221*
Drypetes, **K** *26*, *98*, **P** *226*, *227*
Duabanga, **K** *74*, **F** *136*, *140*, *142*, *158*, **P** *209*
Duguetia, **K** *37*, **F** *154*, *162*, **P** *178*
Dunalia, **K** *46*, **P** *258*
Durio, **K** *106*, **F** *136*, *162*, **P** *221*
Duroia, **P** *265*
Dussia, **F** *156*
Dysoxylum, **K** *79*, **F** *130*, **P** *240*

EBENACEAE, **K** *38*, *40*, *46*, **F** *134*, *140*, *142*, *146*, *162*, **P** *203*

Ecclinusa, **K** *14,*, *31*, **F** *156*
Edgeworthia, **P** *246*
ELAEAGNACEAE, **K** *48*, *89*, **F** *125*, *162*, **P** *171*
Elaeagnus, **K** *48*, *89*, **F** *125*
Elaeis, **K** *60*
ELAEOCARPACEAE, **K** *28*, *29*, *32*, *48*, *66*, **F** *126*, *140*, *152*, *154*, **P** *221*
Elaeocarpus, **K** *28*, *29*, *31*, *32*, *48*, **F** *136*, **P** *221*
Elaeophora, **F** *160*
Elateriospermum, **K** *102*, **P** *226*, *227*
Ellipanthus, **F** *152*
Elmerrillia, **K** *19*, **P** *179*
Embelia, **K** *90*
Emmotum, **K** *37*, **P** *253*
Endlicheria, **F** *136*
Endospermum, **K** *103*, **P** *227*
Engelhardia, **K** *57*
Enkleia, **K** *93*
Ellipanthus, **K** *39*
Elvasia, **P** *214*
Entada, **K** *96*, *119*, **F** *132*
Entandophragma, **K** *114*, **F** *136*, **P** *240*
Enterolobium, **K** *117*, *120*
EPACRIDACEAE, **F** *125*
Epiphyllum, **K** *85*, **F** *148*
Epiprinus, **F** *162*
EREMOLEPIDACEAE, **K** *90*, **F** *124*
Erica, **P** *203*
ERICACEAE, **K** *36*, *39*, *45*, *47*, *89*, **F** *144*, *154*, **P** *203*
Eriotheca, **K** *109*
Erismadelphus, **K** *65*
Ernodea, **P** *265*
Erycibe, **P** *258*
Eryobotrya, **K** *28*, **P** *191*
Erythrina, **K** *117*, *118*, **P** *231*
Erythrochiton, **K** *48*, **F** *152*
ERYTHROPALACEAE, **F** *124*, **P** *171*
Erythropalum, **K** *83*, **P** *171*
Erythrophleum, **F** *160*
ERYTHROXYLACEAE, **K** *22*, **F** *142*, *144*, *148*, *154*, *158*,

P 235
Erythroxylum, *K* 22, *F* 142, 144, 154, *P* 235
Eschweilera, *K* 19, 37, 40, *F* 128, 140, *P* 208
Esenbeckia, *K* 42, 44, 110, *P* 241
Euadenia, *K* 55, *P* 215
Eucalyptus, *K* 49, 72, *F* 126, 134, 154, *P* 208
Eucarya, *F* 125
Euceraea, *K* 33
Euclea, 203
Eugenia, *K* 72, 74, *F* 136, 138, 140, 144, 154, 158, *P* 208
Euphorbia, *K* 16, 61, 105, *P* 227
EUPHORBIACEAE, *K* 13, 14, 13, 16, 17, 20, 22, 24, 25, 27, 28, 29, 30, 31, 32, 33, 34, 35, 36, 38, 45, 48, 49, 50, 55, 61, 65, 66, 79, 80, 82, 83, 98, *F* 124, 125, 126, 128, 134, 136, 138, 144, 148, 150, 152, 154, 158, 160, 162, *P* 226, 227
Euphronia, *K* 25
EUPOMATIACEAE, *F* 124
EUPTELEACEAE, *F* 124
Euroschinus, *P* 241
Eurycoma, *K* 112, *P* 240
Eusideroxylon, *K* 39, *F* 162, *P* 173
Euterpe, *K* 60, *P* 172
Euthemis, *K* 32
Evodia, *K* 72, 110, *F* 152, *P* 241
Evodianthus, *K* 59, 96
Exbucklandia, *K* 18, *F* 146, 148, *P* 191
Excoecaria, *K* 105, *F* 134, *P* 227

FAGACEAE, *K* 22,32, *F* 125, 126, 128, 134, 142, 144, 146, 156, *P* 191
Fagus, *F* 156, *P* 191
Fagara, *K* 54, 56, *F* 150, 152, *P* 241

Fagraea, *K* 63, 68, 93, *F* 134, 136, 148, 152, 158, *P* 264
Faramea, *F* 142, *P* 265
Ficus, *K* 13, 61, 80, 81, *F* 126, 136, 158, 160, *P* 196
Filicium,, *F* 150, *P* 240
Firmiana, *K* 107, 108, *P* 221
Flacourtia, *K* 31, *F* 144, *P* 215
FLACOURTIACEAE, *K* 20, 21, 23, 24, 25, 26, 28, 29, 30, 31, 32, 33, 35, 36, 47, *F* 126, 128, 130, 144, 148, 152, 158, 160, 162, *P* 215, 220
Flemingia, *K* 20, 115
Flindersia, *K* 78, 110
Flueggea, *K* 33, 100
Foetidia, *K* 45
FOETIDIACEAE, *F* 124
Fontanesia, *K* 76
Forsteronia, *K* 80, *F* 162
Forsythia, *K* 75
Fraxinus, *K* 79, *P* 259
Fremontodendron, *K* 107
Freycinetia, *K* 89, *F* 140
Froesia, *K* 78
Fuchsia, *K* 73
Funifera, *K* 45
Funtumia, *K* 62, *F* 138
Freziera, *K* 40, *P* 202

Gaertnera, *K* 63, *P* 265
Gagnebina, *K* 117
Galearia, *K* 98
Garcinia, *K* 61, 62, 81, *F* 136, 148, 156, 158, *P* 214
Gardenia, *P* 265
Gardneria, *K* 92
GARRYACEAE, *K* 68
Garuga, *K* 110
Gaylussacia, *K* 45
Geissois, *K* 63, 78
Geissospermum, *K* 16, *F* 138, *P* 264
GELSEMIACEAE, *P* 171, , see LOGANIACEAE
Gelsenium, *K* 92
Genipa, *F* 136, *P* 265

GENTIANACEAE, *K* 68, *F* 140, 148, *P* 264
Geonoma, *K* 59, *P* 172
GERANIACEAE, *P* 171
Gigantochloa, *K* 60
Gilbertiodendron, *K* 116, *F* 160
Gillietodendron, *P* 231
Gironniera, *K* 18, 19
Givotia, *K* 103
Gleditsia, *K* 119
Gliricida, *K* 118
Glochidion, *K* 98, *P* 227
Glossocalyx, *K* 74, *F* 136, 142
Gluta, *K* 111, 112, *P* 241
Glycydendron, *K* 99, 104, *F* 160, *P* 227
Gmelina, *K* 67, 71
GNETACEAE, *K* 71, 80, 91, *F* 130, 140, 144
Gnetum, *K* 71, 80, 91, *F* 140, 158
Gochnatia, *K* 53, *P* 253
Godmania, *K* 79, *P* 259
Godoya, *K* 54
Gomphandra, *K* 40, *F* 140, *P* 253
Gonocaryum, *K* 40, *F* 130, *P* 253
Gonystylus, *K* 37, *F* 140, 158, *P* 246
GOODENIACEAE, *K* 43, *F* 154
Gordonia, *K* 43, 53, *P* 202
Gossypium, *K* 107, *F* 136
Gouania, *K* 83, *F* 132, 156, *P* 197
Goupia, *K* 24, 26, *F* 162, *P* 171
GOUPIACEAE, *K* 24, 26, *F* 124, 162, *P* 171
Grevea, *K* 76, *P* 171
Grevillea, *K* 67, *P* 185
Grewia, *K* 84, 106, 107, *F* 158
Greyia, *P* 191
Greyiaceae, *P* 191
Grias, *K* 45
Griffonia, *K* 82, 115
Grossera, *K* 102, *P* 227
GROSSULARIACEAE, *F* 124, 125, *P* 191

Guaiacum, **K** 78, **P** 191
Guarea, **K** 111, 114, **F** 150, **P** 240
Guatteria, **F** 128, **P** 178
Guazuma, **K** 106, **P** 221
Gustavia, **K** 45, **P** 208
Gymnocladus, **K** 121, **F** 146
Gynotroches, **K** 64
Gyranthera, **K** 109
Gyrocarpus, **K** 36, **P** 173

Haematoxylum, **K** 118
Hagenia, **K** 54
HAMAMELIDACEAE, **K** 18, 19, 24, 33, 34, 36, 44, 51, **F** 124, 142, 146, 148, **P** 171, 191
Hamamelis, **K** 24
Harpullia, **K** 112, 113
Harrisonia, **K** 112
Harungana, **K** 61, **F** 154, 158, **P** 214
Hasseltia, **K** 30, **F** 158, 160, 162
Hebepetalum, **K** 32, **P** 235
Hedera, **K** 83, **P** 252
Hedyosmum, **K** 63, **F** 125, **P** 179
Heisteria, **K** 39, **F** 140, **P** 190
Helicia, **K** 51, **P** 185
Heliciopsis, **P** 185
Helicteres, **K** 106, 107, **P** 221
Heliocarpus, **K** 107
Hemiscolopia, **F** 162
Henriquezia, **P** 265
Heritiera, **K** 106, 108, **F** 162, **P** 221
Hernandia, **K** 36, **F** 144, 150, 158, **P** 173
HERNANDIACEAE, **K** 36, 82, 83, 95, 96, **F** 132, 144, 150, 152, 158, **P** 173
Heteropsis, **F** 146
Heteropteris, **K** 89, 92
Hevea, **K** 55, 103, **F** 134, 136, 150, **P** 226, 227
Hibbertia, **K** 85
Hibiscus, **K** 27, 107, **F** 158, **P** 221

Hieronyma, **K** 101, **P** 227
Hildegardia, **K** 107
Hillia, **K** 93
HIMANTANDRACEAE, **F** 124
Himatanthus, **K** 16, **F** 126, 158, **P** 264
HIPPOCASTANACEAE, see SAPINDACEAE
HIPPOCRATEACEAE: see CELASTRACEAE
Hippomane, **K** 104
Hiptage, **K** 91
Hiraea, **F** 148
Hirtella, **K** 25, **F** 160, **P** 234
Holigarna, **K** 111
Homalium, **K** 25, **P** 215
Hopea, **K** 20, 23
HOPLESTIGMATACEAE, **F** 124
Horsfieldia, **P** 179
HUACEAE, **K** 25, **F** 124, 162
Huertea, **K** 54
Hugonia, **K** 87, **F** 132, **P** 231
HUGONIACEAE
Humbertiodendron, **F** 125
Humboldtiella, **K** 116
Humiria, **K** 52, **P** 235
HUMIRIACEAE, **K** 26, 40, 52, **F** 125, 126, 152, 154, 156, 158, **P** 235
Hura, **K** 14, 104, **F** 126, **P** 227
Hybanthus, **K** 24
Hydnocarpus, **K** 20, 24, 28, **F** 158, **P** 215
Hydrangea, **K** 69, **F** 126, 154
HYDRANGEACEAE, **K** 69, **F** 126, 154
HYDROPHYLLACEAE, **K** 46
Hymenaea, **K** 116, **P** 231
Hymenocardia, **K** 101, **F** 162
HYMENOCARDIACEAE, **F** 125
Hymenolobium, **K** 117, **F** 150
Hyophorbe, **P** 172

ICACINACEAE, **K** 37, 41, 46, 52, 76, 82, 86, 87, 89, 91, **F** 130, 132, 134, 138, 140, 152, **P** 171, 253

Idesia, **K** 30
Ilex, **K** 30, 33, 52, **P** 171, 252
ILLICIACEAE, **K** 44, **F** 125, 140, **P** 173
Illicium, **K** 44, **F** 140, **P** 173
Illigera, **K** 95, 96, **F** 132, **P** 173
Indorouchera, **K** 88, **F** 132
Inga, **K** 116, **F** 128, 138, 146, 160, **P** 230
Inocarpus, **K** 13, 116, **P** 230
Intsia, **P** 230
Iodes, **K** 91, **F** 130, 132, **P** 253
Ipomoea, **K** 80, 82, **P** 258
Iriartea, **P** 172
Irvingia, **K** 18, **F** 125, **P** 235
IRVINGIACEAE, **K** 18, **F** 125, 148, 154, **P** 235
Iryanthera, **K** 37, **P** 179
Isertia, **P** 265
Isolona, **P** 178
Isoptera, **K** 23, **F** 156
ITEACEAE, **F** 125
IXONANTHACEAE, **K** 23, 31, 32, 51, **F** 125, 126, 130, 152, 154, **P** 235
Ixonanthes, **K** 31, 32, **F** 126, **P** 235
Ixora, **F** 126, 148, **P** 265

Jacaranda, **K** 79, **P** 259
Jacquinia, **K** 45, 47, **F** 140, **P** 202
Jasminum, **K** 76, 92, 93, 94, 95, **F** 132, 136, **P** 259
Jatropha, **K** 17, 103, **F** 160, **P** 226, 227
Jaundea, **F** 130
Jodina, **K** 52, **F** 125
Johannesteijsmannia, **K** 59
JUGLANDACEAE, **K** 57, **F** 125

Kadsura, **K** 86
Kaliphora, **K** 39
KALIPHORACEAE, **K** 39
Kibara, **P** 178
Kielmeyera, **K** 16
Kiggelaria, **K** 36
Kingiodendron, **K** 116

KIRKIACEAE, *F 124*
Kleinhovia, *K 107*
Knema, *P 179*
Koompassia, *P 230*
Kopsia, *K 61*
Krugiodendron, *K 65*

Labordia, *P 264*
Lacistema, *K 24*
LACISTEMATACEAE, see
FLACOURTIACEAE, *K 24*,
F 162, *P 171*
LACTORIDACEAE, *F 124*
Lacunaria, *P 214*
Laetia, *K 21, 23*, *F 130, 144,
158*
Lafoensia, *K 73*
Lagerstroemia, *K 65, 73*,
F 144, *P 209*
Laguncularia, *K 72*, *F 136*
Lamanonia, *K 64*
LAMIACEAE, *K 67*
Landolphia, *K 80*, *F 132*, *P 264*
Lansium, *P 240*
Laplacea, *K 52*, *P 202*
Laportea, *P 196*
LARDIZABALACEAE, *K 95*,
F 132, *P 184*
Lasianthera, *K 41*, *F 134*
Lasianthus, *F 134*, *P 265*
Lasiodiscus, *K 64*, *F 162*,
P 197
Lasiosiphon, *K 44*, *P 246*
LAURACEAE, *K 34, 35, 39, 42,
43, 51, 53, 71, 75*, *F 125, 130,
134, 136, 140, 142, 144, 146,
156, 158, 162*, *P 173*
Leandra, *K 91*
Lecointea, *K 24, 116*
LECYTHIDACEAE, *K 19, 31,
37, 38, 40, 45*, *F 124, 128, 140,
144, 154, 160*, *P 208*
Lecythis, *K 37, 40*, *P 208*
Leea, *K 18, 27, 54*, *F 138, 148*,
P 197
LEEACEAE, *K 18, 27, 54*,
F 125, 138, 148, 152, 156,
P 197

LEGUMINOSAE, *K 13, 20, 23,
24, 52, 54, 55, 56, 78, 80, 82,
83, 88, 95, 96, 97, 115*, *F 126,
128, 130, 132, 138, 140, 142,
144,146, 148, 150, 152, 154,
156, 160*, *P 171, 230, 231*
Leonardoxa, *P 231*
Leonia, *K 26, 32*, *P 220*
LEPIDOBOTRYACEAE, *F 124*
Lepidotrichilia, *K 113*
Lepionurus, *K 39*
Lepisanthes, *K 113*, *F 150*,
P 240
Leptaulus, *K 41*, *P 171, 253*
Leptolaena, *K 18*
Leretia, *K 86*
Leucadendron, *K 49*
Leucosyke, *K 19*
Leucothoe, *P 203*
Licania, *K 21, 22, 25, 41*,
F 148, 160, *P 234*
Licaria, *K 51, 53*
Licuala, *P 172*
Ligustrum, *F 154*
LINACEAE, *K 32, 87, 88*,
F 125, 132, 152, 154, 156,
P 235
Lindackeria, *K 29*, *P 215*
Linociera, *K 76*
Liquidambar, *K 31*
Lithocarpus, *K 22*, *F 146, 156*,
P 191
Litsea, *K 51*, *F 142*
Livistona, *K 60*
Logania, *K 64*, *F 148*, *P 264*
LOGANIACEAE, *K 63, 64, 67,
68, 69, 92, 93, 94*, *F 130, 132,
134, 138, 148, 152, 158*,
P 264
Lonchocarpus, *K 115, 118*,
F 146, *P 230*
Lonicera, *K 91*, *F 126, 146*
Lophira, *K 32*, *P 214*
Lophopetalum, *K 66*, *F 148*
LOPHOPYXIDACEAE, *K 87*,
F 124, 132, *P 171*
Lophopyxis, *K 87*, *F 132*, *P 171*
Lophostoma, *K 75*

Loranthaceae, *P 190*
Loranthus, *P 190*
Loxopterygium, *K 112*
Lozanella, *K 63, 66*
Ludia, *K 47*, *P 215*
Ludovia, *K 59*
Ludwigia, *K 53, 73*
Luehea, *K 106*
Lueheopsis, *K 106*
Lumnitzera, *K 53*
Luvunga, *K 97*
Lycianthes, *K 89*, *P 258*
LYTHRACEAE, *K 65, 69, 73,
74*, *F 126, 144*, *P 209*

Maba, *K 40*
Mabea, *K 105*, *F 154, 160*,
P 227
Macadamia, *K 46*
Macaranga, *K 27, 29, 32, 101*,
F 150, 158, 160, *P 226, 227*
Machaerium, *K 80, 97, 115*,
F 130, 132, 156
Maclura, *K 13*
Macoubea, *F 158*
Macphersonia, *K 113*, *P 240*
Macrolobium, *K 116*, *P 231*
Madhuca, *P 203*
Maesa, *K 49*, *F 136, 156*, *P 202*
Maesobotrya, *K 99*
Maesopsis, *F 148*, *P 197*
Magnolia, *K 19, 27*, *F 136,
148, 154*, *P 179*
MAGNOLIACEAE, *K 19, 27*,
F 125, 148, 154, 156, 158,
P 179
Mahonia, *K 56*, *P 184*
Mahurea, *K 13, 32*, *P 214*
Maingaya, *K 19*, *P 191*
Mallotus, *K 29, 66, 101*, *F 158*,
P 227
Malpighia, *F 146*, *P 246*
MALPIGHIACEAE, *K 30, 61,
63, 65, 69, 71, 80, 86, 91, 92,
93*, *F 126, 132, 140, 146, 148,
158, 160, 162*, *P 246*
MALVACEAE, *K 19, 27, 28,
55*, *F 128, 152, 158*, *P 171, 221*
Mammea, *K 62*, *P 214*

Manettia, *K 91*
Mangifera, *K 111*, *F 136*, *P 241*
Manihot, *K 14, 103*, *F 150*, *P 226, 227*
Manilkara, *K 16*, *F 134, 136*, *P 203*
Maniltoa, *K 116*
Manniophyton, *K 80, 83, 102*, *P 227*
Manotes, *K 39, 80*
Mansonia, *K 108*
Maprounea, *K 104*, *P 227*
Marcgravia, *K 89*, *P 202*
MARCGRAVIACEAE, *K 89*, *F 124, 154*, *P 202*
Mareya, *K 29, 100*, *P 227*
Marila, *K 68*
Maripa, *P 258*
Markea, *K 89*, *P 258*
Marquesia, *K 21*, *P 221*
Mastixia, *K 46*, *P 252*
Matayba, *P 240*
Matisia, *K 107*
Matthaea, *F 138*, *P 178*
Mauritia, *K 60*, *P 172*
Maytenus, *K 26, 32*, *P 235*
Medinilla, *F 158*, *P 209*
MEDUSAGYNACEAE, *F 124*
MEDUSANDRACEAE, *F 124*
Medusanthera, *P 253*
MELANOPHYLLACEAE, *K 47*
Melanophyllum, *K 47*
MELASTOMATACEAE, *K 67, 67, 68, 69, 71, 73, 76, 91, 94*, *F 126, 138, 140, 142, 148, 154, 156, 158*, *P 209*
Melia, *K 111*
MELIACEAE, *K 40, 43, 79, 111, 112, 113, 114*, *F 126, 130, 134, 142, 144, 146, 150, 152, 162*, *P 240*
MELIANTHACEAE, *K 54, 110*, *F 124, 152*, *P 191*
Melianthus, *K 54, 110*, *P 191*
Melicoccus, *K 113*, *F 126*
Meliosma, *K 40, 46, 57*, *P 171, 184*

MELIOSMACEAE, *K 40, 46, 57*, *F 124, 125*, *P 171, 184*
Melochia, *K 107*
Memecylon, *K 68, 69, 73, 76*, *P 209*
Mendoncia, *K 91*, *F 125*
MENDONCIACEAE, *F 125*, *P 171*
MENISPERMACEAE, *K 34, 82, 84, 95*, *F 130, 144, 150, 152, 156*, *P, 184*
Merremia, *K 95*
Mesua, *K 74*, *P 214*
Metrodorea, *K 72, 111*, *F 148*
Metroxylon, F 138
Mezoneuron, *K 121*
Michelia, *P 179*
Miconia, *K 71*, *F 158*, *P 209*
Micrandra, *K 103*, *F 160*
Microdesmis, *K 98*, *P 227*
Micropholis, *K 16*, *F 156*, *P 203*
Mikania, *K 91*, *F 128, 132*, *P 253*
Millettia, *K 96, 117*, *F 130*
Millingtonia, *K 79*
Mimosa, *K 117, 119, 120*
Mimusops, *K 14, 16*, *P 203*
Minquartia, *P 190*
Mirabilis, *K 67*, *F 144*
Mischocarpus, *K 113*, *F 162*
Mischodon, *K 65, 100*
Mollia, *K 106*
Mollinedia, *K 74*, *P 178*
Moniera, *P 241*
MONIMIACEAE, *K 72, 74, 75, 94*, *F 125, 128, 130, 138, 140, 142, 156, 158*
Monodora, *F 138*, *P 178*
MONOTACEAE, *F 124*
Monotes, *K 21*, *F 160*, *P 221*
Montezuma, *K 107*
MONTINIACEAE, *K 76*, *F 124*
Moronobea, *P 214*
Morisonia, *K 45, 48, 49*, *P 215*
Mora, *K 119*
MORACEAE, *K 13, 14, 23, 27, 61, 80, 81*, *F 126, 128, 142, 148, 150, 156, 158, 160*, *P 196*

Morinda, *P 265*
Moringa, *K 56*, *F 160*
MORINGACEAE, *K 56*, *F 125, 160*
Mostuea, *P 171, 264*
Mouriri, *K 68, 69, 73*, *F 138, 140*, *P 209*
Moutabea, *K 86*, *P 247*
Mucuna, *K 80, 95, 117*, *P 230, 231*
Muntingia, *K 19*
MUNTIGIACEAE, *K 19*
Murraya, *P 241*
Musanga, *K 14, 27*
Mussaenda, *P 265*
Myllanthus, *K 79, 110*
MYOPORACEAE, *K 49*
Myoporum, *K 49*
Myrceugenia, *K 72*
Myrcia, *P 208*
Myrica, *K 49*, *F 156*
MYRICACEAE, *K 49*, *F 156*
MYRISTICACEAE, *K 15, 37*, *F 134, 136, 140, 146, 156, 158*, *P, 179*
Myrocarpus, *P 231*
Myrospermum, *P 231*
Myroxylon, *K 115*, *F 146*, *P 231*
MYRSINACEAE, *K 38, 43, 44, 48, 49, 90*, *F 144, 156*, *P 202*
MYRTACEAE, *K 17, 44, 49, 71, 72, 73, 74, 76*, *F 126, 138, 140, 144, 154, 158, 160*, *P 208*
Mystroxylon, *K 33*
Myxopyrum, *K 92*

Napoleonaea, *K 37, 38*, *F 136, 160*, *P 208*
Natsiatum, *K 82*
Nauclea, *P 265*
Nealchornea, *F 160*
Nectandra, *P 172*
Neea, *K 76*, *P 190*
Neesia, *K 108*, *P 221*
Nemuaron, *P 178*
Neocouma, *F 158*
Neoscortechinia, *K 100*
Neosprucea, *P 215*

Neostachyanthus, **K** 86
Nephelium, **K** 113, **F** 150, **P** 240
Neuburgia, **K** 64, **F** 148
Neuropeltis, **K** 86, **F** 130, **P** 258
Newbouldia, **F** 160
Newtonia, **K** 117
Norantea, **K** 89
Noronhia, **K** 76
Norrisia, **K** 64
Nothopora, **P** 203
Notobuxus, **K** 75
Nuxia, **K** 69, **P** 171
Nuytsia, **P** 190
NYCTAGINACEAE, **K** 51, 67, 73, 76, 94, **F** 128, 138, 144, 146, **P** 190
Nyctanthes, **F** 162

Ochanostachys, **K** 40, **P** 190
Ochna, **K** 22, **F** 158, **P** 214
OCHNACEAE, **K** 19, 20, 22, 31, 32, 54, **F** 124, 142, 154, 158, **P** 214
Ochroma, **K** 107, **F** 150, 158, **P** 221
Ochthocosmus, **K** 23, 31, **F** 130
Ocotea, **K** 42, 53, **F** 134, 136, **P** 173
Octoknema, **K** 38, **P** 190
OCTOKNEMACEAE, **F** 124
Octomeles, **K** 35, **F** 144, 162, **P** 171, 220
Odontadenia, **F** 148, **P** 264
Odyendea, **K** 112, **P** 240
OLACACEAE, **K** 15, 38, 39, 41, 42, 83, 88, **F** 124, 140, 144, 152, **P** 190
Oldfieldia, **K** 79
Olea, **F** 144
OLEACEAE, **K** 75, 76, 77, 79, 92, 93, 94, 95, **F** 126, 132, 134, 144, 154, 162, **P** 259
Olinia, **K** 74, **F** 144
OLINIACEAE, **K** 74, **F** 124, 144
Omalanthus, **K** 104, **F** 160

Omphalea, **K** 82, 102, **F** 154, 160, 162, **P** 227
ONAGRACEAE, **K** 53, 73, **F** 126
Oncostemum, **K** 48, 49, **F** 144, **P** 202
ONCOTHECACEAE, **F** 124
Ophiocaryon, **K** 57
Opilia, **K** 39
OPILIACEAE, **K** 39, 47, **P** 190
Oreomunnea, **K** 57
Ormocarpopsis, **K** 115
Ormosia, **K** 118, **F** 146
Osmanthus, **K** 76, **F** 162
Ostodes, **K** 102
Ouratea, **K** 19, 20, 22, **F** 142, 158, **P** 214
OXALIDACEAE, **K** 56, 57, 58, 89, **F** 124, 152, 158, **P** 234
Oxalis, **K** 58
Ozoroa, **K** 50, 111, **F** 156, **P** 241

Pachira, **K** 109
Pachyelasma, **K** 121
Pachylobus, **K** 113, **F** 150, **P** 241
Pachystroma, **K** 105
Pacouria, **F** 132, **P** 264
Paederia, **K** 91
Pagamea, **P** 265
Palaquium, **K** 14, **F** 156, **P** 203
Palicourea, **P** 265
Palmeria, **K** 94
Pakaraimea, **K** 30, **F** 160
Pancheria, **P** 247
Pandorea, **K** 95
Panda, **K** 98
PANDACEAE, **K** 24, 98, **F** 125, **P** 227
PANDANACEAE, **K** 59, 89, **F** 125, 140, 152
Pandanus, **K** 59, **F** 136, 140
Pangium, **K** 35, **F** 128, 158, **P** 220
Panopsis, **K** 46, 75, **P** 185
PARACRYPHIACEAE, **F** 124
Paranephelium, **K** 79
Parashorea, **K** 32, **F** 156

Parinari, **K** 22, **F** 160, **P** 234
Parkia, **K** 121, **F** 160, **P** 230
Parkinsonia, **K** 121
Parmentiera, **K** 79, **F** 134
Parsonsia, **K** 80
Parvatia, **K** 95, **P** 184
Passiflora, **K** 30, 83, 87, 96, 132, 160, **P** 215
PASSIFLORACEAE, **K** 22, 30, 83, 87, 87, 96, **F** 126, 132, 144, 160, 162
Paullinia, **K** 54, 81, 96, **F** 132, **P** 240
Pausandra, **K** 14, 102, **F** 160
Payena, **K** 14, **F** 158
Paypayrola, **K** 28, 31, **P** 220
Pedilanthus, **K** 105, **P** 227
Pellacalyx, **K** 64, **P** 247
Pelliciera, **K** 47, **P** 202
PELLICIERACEAE, **K** 47, **F** 124, 154, **P** 202
Peltogyne, **K** 116, **P** 231
Peltophorum, **K** 121, **F** 146
Pemphis, **K** 73
Pentace, **P** 221
Pentaclethra, **K** 121, **F** 146, **P** 230, 231
PENTADIPLANDRACEAE, **F** 124
PENTAPHYLACACEAE, **F** 124
Pentaspadon, **K** 111
Pera, **K** 99, 100, **F** 160
PERACEAE, **F** 124
Perebea, **K** 13, **P** 196
Pereskia, **K** 42
PERIDISCACEAE, **K** 20, **F** 124
Peridiscus, **K** 20
Pernettya, **P** 203
Peronema, **K** 78, **F** 150, **P** 259
Perriera, **K** 112
Persea, **F** 136 Persea, **P** 173
Petersianthus, **K** 45
Petiveria, **P** 190
Petrea, **K** 92, **F** 132, **P** 259
Phaleria, **K** 75, **P** 246
PHELLINACEAE, **F** 124
Phoenix, **K** 60, **P** 172
Phoradendron, **K** 94
Phthirusa, **P** 190

Phyllanthus, **K** *98, 99,* **F** *126, 136, 158,* **P** *226, 227*
Phyllobotryon, **K** *31*
Physena, **K** *41*
PHYSENACEAE, **K** *41*
Phytocrene, **K** *82,* **F** *132,* **P** *253*
Phytolacca, **F** *126,* **P** *190*
PHYTOLACCACEAE, **K** *88,* **F** *126, 146,* **P** *190*
Picralima, **K** *62,* **P** *171*
Picramnia, **K** *56, 112,* **F** *152*
PICRAMNIACEAE, **K** *112,* **F** *124, 152*
Picrasma, **K** *110, 112,* **F** *150,* **P** *240*
PICRODENDRACEAE, **K** *58, 98,* **F** *124*
Picrodendron, **K** *58*
Pilea, **K** *93,* **F** *142*
Pilocarpus, **K** *44,* **F** *140*
Pimeleodendron, **K** *104*
Pimenta, **K** *72*
PINACEAE, **K** *59,* **F** *125*
Pinus, **K** *59*
Pinzona, **K** *85*
Piper, **K** *26, 34, 38, 89,* **F** *144, 146,* **P** *179*
PIPERACEAE, **K** *26, 34, 38, 89,* **F** *144, 146, 152, 156,* **P** *179*
Piptadenia, **K** *117, 119, 120,* **F** *134,* **P** *230*
Piptadeniastrum, **K** *120*
Piptocarpha, **K** *89*
Piranhea, **K** *55, 101*
Pisonia, **K** *73, 76, 94,* **F** *128, 138,* **P** *190*
Pithecellobium, **K** *121* **F** *130, 146, 150, 160,* **P** *230, 231*
PITTOSPORACEAE, **K** *17, 45, 46, 47,* **F** *125, 138, 140,* **P** *252*
Pittosporum, **K** *17, 45, 46, 47,* **F** *125, 138, 140,* **P** *252*
Piscidia, **F** *156*
Planchonella, **P** *203*
Planchonia, **P** *208*
PLATANACEAE, **K** *18,* **F** *128, 148, 154,* **P** *171*
Platanus, **K** *18,* **F** *148*
Platonia, **K** *62,* **P** *214*

Platycarpum, **P** *265*
Platymiscium, **K** *78, 117*
Plukenetia, **K** *101,* **P** *227*
Plumeria, **K** *16,* **P** *264*
Ploiarium, **K** *43,* **F** *134, 150,* **P** *202*
Pochota, **K** *109*
PODOCARPACEAE, **K** *59,* **F** *136, 152, 156*
Podocarpus, **K** *59*
Poecilandra, **K** *31*
Poecilanthe, **K** *115*
Poeciloneuron, **K** *77*
Poeppigia, **K** *116*
Pogonophora, **K** *100*
Pogonopus, **P** *265*
Poikilospermum, **P** *196*
POLEMONIACEAE, **K** *96,* **F** *124, 132*
Polyalthia, **P** *178*
POLYGALACEAE, **K** *22, 38, 47, 51, 86,* **F** *124, 132, 154, 160,* **P** *247*
POLYGONACEAE, **K** *18, 27, 85, 87, 88,* **F** *130, 148, 154,* **P** *185*
Polylepis, **P** *191*
Polyscias, **P** *252*
Pometia, **F** *126, 150,* **P** *240*
Porophyllum, **K** *73,* **P** *253*
Potalia, **K** *67,* **F** *148*
Potentilla, **P** *191*
Pothomorphe, **F** *156*
Pourouma, **K** *13*
Pouteria, **K** *16,* **F** *158,* **P** *203*
Pradosia, **K** *62*
Premna, **K** *71, 75*
Prionostema, **K** *80*
Pritchardia, **P** *172*
PROTEACEAE, **K** *46, 51, 57, 75,* **F** *128, 142,* **P** *171, 185*
Protium, **K** *51, 111,* **P** *241*
Protomegabaria, **K** *99*
PRUNACEAE, **F** *144, 154, 158, 160,* **P** *234*
Prunus, **K** *21, 23, 38,* **F** *128, 144, 154, 160,* **P** *171, 234*
Pseudobombax, **K** *108*
Pseudolmedia, **K** *13,* **P** *196*

Pseudopiptadenia, **K** *121*
Pseudoprosopis, **K** *96, 120*
Pseudosamanea, **K** *120*
Psidium, **K** *72,* **F** *140*
Psittacanthus, **P** *190*
Psychotria, **P** *265*
Pterocarpus, **K** *115, 117,* **P** *230, 231*
Pterocymbium, **K** *108,* **P** *221*
Pterolobium, **P** *230*
Pterospermum, **K** *106,* **F** *128*
Pterygota, **P** *221*
Ptychopyxis, **K** *99,* **F** *158, 160,* **P** *226, 227*
Putranjiva, **K** *99*
PUTRANJIVACEAE, **K** *98, 99*
Pycnanthus, **P** *179*
Pycnocoma, **K** *99*
Pyrenacantha, **K** *82, 86, 87*
Pyrostegia, **F** *162*

Qualea, **K** *65,* **F** *154, 156, 160,* **P** *209*
Quararibea, **K** *107,* **F** *150,* **P** *221*
Quassia, **K** *112,* **P** *240*
Quercus, **K** *32,* **F** *134, 136, 140, 142, 146,* **P** *191*
Quiina, **K** *66,* **F** *138,* **P** *214*
QUIINACEAE, **K** *66, 78,* **F** *124, 138, 142, 148,* **P** *214*

Randia, **F** *144,* **P** *265*
RANUNCULACEAE, **K** *95,* **F** *132,* **P** *184*
Rapanea, **K** *44, 48, 49*
Rauvolfia, **K** *62,* **P** *264*
Rauwenhoffia, **K** *85*
RHABDODENDRACEAE, **K** *49,* **F** *124,* **P** *171*
Rhabdodendron, **K** *49*
RHAMNACEAE, **K** *24, 30, 64, 65, 66, 83, 86, 87, 94,* **F** *132, 140, 148, 154, 156, 160, 162,* **P** *197*
Rhaphiostylis, **K** *89,* **F** *132, 138*
Rhapis, **P** *172*
Rhaptopetalum, **K** *40*
Rheedia, **K** *61,* **F** *156, 158,*

P 214
Rhipsalis, **K** *85*, **F** *124*
Rhizophora, **K** *64*, **F** *136*, **P** *247*
RHIZOPHORACEAE, **K** *64*,
F *128, 134, 140, 148*, **P** *171,
247*
Rhodamnia, **K** *71*
Rhododendron, **K** *45*, **P** *203*
Rhodoleia, **K** *34, 44*, **P** *191*
RHODOLEIACEAE, F *124*
Rhodomyrtus, **P** *208*
Rhus, **K** *112, 113*, **P** *241*
Rhyticaryum, **F** *140*, **P** *253*
Ribes, **P** *191*
Richeria, **K** *28, 99*, **P** *226, 227*
Ricinodendron, **K** *103*, **P** *227*
Ricinus, **K** *101*
Rinorea, **K** *23, 25, 26, 32, 65, 66*, **F** *134*, **P** *220*
Ripogonum, **K** *92*
Robinia, **F** *148*
Rollinia, **P** *178*
Rondeletia, **P** *265*
Rothmania, **P** *265*
Roucheria, **K** *32*, **P** *231*
ROSACEAE, **K** *21, 23, 38, 54*,
P *191, 234*
Roupala, **K** *46, 51, 57*, **F** *142*,
P *185*
Roureopsis, **P** *234*
Roystonea, **K** *60*
RUBIACEAE, **K** *61, 63, 64, 91,
93*, **F** *126, 130, 132, 134, 140,
142, 144, 148, 154, 156, 158,
162*, **P** *171, 265*
Rudgea, **P** *265*
Ruellia, **F** *140*, **P** *259*
Ruizterania, **K** *74*
Ruprechtia, **K** *18*
RUTACEAE, **K** *42, 44, 48, 54,
56, 57, 72, 78, 79, 97, 110, 111*,
F *134, 138, 140, 144, 148, 150,
152, 160*, **P** *241*
Ryania, **K** *2*, **P** *215*
Ryparosa, **K** *48*, **F** *152*, **P** *215*

Sabal, **P** *172*
Sabia, **K** *86*, **P** *171*
SABIACEAE, **K** *86*, **F** *124, 125*,
P *171*
Sabicea, **K** *93*
Sacoglottis, **K** *26, 40*, **P** *235*
Sagotia, **K** *102*, **P** *226*
Salacia, **K** *66, 80, 93*, **F** *130,
132, 148, 158*, **P** *235*
Samadera, **K** *38, 110*, **P** *240*
Samanea, **K** *120*
Sambucus, **K** *79*, **F** *148, 154*
Sandoricum, **K** *113*, **P** *240*
SANTALACEAE, **K** *52, 77*,
F *125*
Santalum, **K** *77*, **F** *125*
Santiria, **P** *241*
SAPINDACEAE, **K** *49, 52, 54,
81, 71, 79, 96, 110, 112, 113*,
F *126, 132, 134, 138, 142, 144,
150, 152*, **P** *240*
Sapium, **K** *14, 104*, **F** *160*,
P *226*
SAPOTACEAE, **K** *14, 15, 16,
61, 62, 81*, **F** *124, 126, 128,
134, 144, 146, 154, 156, 158*,
P *203*
Saraca, **F** *148, 150*, **P** *231*
Sarcandra, **K** *67*, **P** *179*
Sarcolaena, **K** *18*
SARCOLAENACEAE, **K** *18*,
F *124, 148, 154*
Sarcosperma, **K** *61*
SARCOSPERMATACEAE,
F *124*
Sarcostigma, **K** *86*
Sarcotheca, **K** *57*, **F** *158*, **P** *234*
Satyria, **K** *36, 89*
Saurauia, **K** *43, 50*, **F** *125, 156,
162*, **P** *202*
Sauropus, **K** *98*
Sauvagesia, **P** *214*
Savia, **K** *25, 98*, **P** *226, 227*
Scaevola, **K** *43*, **F** *154*
Scaphium, **P** *221*
Scaphocalyx, **P** *215*
Scaphopetalum, **K** *106*
Schaefferia, **K** *46*
Scheelea, **P** *172*
Schefflera, **K** *57*, **F** *138, 156*,
P *252*

Schinus, **K** *112*
Schisandra, **K** *86*, **F** *146*
SCHISANDRACEAE, **K** *86*,
F *125, 146*
Schizolobium, **K** *121*
Schizomeria, **K** *64*
Schlegelia, **K** *94*
Schleichera, **K** *113*, **P** *240*
Schoutenia, **P** *221*
Schrebera, **K** *79*
Schumanniophyton, **P** *265*
Sciadodendron, **K** *56*
Sclerolobium, **K** *119*, **F** *144*,
P *230*
Scolopia, **K** *25*, **F** *144, 158*
Scorodocarpus, **P** *190*
Scottelia, **F** *162*
Scutia, **K** *66, 94*, **P** *197*
SCYPHOSTEGIACEAE, **F** *124*
SCYTOPETALACEAE, **F** *124*,
see LECYTHIDACEAE
Scytopetalum, **K** *40*
Sebastiana, **K** *102*
Securidaca, **K** *22, 38, 86*, **F** *160*
Seguieria, **K** *88*, **F** *146*, **P** *190*
Senefeldera, **K** *105*, **P** *227*
Senna, **K** *116, 119*, **F** *130, 160*,
P *231*
Serianthes, **K** *120*
Sericolea, **K** *66*
Serjania, **K** *96*, **P** *240*
Sesbania, **K** *116, 118*
Shorea, **K** *13, 20, 21, 22, 23*,
F *156, 162*, **P** *221*
Sideroxylon, **K** *16*, **F** *126, 144*
Simaba, **K** *112, 114*,
F *150*, **P** *240*
SIMAROUBACEAE, **K** *38, 52,
56, 110, 112, 113, 114*, **F** *124,
125, 134, 144, 150, 152, 162*, **P**
240
Simarouba, **K** *112, 114*, **F** *136*,
P *240*
Sindora, **P** *231*
Siparuna, **K** *75*, **F** *140*, **P** *178*
Siphonodon, **K** *26*
SIPHONODONTACEAE,
F *124*

Sloanea, **K** *29, 48,* **F** *126, 140, 152,* **P** *221*
Smeathmannia, **F** *162*
SMILACACEAE, **K** *83, 87,* **F** *132*
Smilax, **K** *83, 87,* **F** *132, 158*
Socratea, **P** *172*
SOLANACEAE, **K** *42, 46, 53, 57, 87, 89,* **F** *132, 144, 146, 162,* **P** *258*
Solandra, **K** *89*
Solanum, **K** *46, 87,* **F** *132, 146,* **P** *258*
Sonerila, **F** *142,* **P** *209*
Sonneratia, **K** *69,* **F** *125, 160,* **P** *209*
SONNERATIACEAE, see *LYTHRACEAE*
Sophora, **K** *118,* **F** *150*
Sorbus, **K** *24, 54*
Sorindeia, **K** *112, 113,* **F** *158*
Sorocea, **K** *196*
Soulamea, **K** *52*
Spachea, **K** *61, 63*
Sparattanthelium, **K** *82, 83,* **F** *158,* **P** *173*
Sparmannia, **P** *221*
Spathelia, **F** *138,* **P** *241*
Spathodea, **K** *79,* **P** *259*
Spiraea, **F** *154,* **P** *191*
Spondianthus, **K** *99*
Spondias, **K** *113,* **F** *158,* **P** *241*
STAPHYLEACEAE, **K** *54, 78,* **F** *154,* **P** *191*
Stenocarpus, **P** *185*
Stephania, **P** *184*
Sterculia, **K** *108,* **F** *128, 136, 142, 158, 162,* **P** *221*
STERCULIACEAE, **K** *18, 82,* **F** *128, 130, 136, 140, 148, 150, 152, 158, 160, 162,* **P** *171, 221*
Sterigmapetalum, **F** *128,* **P** *247*
Steriphoma, **K** *29,* **P** *215*
Stifftia, **P** *253*
Stigmaphyllon, **F** *160,* **P** *246*
Stixis, **F** *152,* **P** *215*
Stomatocalyx, **P** *227*
STRASBURGERIACEAE, **F** *124*

Streblus, **K** *13, 81,* **P** *196*
Strephonema, **K** *50*
Strombosia, **K** *39,* **P** *190*
Strophioblachia, **K** *36, 103*
Strychnos, **K** *65, 67, 93, 94,* **F** *130, 132, 138, 158,* **P** *264*
Stryphnodendron, **K** *120*
STYLOCERATACEAE, **F** *124*
Styrax, **K** *37, 48,* **F** *125, 152*
STYRACACEAE, **K** *37, 48,* **F** *125, 128, 162*
Sumbaviopsis, **K** *100*
Suriana, **K** *52*
SURIANACEAE, **K** *52*
Swartzia, **K** *115,* **F** *138, 140, 146, 150,* **P** *230, 231*
Swietenia, **K** *114,* **P** *241*
Syagrus, **P** *172*
Symphonia, **K** *61, 62,* **F** *136,* **P** *214*
Symingtonia, **P** *191*
SYMPLOCACEAE, **K** *52, 53,* **F** *125, 158*
Symplocos, **K** *52, 53,* **F** *125*
Synadenium, **K** *105*
Syzygium: see *Eugenia*

Tabebuia, **K** *74, 79,* **F** *146, 152, 162,* **P** *259*
Tabernaemontana, **K** *62,* **P** *264*
Tachia, **K** *68,* **F** *140,* **P** *264*
Tachigali, **K** *119,* **P** *231*
Talauma, **P** *179*
Talisia, **F** *138,* **P** *231*
Tambourissa, **F** *130,* **P** *178*
Tapeinosperma, **K** *44,* **P** *202*
Tapirira, **K** *112*
Tapura, **K** *21, 25,* **P** *246*
Taralea, **K** *78, 117*
Tecoma, **K** *79*
Tectona, **K** *69*
Teijsmanniodendron, **K** *74, 79,* **F** *152*
Tephrosia, **K** *118*
Terminalia, **K** *43, 72, 73,* **F** *134, 136,* **P** *208*
Ternstroemia, **K** *47, 50,* **F** *125,* **P** *202*
Tessarandra, **K** *75*

Tessmannia,, **F** *150*
Tetracera, **K** *85*
Tetractomia, **K** *111,* **F** *152,* **P** *241*
Tetragastris, **K** *111, 112*
TETRAMELACEAE, **K** *35,* **F** *124, 162,* **P** *171, 220*
Tetrameles, **K** *35,* **P** *171, 220*
TETRAMERISTACEAE, **F** *125*
Tetrapleura, **K** *120*
Tetrastigma, **K** *19, 83, 96,* **P** *197*
Tetrorchidium, **K** *66, 104,* **P** *227*
THEACEAE, **K** *39, 40, 43, 47, 50, 52, 53,* **F** *125, 154, 158, 162,* **P** *171, 202*
Thecacoris, **K** *98*
Theobroma, **K** *106,* **F** *128,* **F** *136, 140*
Theophrasta, **K** *45,* **P** *202*
THEOPHRASTACEAE, **K** *45, 47,* **F** *124, 140,* **P** *202*
Thunbergia, **K** *91,* **F** *125,* **P** *259*
THUNBERGIACEAE, **F** *125*
Thuja, **F** *134*
THYMELAEACEAE, **K** *37, 45, 48, 75, 89, 93,* **F** *128, 132, 138, 140, 156, 158, 162,* **P** *171, 246*
Thyrsodium, **K** *112, 113*
Tilia, **F** *162*
TILIACEAE, **K** *84,* **F** *128, 152, 158, 162,* **P** *171, 221*
Tiliacora, **K** *84*
Tinospora, **P** *184*
Tipuana, **K** *117*
Tococa, **P** *209*
Toddalia, **K** *44, 110*
Toechima, **K** *113,* **F** *162,* **P** *240*
Toona, **K** *114,* **P** *240*
Topobea, **K** *91*
Toulicia, **K** *113,* **F** *150,* **P** *240*
Tournefortia, **K** *90*
Touroulia, **K** *78*
Tovaria, **K** *55*
TOVARIACEAE, **K** *55*
Tovomita, **K** *68, 70,* **F** *158,* **P** *214*
Tragia, **K** *101,* **P** *227*

Trattinickia, **K** *111*, **P** *241*
Trema, **K** *19, 23*, **P** *196*
Trevesia, **P** *252*
Trichilia, **K** *111, 113, 114*, **F** *150*, **P** *240*
Trichoscypha, **K** *112*
Trigonia, **K** *64, 91*, **F** *148*
TRIGONIACEAE, **K** *25, 64, 91*, **F** *125, 148, 162*
Trigoniastrum, **K** *25*
Trigonobalanus, **F** *125*
Trigonostemon, **K** *102*, **P** *227*
TRIMENIACEAE, **F** *124*
Triphasia, **P** *241*
Triplaris, **K** *18*
Triplochiton, **K** *18, 106*, **F** *148, 150, 158*, **P** *221*
Tristania, **K** *17, 44*
Triumfetta, **K** *107*, **F** *162*
Trophis, **K** *13, 81*, **P** *196*
Turnera, K *50*
TURNERACEAE, **K** *50*
Turpinia, **K** *78*, **F** *154*, **P** *191*
Turraea, **K** *40*, **F** *144, 162*, **P** *240*

Uapaca, **K** *32, 99*, **F** *152*, **P** *226, 227*
UAPACACEAE, **F** *124*
Ugni, **P** *208*
ULMACEAE, **K** *18, 19, 23, 63, 66, 84*, **F** *140, 156, 158*, **P** *196*
Uncaria, **K** *93*, **F** *130, 132*, **P** *265*
Urera, **K** *28, 84*, **F** *148*, **P** *196*
URTICACEAE, **K** *19, 28, 66, 84, 93*, **F** *126, 128, 142, 148, 158*, **P** *196*
Uvaria, **K** *85*, **F** *132*, **P** *178*
Uvariodendron, **P** *178*

Vaitarea, **K** *118*
Vantanea, **K** *40*, **P** *235*
Vateria, **K** *20,*
Vatica, **K** *21, 28*
Vaupesia, **K** *102*
Vavaea, **K** *43, 111*, **P** *240*
Veitchia, **K** *60*, **P** *172*
Vellosiaceae, **F** *125*

Ventilago, **K** *24, 86, 87*, **F** *132, 156*
Vepris, **F** *150*, **P** *241*
VERBENACEAE, **K** *67, 68, 69, 72, 74, 75, 76, 78, 79, 92, 93*, **F** *132, 140, 144, 146, 148, 150, 152, 160, 162*, **P** *259*
Verbesina, **K** *69*, **P** *253*
Vernonia, **P** *253*
Vesselowskya, **K** *78*
Viburnum, **K** *67, 69, 73*, **F** *154*
Villaresia, **K** *52*
VIOLACEAE, **K** *23, 24, 25, 26, 28, 31, 32, 65, 66*, **F** *134, 158*, *P 220*
Virola, **P** *179*
VISCACEAE, **K** *94*, **F** *152*
Viscum, **K** *94*, **P** *190*
Vismia, **K** *61, 68, 81*, **F** *140, 154, 158, 162*, **P** *214*
VITACEAE, **K** *19, 83, 96*, **F** *130, 132, 144, 162*, **P** *171, 197*
Vitex, **K** *79*, **F** *152*, **P** *259*
Vochysia, **K** *65*, **F** *140, 154*, **P** *171, 209*
VOCHYSIACEAE, **K** *61, 65, 74, 77*, **F** *125, 140, 142, 154, 156, 160*, **P** *209*
Vouacapoua, **K** *118*, **F** *160*, **P** *230*

Walsura, **K** *111*, **F** *152*, **P** *240*
Weinmannia, **K** *78*, **F** *148, 150*, **P** *247*
Wigandia, **K** *46*
Willughbea, **F** *132*, **P** *264*
WINTERACEAE, **K** *44*, **F** *125*
Witheringia, **P** *258*
Woodfordia, **K** *69*

XANTHOPHYLLACEAE, **K** *38*, **F** *124, 160*, **P** *247*
Xanthophyllum, **K** *38*, **F** *160*, **P** *247*
Xerospermum, **P** *240*
Xylia, **K** *120*
Xylopia, **F** *146, 154*, **P** *178*
Xylosma, **F** *144*

Xymalos, **K** *72*, **P** *178*
Ximenia, **K** *42*

Yucca, **K** *59*

Zanthoxylum, see *Fagara*
Zapoteca, **K** *117*
Zinowiewia, **K** *66*
Ziziphus, **K** *24*, **F** *140, 148, 154, 156*, **P** *197*
Zollernia, **K** *24, 116*, **P** *230*
Zollingeria, **P** *240*
Zygia, **K** *121*
Zygogynum, **K** *44*
ZYGOPHYLLACEAE, **K** *78*, **F** *124, 152*, **P** *171, 191*

Captions to photos

Numbers refer to corresponding plates

1.1 *Korthalsia sp.* spiny liana at forest edge, Manusela national Park, Seram, Moluccas Islands.
1.2 *Veitchia merrillii*, in cultivation, Pointe-à-Pitre, Guadeloupe.
1.3 *Salacca edulis*, market of Bukittinggi, Sumatra.
1.4 *Gaussia principes*, inflorescence and its bract, Marnier Lapostolle botanical garden, Saint-Jean-Cap-Ferrat, France.
1.5 *Euterpe precatoria*, sclerenchymatous inclusions in wood, El Salto, Caura, Venezuela.
1.6 *Licuala sp.*, forest undergrowth, Pasir Mayang, Sumatra.
1.7 *Maximilliana regia*, Trapichote, upper Orinoco, Vevezuela.

2.1 *Hernandia sonora* (Hernandiaceae), inflated bracts, Isle of Ambon, Moluccas.
2.2 *Sparattanthelium tupiniquinorum*, climbing with short shoots, Caura, Venezuelan Guiana.
2.3 *Canella winterana* (Canellaceae), capsules, seeds embeded in mucilage, Guadeloupe, West Indies.
2.4 *Cinnamomum camphora* (Lauraceae) an Asiatic species with leaves grouped at distal ends of modules, Marnier Lapostolle botanical garden, Saint-Jean-Cap-Ferrat, France.
2.5 An unidentified genus of a Lauraceae, outer bark lenticellate, inner bark with sclerenchy-matous orange inclusions.
2.6 *Litsea umbellata*, stouts and short peduncles supporting berries, Kebun Raya botanical garden, Java.
2.7 *Ocotea sp.*, fruit and its cupular receptacle, Lago do Peri, Isle of Santa Catarina, Brasil.
2.8 *Ocotea lanceolata*, inflorescence, Isle of Santa Catarina, Brasil.

3.1 *Ephippiandra madagascarensis* (Monimiaceae), trunk module bendind abruptly (Mangenot's model), Montagne d'Ambre, Madagascar.
3.2 *Kibara coriacea*, ripening fruits green, then black supported by stout peduncles, Kebun Raya botanical garden, Java
3.3 *Xylopia frutescens*, plagiotropic twig and berries, Cayenne, French Guiana
3.4 *Duguetia sp.* (Annonaceae), syncarpous compound fruits, Caura, Venezuelan Guiana
3.5 *Polyalthia sp.*, apocarpous compound fruits, Montagne d'Ambre, Madagascar

4.1 *Horsfieldia irya* (Myristicaceae), large and fruiting branch hold by the author and Daniel Atuany, Manusela National Park, Seram, Moluccas Islands.
4.2 An unidentified Myristicaceae, abundant red exsudate, Pasir Mayang, Sumatra.
4.3 *Myristica malabarica*, capsule, seed embedded in a red arilla, Kerala, India.
4.4 *Virola surinamensis*, rufescent young twig, trail of Saint Elie, french Guiana.
4.5 *Magnolia hypoleuca* (Magnoliaceae), fruit syncarpous, Geneva botanical garden, Switzerland.
4.6 *Piper aduncum* (Piperaceae), modules ending into a spike, trail of Saint Elie, French Guiana.

5.1 *Epinetrum sp.* (Menispermaceae), drupes, reserve of Campo, Cameroon.
5.2 *Hyperbaena sp.* petiole geniculate at both ends, Isle of santa Catarina, Brasil.
5.3 An unidentified Menispermaceae with its typical trinerved leaves, Upata, Venezuela.
5.4 *Aristolochia gigantea* (Aristolochiaceae), an American species, Marnier Lapostolle botanical garden, Saint-Jean-Cap-Ferrat, France.
5.5 *Pararistolochia sp.*, large, fleshy? capsule, reserve of Campo, South Cameroon.
5.6 *Meliosma* cf. *simplicifolia* (Meliosmaceae), treelet conforming to Roux'model, Pasir Mayang, Sumatra.

5.7 *Akebia quinata* (Lardizabalaceae), flowers with ovaries secreting large amounts of mucilage, in Lausanne botanical garden, Switzerland.

6.1 *Panopsis rubesens* (Proteacease), pseudodistichous phyllotaxy, Caura, Venezuelan Guiana.
6.2 *Roupala montana*, spike with fragrant flowers, near Ciudad Bolivar, Venezuelan Guiana.
6.3 *Roupala montana*, furrowed external wood, near Ciudad Bolivar, Venezuela.
6.4 *Antigonon sp.* (Polygonaceae), in cultivation, Florianopolis, Brasil.
6.5 *Coccoloba sp.*, liana with short shoots bearing flush of red leaves, Imataca, Venezuelan Guiana.
6.6 *Cocccoloba sp.*, red, somewhat fleshy small fruits, waterside of the Orinoco, near Ciudad Bolivar, Venezuelan.
6.7 *Dillenia indica* (Dilleniaceae), fruits consisting by fleshy bracts, Kebun Raya botanical garden, Java.
6.8 *Davilla rugosa*, the yellow petals are quickly caducous, Isle of Santa Catarina, Brasil.

7.1 *Neea cf. ferruginea* (Nyctaginaceae), inflorescences rufescent, cymoses, near Upata, Venezuelan Guiana.
7.2 *Guapira fragrans*, Modular architecture, Guadeloupe, West Indies.
7.3 An unidentified Nyctaginaceae, furrowed external wood, parque de Las Lloviznas, Puerto Ordaz, Venezuelan Guiana.
7.4 *Phytolacca dioica* (Phytolaccaceae), a large tree with a basal sprout, trail of Saint Elie, French Guiana.
7.5 *Phytolacca americana*, an herb bearing purple berries, Lausanne botanical garden, Switzerland.
7.6 *Heisteria sp.* (Olacaceae), orange drupe, reserve of Campo, South Cameroon.
7.7 *Heisteria sp.*, black drupe surouded by a coloured calyx, trail of Saint Elie, French Guiana.

8.1 *Rhodoleia championii* (Hamamelidaceae), flowers with yellowish bracts and pink corolla, Lembah Harau, Sumatra.
8.2 *Lithocarpus sp.* (Fagaceae), trunk with eye-marks, Unggan valley, Sumatra.
8.3 *Lithocarpus sundaicus*, fruits embedded by a cupule, note the three different kind of ribs on cupules: left and right spirals, and concentric circles! Kebun Raya botanical garden, Java.
8.4 *Guaiacum officinale* (Zygophyllaceae), bright capsules and seeds, jardín botanico del Orinoco, Ciudad Bolivar, Venezuela.
8.5 *Maingaya malayana* (Hamamelidaceae), trunk and lenticels, Forest Research Institute of Malaysia, Kepong.
8.6 *Exbucklandia populnea*, a pioneer tree, Bukittingi, Sumatra.

9.1 *Celtis phillipensis* (Ulmaceae), Piper-like leaves, Manusela National Park, Seram, Moluccas Islands.
9.2 *Laportea sp.* (Urticaceae), Montagne des Français, Madagascar.
9.3 *Artocarpus integrifolia* (jak fruit), trunciflory, in cultivation, Venezuelan Guiana.
9.4 *Ficus auriculata* (Moraceae), an African trunciflorous species, Marnier Lapostolle botanical garden, Saint-Jean-Cap-Ferrat, France.
9.5 *Cecropia obtusa*, stipular annular scars and mullerian bodies, Gran Sabana, Venezuela.
9.6 *Celtis sp.* (Ulmaceae), lying trunk exhibiting radicular wood issuing from the erect reiterated complexes, Montagne d'Ambre, Madagascar.
9.7 *Ulmus glabra*, samaras, Geneva, Switzerland.

10.1 *Maesopsis eminii* (Rhamnaceae), an East African species conforming to Roux's model, Forest Research Institute of Malaysia, Kepong.
10.2 *Maesopsis eminii*, alternating ramification on a plagiotropic branch.
10.3 *Ventilago sp.*, liana, Seram, Indonesia.
10.4 *Colubrina glandulosa*, leaves base bears two glands, Imataca, Venezuelan Guiana.

10.5 Leea sp. (Leeaceae), shrub with cymose and bright red inflorescence, Montagne d'Ambre, Madagascar.
10.6 Leea sp., a small tree with spiny trunk, Manusela National Park, Seram, Moluccas Islands.

11.1 Laplacea sp. (Theaceae), shortly petiolate leaves, Isle of Santa Catarina, Brasil.
11.2 Ternstroemia sp., shrub with coriaceous leaves, Gran Sabana, Venezuelan Guiana.
11.3 Saurauia sp. (Actinidiaceae), large, hairy leaves, Bukittinggi, Sumatra.
11.4 Rapanea sp. (Myrsinaceae), purple arillas, Isle of Santa Catarina, Brasil.
11.5 Clavija guianensis, (Theophrastaceae), fragrant orange coloured flowers, Gurí, Venezuelan Guiana.
11.6 Marcgravia umbellata (Marcgraviaceae) flowers, Guadeloupe, West Indies.
11.7 Marcgravia umbellata, fruits, Guadeloupe, West Indies.

12.1 Bejaria sp. (Ericaceae), slightly zygomorphic flowers, Gran sabana, Venezuelan Guiana.
12.2 Satyria sp., epiphytic shrub, Sierra de Lema, Venezuelan Guiana.
12.3 Vaccinium euryanthum, corolla quickly caducous, Gran sabana, Venezuelan Guiana.
12.4 Manilkara sp., flowering and fruiting, twig sympodial by apposition, in cultivation, Puerto Ordaz, Venezuela.
12.5 Micropholis sp. (Sapotaceae), typical creamy latex of the family, trail of Saint Elie, French Guiana.
12.6 Pouteria sp., berry and typical seeds with large hilum, Imataca, Venezuelan Guiana.
12.7 Diospyros sp. (Ebenaceae), fruit with four persisting sepals, Imataca, Venezuelan Guiana.

13.1 Eschweilera cf. tenuifolia (Lecythidaceae), low Orinoco, Venezuelan Guiana.
13.2 Couroupita guianensis, large trunciflorous inflorescence, a Guianan species at Kuala Lumpur, Malaysia.
13.3 Barringtonia sp., indehiscent woody capsules, Manusela National Park, Seram, Moluccas Islands.
13.4 Combretum sp. (Combretaceae), bottle brush inflorescence, petit Saut, French Guiana.
13.5 Laguncularia racemosa, fruting in mangrove, Isle of Santa Catarina, Brasil.
13.6 Terminalia subspathula, crown shyness, botanical garden of Singapore.
13.7 Eugenia (or *Syzygium*) *malaccensis* (Myrtaceae), large flowers and fruits, an Asiatic species in cultivation, Venezuelan Guiana.

14.1 Lafoensia pacari (Lythraceae), brown capsules, Isle of Santa Catarina, Brasil.
14.2 Erisma uncinatum (Vochysiaceae), the reiterate tree and a winged seed, Imataca, Venezuelan Guiana.
14.3 Vochysia sp., crown shyness, Imataca, Venezuelan Guiana.
14.4 Maieta guianensis (Melastomataceae), petiole inhabitating ants, Gran Sabana, Venezuelan Guiana.
14.5 Mouriri cf. crassifolia, Piste de Saint Elie, French Guiana.
14.6 Medinilla sp., Montagne d'Ambre, Madagascar.
14.7 Miconia radulaefolia, Imataca, Venezuelan Guiana.
14.8 Miconia sp., white flower and *Miconia radulaefolia*, pink peduncles and pale blue fruits, Imataca, Venezuelan Guiana.

15.1 Garcinia sp. (Clusiaceae), typical showy flowers of this family, Ambon, Moluccas Islands.
15.2 Clusia fockeana, coastal savanna, French Guiana.
15.3 Clusia sp., abundant yellow resinous latex in fruits, Imataca, Venezuelan Guiana.
15.4 Ochna kirkii (Ochnaceae), fruits separating into free black drup-like mericarps on a fleshy, enlarged red torus, an African species, Kebun Raya botanical garden, Java.
15.5 Ouratea sp., inflorescence terminal, near Upata, Venezuelan Guiana.

15.6 *Quiina sp.* (Quiinaceae), orange berries, Imataca, Venezuelan Guiana.
16.1 *Casearia sp.* (Flacourtiaceae), capsule and orange arillas, near Upata, Venezuelan Guiana.
16.2 *Capparis cynophallophora* (Capparidaceae), long gynosteme, Guadeloupe, West Indies.
16.3 *Steriphoma paradoxum*, orange flower bud, open flower and young fruit, Gurí, Venezuelan Guiana.
16.4 *Passiflora sp.* (Passifloraceae), delta-shaped leaves, near Upata, Venezuelan Guiana.
16.5 *Passiflora sp.*, glandular bracts, near Upata, Venezuelan Guiana.
16.6 *Passiflora glandulosa*, trail of Saint Elie, French Guiana.
17.1 *Hybanthus phyllanthoides* (Violaceae), zygomorphic flowers, near Upata, Venezuelan Guiana.
17.2 *Cochlospermum vitifolium* (Bixaceae), small tree flowering after leaf fall, savannas of Venezuela.
17.3 *Rinorea riana* (Violaceae), capsule, placentation parietal, trail of Saint Elie, French Guiana.
17.4 *Muntingia calabura* (Muntigiaceae), an American species, Kovallam, Kerala, India.
17.5 *Bixa orellana*, spiny capsules with numerous red seeds, trail of Saint Elie, French Guiana.
17.6 *Paypayrola longifolia* (Violaceae), Bochinche, Venezuelan Guiana.
18.1 *Elaeocarpus angustifolius* (Elaeocarpaceae), old leaves turning red, drupes blue, Kebun Raya botanical garden, Java.
18.2 *Helicteres guazumaefolia*, (Malvaceae s.l.), zygomorphic flowers with androgynosteme, Caroni, Venezuelan Guiana.
18.3 *Herrania lemniscata*, trunk bearing fruit, Imataca, Venezuelan Guiana.
18.4 *Apeiba echinata*, axillary yellow flowers, Gurí, Venezuelan Guiana.
18.5 *Ceiba pentandra*, spiny green branches, Cayenne, French Guiana.
18.6 *Catostemma commune*, orange capsules with red arilla, Imataca, Venezuelan Guiana.
18.7 *Durio zibethinus*, large capsules with fleshy arilla, elephants hap to eat the fruit in pristine forests, Sumatra.
18.8 *Grewia sp.*, columnar stamens, Kerala, India.
19.1 *Maprounea guianensis* (Euphorbiaceae), treelet with continuous branching, Imataca, Venezuelan Guiana.
19.2 *Phyllanthus emblica*, flowering on old units of extension, an Asiatic species, Delta Amacuro, Venezuela.
19.3 *Maesobotrya sp.*, fruits, trunciflory, reserve of Campo, South Cameroon.
19.4 *Baccaurea sp.*, inflorescences, trunciflory, Pasir Mayang, Sumatra.
19.5 *Dalechampia scandens*, resin producing inflorescence of reduced male and female flowers, near Upata, Venezuelan Guiana.
19.6 *Pedilanthus tithymaloides*, strongly differentiated flowers, Guadeloupe, West Indies.
20.1 *Delonix regia* (Leguminosae), unguiculate petals, a Malagasian species, Delta Amacuro, Venezuela.
20.2 *Campsiandra sp.* contrasting red and white flowers, low Orinoco, Venezuelan Guiana.
20.3 *Brownea sp.*, red capitate inflorescence, FIBV botanical garden, Caracas, Venezuela.
20.4 *Senna quinquangulata*, stamens opening by a pore, Imataca, Venezuelan Guiana.
20.5 *Adenanthera pavonina*, inflorescence in spike, Forest Research Institute of Malaysia, Kepong.
20.6 *Entada polystachya*, riverine shrub with desarticulating fruits, Sinamari, French Guiana.
20.7 *Pentaclethra macroloba*, fibrous bark, Imataca, Venezuelan Guiana.
20.8 *Mucuna urens*, twining lianas with drooping inflorescences, near Upata, Venezuela.
20.9 *Coumarouna odorata*, pollinated by a bee, winged rachis, jardín botanico del Orinoco, Ciudad Bolivar, Venezuela.

21.1 *Couepia ovatifolia* (Chrysobalanaceae), petals early caducous, jardín botanico del Orinoco, Ciudad Bolivar, Venezuela.
21.2 *Hirtella sp.*, Caura, Venezuelan Guiana.
21.3 *Connarus venezuelensis* (Connaraceae), orange capsules and black seeds, near Upata, Venezuela.
21.4 *Agelaea sp.*, liana with trunk bearing fruits, Montagne d'Ambre, Madagascar.
21.5 *Averrhoa bilimbi* (blimbing, Oxalidaceae), trunk bearing fruits, an Asiatic shrub, FIBV botanical garden, Caracas.

22.1 *Prionostema aspera* (Celastraceae), thigmonastic twig, Imataca, Venezuelan Guiana.
22.2 *Maytenus elliptica*, pink bark, Guadeloupe, West Indies.
22.3 *Salacia sp.*, tronciflory, reserve of Campo, South Cameroon.
22.4 *Irvingia smithii* (Irvingiaceae), flush of young pink leaves, reserve of Campo, South Cameroon.
22.5 *Erythroxylum sp.* (Erythroxylaceae), orange coloured bark, near Upata, Venezuela.
22.6 *Erythroxylum sp.*, strong rhythmic growth and scaly stipules, near Upata, Venezuela.
22.7 *Erythroxylum novogranatense*, an American species, Kebun Raya botanical garden, Java.

23.1 *Sapindus saponaria* (Sapindaceae), drupes, Imataca, Venezuelan Guiana.
23.2 *Alophyllus sp.*, red capsule, Montagne d'Ambre, Madagascar.
23.3 *Xerospermum sp.*, spiny capsules, arillas yellow, Pasir Mayang, Sumatra.
23.4 *Cardiospermum sp.*, inflated calyx, Imataca, Venezuelan Guiana.
23.5 *Quassia amara* (Simaroubaceae), red and slightly zygomorphic flowers pollinated by humingbirds, Imataca, Venezuelan Guiana.
23.6 *Guarea sp.* (Meliaceae), twig without leaves, but bearing flowers, Isle of Santa Catarina, Brasil.
23.7 *Dysoxylum sp.*, orange capsule, Manusela National Park, Seram, Moluccas Islands.
23.8 *Turraea sericea*, flowers pollinated by bats, Montagne des Français, Madagascar.

24.1 *Anacardium occidentale* (Anacardiaceae), fleshy and hypertrophied peduncle bearing a nut, in a savanna near Ciudad Bolivar, Venezuela.
24.2 *Trychoscypha sp.*, ripening drupes, purple inside, reserve of Campo, South Cameroon.
24.3 *Esenbeckia grandiflora* (Rutaceae), spiny capsule, Isle of Santa Catarina, Brasil.
24.4 *Protium unifoliolatum* (Burseraceae), red capsules, arillas white, Caura, Venezuelan Guiana.
24.5 *Fagara sp.* (Rutaceae), rhythmic growth of a corky spine on trunk, Las Lloviznas, Puerto Ordaz, Venezuela.
24.6 *Angostura trifoliata* has a bitter bark, Venezuelan Guiana.
24.7 *Citrus medica*, the true lemon, Marnier Lapostolle garden, Saint-Jean-Cap-Ferrat, France.

25.1 *Dicranolepis cf. persei* (Thymeleaceae), liana, drupe orange, reserve of Campo, Cameroon.
25.2 *Lasiosiphon cf. decaryi*, Leeuwenberg's model, Montagne des Français, Madagascar.
25.3 *Dichapetalum cf. pedunculatum* (Dichapetalaceae), infrutescences of hairy drupes, Gran Sabana, Venezuela.
25.4 *Stigmaphyllum sp.* (Malpighiaceae), unguiculate petals, leaves slightly crenate (unfrequent for Malpighiaceae), Isle of Santa Catarina, Brasil.
25.5 *Banisteriopsis sp.*, infrutescences of schizocarps separating into three samaras, Rio Candelaria, Venezuelan Guiana.
25.6 *Byrsonima crassifolia*, thick bark adapted to savannas' fires, near Upata, Venezuela.

26.1 *Securidaca sp.* (Polygalaceae), papilionoid flowers, Imataca, Venezuelan Guiana.
26.2 *Casipourea guianensis* (Rhizophoraceae), laciniate petals, Guadeloupe, West Indies.
26.3 *Casipourea guianensis*, trunk lenticelate with eye marks, El Buey, Venezuelan Guiana.
26.4 *Blepharistemma sp.*, lenticelate trunk, Kerala Research Forest Institute, India.
26.5 *cf. Carallia sp.*, furrowed external wood, Manusela National Park, Seram, Moluccas Islands.

26.6 *Rhizophora racemosa*, in mangrove, French Guiana.
27.1 *Ilex aquifolium* (Aquifoliaceae), two different stages of ripening on a same extension unit, Cully, Switzerland.
27.2 *Alangium grifithii* (Alangiaceae), orange-coloured young leaves, Forest Research Institute of Malaysia, Kuala Lumpur.
27.3 *Brassaia actinophylla* (Araliaceae), large inflorescence, Singapore.
27.4 *Polyscias sp.*, large foliar scars, Montagne d'Ambre, Madagascar.
27.5 *Anisophyllea disticha*, pseudodistichous phyllotaxy, Pasir Mayang, Sumatra.

28.1 *Desmostachys sp.* (Icacinaceae), twining liana, Montagne d'Ambre, Madagascar.
28.2 *Dendrobangia boliviana*, fibrous wood, slightly aromatic,
28.3 *Lasianthera africana*, inflorescence opposite to leaf, reserve of Campo, South Cameroon.
28.4 *Gochnatia calophylla* (Asteraceae), fissured bark adapted to fire, near Upata, Venezuela.
28.5 *Gongylolepis cf. pedunculata*, shrub with coriaceous leaves, Gran Sabana, Venezuelan Guiana.
28.6 *Centauropsis rhaponticoides*, a Vernonieae, Montagne d'Ambre, Madagascar.

29.1 *Operculina sp.*, (Convolvulaceae), capsules surrounded by papery bracts, Upata, Venezuela.
29.2 *Bourreria cumanensis*, (Borraginaceae), samaroid drupes, near Upata, Venezuela.
29.3 *Tournefortia sp.*, compact scorpioid inflorescences, Imataca, Venezuelan Guiana.
29.4 *Cordia sebestana*, corolla with six petals are unusual in Borraginaceae, jardín botanico del Orinoco, Venezuela.
29.5 *Lycianthes sp.*, (Solanaceae), sarmentous liana, reflexed sepals persisting on fruits, Caura, Venezuelan Guiana.
29.6 *Solanum wrightii*, spiny capsules, near Upata, Venezuelan Guiana.
29.7 *Cyphomandra sp.*, drooping fruits, Moró do Cará, Santa Catarina, Brasil.

30.1 *Tabebuia chrysantha* (Bignoniaceae), flowering and fruting tree, Upata, Venezuela.
30.2 *Jasminum sp.* (Oleaceae), cymose axillary inflorescences, petals are shed, Montagne d'Ambre, Madagascar.
30.3 *Petrea volubilis* (Verbenaceae), trail of Saint Elie, French Guiana.
30.4 *Anisacanthus secundus* (Acanthaceae), sarmentous liana, near Upata Venezuelan Guiana.
30.5 *Barleria sp.*, stem spiny, Kerala, India.
30.6 *Aphelandra sp.*, Imataca, Venezuela.

31.1 *Potalia amara* (Loganiaceae), Crique Toussaint, French Guiana.
31.2 *Strychnos madagascarensis*, shrub bearing large drupes Ankarana, Madagascar.
31.3 *Aspidosperma sp.* (Apocynaceae), flutted trunk, bilocular capsule and glaucous leaf, Venezuelan Guiana.
31.4 *Peschiera cymosa*, latex in capsules, arillas orange, Gurí, Venezuela Guiana.
31.5 *Strophanthus boivinii*, Montagne des Français, Madagascar.
31.6 *Allamanada cathartica*, trail of Saint Elie, French Guiana.

32.1 *Nauclea subdita* (Rubiaceae), globose inflorescence with protruding styles, Kebun Raya botanical garden, Java.
32.2 *Atractogyne sp.*, sarmentous liana, only one of each two nodes is ramified, reserve of Campo, South Cameroon.
32.3 *Palicourea rigida*, shrub in savanna, near Upata, Venezuelan Guiana.
32.4 *Morinda ctrifolia*, terminal inflorescence, Kovallam, Kerala, India.
32.5 *Lasianthus sp.*, axillary fruits on a plagiotropic twig, Batang Palupuh, Sumatra.
32.6 *Psychotria poeppigiana*, blue fruits and red bracts, Boca de Nichare, Venezuelan Guiana.

1.1 Korthalsia sp.
1.2 Veitchia merrillii
1.3 Salacca edulis
1.4 Gaussia principes
1.5 Euterpe precatoria
1.6 Licuala sp.
1.7 Maximilliana regia

3.1 *Ephippiandra madagascariensis*

2.2 *Sparattanthelium tupiniquinorum*

3.3 *Xylopia frutescens*

3.2 *Kibara coriacea*

3.4 *Duguetia sp.*

3.5 *Polyalthia sp.*

4.1 Horsfieldia irya

4.2 Myristicaceae sp.

4.3 Myristica malabarica

4.4 Virola surinamensis

4.5 Magnolia hypoleuca

4.6 Piper aduncum

5.1 *Epinetrum sp.*

5.2 *Hyperbaena sp.*

5.4 *Aristolochia gigantea*

5.5 *Pararistolochia sp.*

5.3 Menispermaceae

5.6 *Meliosma sp.*

5.7 *Akebia quinata*

7.1 *Neea sp.*

7.2 *Guapira fragrans*

7.3 Nyctaginaceae sp.

7.4 *Phytolacca dioica*

7.5 *Phytolacca americana*

7.6 *Heisteria sp.*

7.7 *Heisteria sp.*

8.1 *Rhodoleia championii*
8.2 *Lithocarpus sp.*
8.3 *Lithocarpus sundaicus*
8.4 *Guaiacum officinale*
8.5 *Maingaya malayana*
8.6 *Exbucklandia populnea*

10.1 *Maesopsis eminii*
10.2 *Maesopsis eminii*
10.3 *Ventilago sp.*
10.4 *Colubrina glandulosa*
10.5 *Leea sp.*
10.6 *Leea sp.*

11.1 Laplacea sp.

11.2 Ternstroemia sp.

11.3 Saurauia sp.

11.6 Marcgravia umbellata

11.4 Rapanea sp.

11.5 Clavija guianensis

11.7 Marcgravia umbellata

14.1 *Lafoensia pacari*
14.2 *Erisma uncinatum*
14.3 *Vochysia sp.*
14.4 *Maieta guianensis*
14.5 *Mouriri sp.*
14.6 *Medinilla sp.*
14.7 *Miconia radulaefolia*
14.8 *Miconia sp.*

15.1 Garcinia sp.

15.2 Clusia fockeana

15.4 Ochna kirkii

15.3 Clusia sp.

15.6 Quiina sp.

15.5 Ouratea sp.

17.1 *Hybanthus phyllanthoides*
17.2 *Cochlospermum vitifolium*
17.3 *Rinorea riana*
17.4 *Muntingia calabura*
17.5 *Bixa orellana*
17.6 *Paypayrola longifolia*

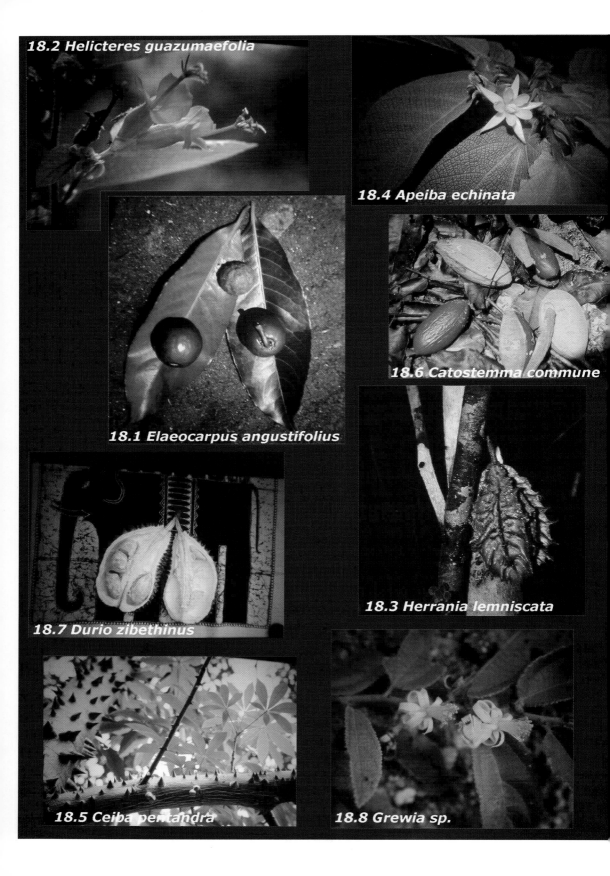

19.1 Maprounea guianensis
19.2 Phyllanthus emblica
19.3 Maesobotrya sp.
19.4 Baccaurea sp.
19.6 Pedilanthus tithymaloides
19.5 Dalechampia scandens

21.1 *Couepia ovatifolia*

21.2 *Hirtella sp.*

21.3 *Connarus venezuelensis*

21.4 *Agelaea sp.*

21.5 *Averrhoa bilimbi*

24.1 *Anacardium occidentale*

24.2 *Trychoscypha sp.*

24.3 *Esenbeckia grandiflora*

24.5 *Fagara sp.*

24.4 *Protium unifoliolatum*

24.6 *Angostura trifoliata*

24.7 *Citrus medica*

25.1 Dicranolepis sp.
25.2 Lasiosiphon sp.
25.4 Stigmaphyllum sp.
25.3 Dichapetalum sp.
25.5 Banisteriopsis sp.
25.6 Byrsonima crassifolia

26.1 *Securidaca sp.*
26.2 *Casipourea guianensis*
26.3 *Casipourea guianensis*
26.4 *Blepharistemma sp.*
26.5 cf. *Carallia sp.*
26.6 *Rhizophora racemosa*

27.1 Ilex aquifolium
27.2 Alangium grifithii
27.3 Brassaia actinophylla
27.4 Polyscias sp.
27.5 Anisophyllea disticha

28.1 *Desmostachys sp.*
28.2 *Dendrobangia boliviana*
28.3 *Lasianthera africana*
28.4 *Gochnatia calophylla*
28.5 *Gongylolepis sp.*
28.6 *Centauropsis rhaponticoides*

30.1 Tabebuia chrysantha
30.2 Jasminum sp.
30.3 Petrea volubilis
30.5 Barleria sp.
30.6 Aphelandra sp.
30.4 Anisacanthus secundus

31.1 Potalia amara

31.3 Aspidosperma sp.

31.4 Peschiera cymosa

31.2 Strychnos madagascarensis

31.5 Strophanthus boivinii

31.6 Allamanada cathartica

32.1 Nauclea subdita
32.2 Atractogyne sp.
32.3 Palicourea rigida
32.4 Morinda ctrifolia
32.5 Lasianthus sp.
32.6 Psychotria poeppigiana